Studies in Classification, Data Analysis, and Knowledge Organization

More information about this series at
http://www.springer.com/series/1564

Isabella Morlini • Tommaso Minerva •
Maurizio Vichi
Editors

Advances in Statistical Models for Data Analysis

Springer

Editors
Isabella Morlini
Department of Economics "Marco Biagi"
University of Modena & Reggio Emilia
Modena, Italy

Tommaso Minerva
Department of Communication and Economics
University of Modena & Reggio Emilia
Reggio Emilia, Italy

Maurizio Vichi
Department of Statistics
University of Rome "La Sapienza"
Roma, Italy

ISSN 1431-8814
Studies in Classification, Data Analysis, and Knowledge Organization
ISBN 978-3-319-17376-4 ISBN 978-3-319-17377-1 (eBook)
DOI 10.1007/978-3-319-17377-1

Library of Congress Control Number: 2015946232

Springer Cham Heidelberg New York Dordrecht London

Printed on acid-free paper

Springer International Publishing AG Switzerland is part of Springer Science+Business Media (www.springer.com)

Preface

This volume contains peer-reviewed selected contributions presented at the 9th biannual meeting of the Classification and Data Analysis Group (CLADAG) of the Italian Statistical Society that took place in Modena from September 18 to September 20, 2013. The conference brought together not only theoretical and applied statisticians working in Italy but also a number of specialists coming from nine different countries and was attended by more than 180 participants, including those who participated in a special session for young researchers. The conference encompassed 122 presentations organised into two plenary talks, two semi-plenary talks, 11 specialized sessions, 11 contributed sessions, eight coordinate sessions and a poster session. The main emphasis on the selection of the plenary and semi-plenary talks and on the call of papers was put on classification, data analysis and multivariate statistics, to fit the mission of CLADAG. However, many chosen contributions regarded related areas like machine learning, Markov models, structural equation models, statistical modelling in economics and finance, education and social sciences and environment. We would like to express our gratitude to all members of the Scientific Program and in particular to the Chair of the committee Francesco Palumbo. We also thank the local organizing committee, the session organizers, the invited speakers, the chairs and the discussants of all specialized sessions. We thank the authors of the contributions in this volume and the referees who spent time in carefully reviewing the papers and giving useful suggestions to the authors for improving theirs papers. We are largely indebted to the referees and to everyone who contributed their work to this volume. Finally, we thank Alice Blank from Springer for the cooperation provided in the publication of this volume.

Modena, Italy — Isabella Morlini
Modena, Italy — Tommaso Minerva
Roma, Italy — Maurizio Vichi
March 2015

Contents

Using the dglars Package to Estimate a Sparse Generalized Linear Model 1
Luigi Augugliaro and Angelo M. Mineo

A Depth Function for Geostatistical Functional Data 9
Antonio Balzanella and Romano Elvira

Robust Clustering of EU Banking Data 17
Jessica Cariboni, Andrea Pagano, Domenico Perrotta, and Francesca Torti

Sovereign Risk and Contagion Effects in the Eurozone: A Bayesian Stochastic Correlation Model 27
Roberto Casarin, Marco Tronzano, and Domenico Sartore

Female Labour Force Participation and Selection Effect: Southern vs Eastern European Countries 35
Rosalia Castellano, Gennaro Punzo, and Antonella Rocca

Asymptotics in Survey Sampling for High Entropy Sampling Designs 45
Pier Luigi Conti and Daniela Marella

A Note on the Use of Recursive Partitioning in Causal Inference 55
Claudio Conversano, Massimo Cannas, and Francesco Mola

Meta-Analysis of Poll Accuracy Measures: A Multilevel Approach 63
Rosario D'Agata and Venera Tomaselli

Families of Parsimonious Finite Mixtures of Regression Models 73
Utkarsh J. Dang and Paul D. McNicholas

Quantile Regression for Clustering and Modeling Data 85
Cristina Davino and Domenico Vistocco

Nonmetric MDS Consensus Community Detection 97
Carlo Drago and Antonio Balzanella

The Performance of the Gradient-Like Influence Measure in Generalized Linear Mixed Models ... 107
Marco Enea and Antonella Plaia

New Flexible Probability Distributions for Ranking Data ... 117
Salvatore Fasola and Mariangela Sciandra

Robust Estimation of Regime Switching Models ... 125
Luigi Grossi and Fany Nan

Incremental Visualization of Categorical Data ... 137
Alfonso Iodice D'Enza and Angelos Markos

A New Proposal for Tree Model Selection and Visualization ... 149
Carmela Iorio, Massimo Aria, and Antonio D'Ambrosio

Object-Oriented Bayesian Network to Deal with Measurement Error in Household Surveys ... 157
Daniela Marella and Paola Vicard

Comparing Fuzzy and Multidimensional Methods to Evaluate Well-Being in European Regions ... 165
Maria Adele Milioli, Lara Berzieri, and Sergio Zani

Cluster Analysis of Three-Way Atmospheric Data ... 177
Isabella Morlini and Stefano Orlandini

Asymmetric CLUster Analysis Based on SKEW-Symmetry: ACLUSKEW ... 191
Akinori Okada and Satoru Yokoyama

Parsimonious Generalized Linear Gaussian Cluster-Weighted Models ... 201
Antonio Punzo and Salvatore Ingrassia

New Perspectives for the *MDC* Index in Social Research Fields ... 211
Emanuela Raffinetti and Pier Alda Ferrari

Clustering Methods for Ordinal Data: A Comparison Between Standard and New Approaches ... 221
Monia Ranalli and Roberto Rocci

Novelty Detection with One-Class Support Vector Machines ... 231
John Shawe-Taylor and Blaž Žličar

Using Discrete-Time Multistate Models to Analyze Students' University Pathways ... 259
Isabella Sulis, Francesca Giambona, and Nicola Tedesco

Using the dglars Package to Estimate a Sparse Generalized Linear Model

Luigi Augugliaro and Angelo M. Mineo

Abstract `dglars` is a publicly available `R` package that implements the method proposed in Augugliaro et al. (J. R. Statist. Soc. B **75**(3), 471–498, 2013) developed to study the sparse structure of a generalized linear model (GLM). This method, called dgLARS, is based on a differential geometrical extension of the least angle regression method. The core of the `dglars` package consists of two algorithms implemented in `Fortran 90` to efficiently compute the solution curve.

Keywords dgLARS • Generalized linear models • Sparse models • Variable selection

1 Introduction

Nowadays, high-dimensional data sets, in which the number of predictors is larger than the sample size, are becoming more and more common. Modern statistical methods developed to cope with this problem are usually based on the idea of using a penalty function to estimate a sparse solution curve embedded in the parameter space and then to find the point that represents the best compromise between sparsity and fit of the model. Recent statistical literature has a great number of contributions devoted to this problem; examples are the L_1-penalty method [8], the SCAD method [5] and the MC+ penalty function [11], among others.

Differently from the methods cited above, Augugliaro et al. [3] propose a new approach based on the differential geometrical representation of a generalized linear model (GLM) which does not require an explicit penalty function. It has been called differential geometric LARS (dgLARS) because it generalizes the geometrical ideas underlying the least angle regression [4]. Using the differential geometric characterization of the classical signed Rao score test statistic, dgLARS gains important theoretical properties that are not shared by other methods. From

L. Augugliaro (✉) • A.M. Mineo
University of Palermo, Viale delle Scienze Ed. 13, 90128 Palermo, Italy
e-mail: luigi.augugliaro@unipa.it; angelo.mineo@unipa.it

I. Morlini et al. (eds.), *Advances in Statistical Models for Data Analysis*,
Studies in Classification, Data Analysis, and Knowledge Organization,
DOI 10.1007/978-3-319-17377-1_1

a computational point of view, the dgLARS method consists in computing the curve implicitly defined by a system of non-linear equations. In [3] this problem is satisfactorily solved by using a predictor–corrector (PC) algorithm, which however has the drawback of becoming intractable when working with thousands of predictors, since in the predictor step of this algorithm the number of arithmetic operations increases as the cube of the number of predictors. To overcome this problem, in [2], the authors propose a much more efficient cyclic coordinate descend (ccd) algorithm, which connects the original dgLARS problem with an iterative reweighted least squares algorithm. In this paper we present the `dglars` package, version 1.0.5, which is available under general public licence (GPL-2) from the Comprehensive `R` Archive Network (CRAN[1]) at http://CRAN.R-project.org/package=dglars.

2 Description of the dglars() and dglars.fit() Functions

The `dglars` package is an `R` [6] package containing a collection of tools related to the dgLARS method. The main functions of this package are `dglars()` and `dglars.fit()`. The first one

```
dglars(formula, family = c("binomial", "poisson"),
    data, subset, contrast = NULL, control = list())
```

is a wrapper function implemented to handle the formula interface usually used in `R` to create the $N \times p$-dimensional design matrix `X` and the N-dimensional response vector `y`. These objects, together with the arguments `family` and `control`, are passed to the function `dglars.fit()`

```
dglars.fit(X, y, family = c("binomial", "poisson"),
    control = list())
```

which is the `R` function used to compute the dgLARS/dgLASSO solution curve. Although in `R` the formula interface is the more familiar way to specify the linear predictor in a GLM, in a high-dimensional setting, this management of the involved model variables can be computationally inefficient. For this reason we recommend using the function `dglars.fit()` directly for simulation studies and real applications in the cases where p is very large.

As we shall see in more detail in the next section, the solution curve is related to the tuning parameter γ which is equal to the absolute value of the Rao score test statistic evaluated along the solution curve (see also [3] for its geometric meaning). From a computational point of view, the parameters used to set up the solution curve are handled by the argument `control` which is a named list defined as follows:

[1]URL: http://CRAN.R-project.org.

```
control = list(algorithm = "pc", method = "dgLASSO",
   np = NULL, g0 = NULL, eps = 1.0e-05, nv = NULL,
   dg_max = 0, nNR = 50, NReps = 1.0e-06, ncrct = 50,
   cf = 0.5, nccd = 1.0e+05)
```

Using the control parameter `algorithm` it is possible to select the algorithm used to fit the dgLARS solution curve, i.e., setting `algorithm = "pc"` (the default setting) the PC algorithm is used, whereas the ccd algorithm is used when `algorithm = "ccd"` is selected. The group of control parameters `method`, `np`, `g0` and `eps`, is composed of those elements that are shared by the two algorithms. The argument `method` is used to choose between the dgLASSO solution curve (`method = "dgLASSO"`) and the dgLARS solution curve (`method = "dgLARS"`), while `np` is used to define the maximum number of points on the solution curve. Since the PC algorithm can compute the step size by a local approximation [3], the number of effective points of the solution curve can be significantly smaller than `np`. In contrast, the ccd algorithm fits the dgLARS solution curve using a multiplicative grid of `np` values of the tuning parameter. The `g0` control parameter is used to define the smallest value of the tuning parameter. By default this parameter is set to $1.0e - 04$ when $p > N$ and to 0.05 otherwise. Finally, `eps` is used for the test of convergence of the two algorithms. When the PC algorithm is used, `eps` is also used to identify a predictor that will be included in the active set, namely when the absolute value of the corresponding Rao score test statistic belongs to $[\gamma - \texttt{eps}; \gamma + \texttt{eps}]$.

The group composed by `nv`, `dg_max`, `nNR`, `NReps`, `ncrct` and `cf` contains the control parameters specific for the PC algorithm. `nv` is used to define the maximum number of predictors included in the model, while `dg_max` is used to fix the step size. Setting `dg_max = 0` (default) the PC algorithm uses the local approximation to compute the γ value to evaluate the inclusion or exclusion of a predictor from the active set. The control parameters `nNR` and `NReps` are used to set the number of steps and to define the convergence of the Newton–Raphson algorithm used in the corrector step. When the Newton–Raphson algorithm does not converge or when there exists a predictor such that the absolute value of the corresponding Rao score test statistic is greater than γ + `eps`, the step size is reduced by the contractor factor `cf`, i.e., $\Delta\gamma = \Delta\gamma \cdot \texttt{cf}$, and then the corrector step is repeated. The control parameter `ncrct` sets the maximum number of attempts for the corrector step. Finally, the parameter `nccd` is used to define the maximum number of steps of the ccd algorithm.

3 An Example of a Logistic Regression Model

To gain more insight on how to use the main functions of the `dglars` package, in this section we study the sparse structure of a logistic regression model applied to a subset of the breast cancer gene deletion/amplification data set obtained by John

Bartlett at the Royal Infirmary, Glasgow [10]. The aim of the study is to identify which genes play a crucial role in the severity of the disease, defined as whether or not the patient dies as a result of breast cancer. The data set contains 52 samples, 29 of which are labelled as deceased due to breast cancer. For each sample, 287 gene deletion/amplification measurements are available. Missing values are imputed using the method proposed in [9].

The R code used for this data set is the following:

```
R> library("dglars")
R> data("breast", package = "dglars")
R> out_dglasso <- dglars(status ~., family = "binomial",
+ data = breast)
```

In this data set, when `status` is equal to `0` it means that the patient is not died of breast cancer, otherwise if `status = 1` that patient is died. `dglars()` returns an S3 class object called `dglars`, which is a list containing a matrix named `beta` used to store the estimated points of the solution curve, the vector `dev` of deviances, the vector `g` containing the sequence of the used values of the tuning parameter and the vector `df` containing the number of non-zero estimated coefficients including the intercept. By default, `dglars()` computes the dgLASSO solution curve; the dgLARS solution curve can be computed using the control parameter `method`, i.e.,

```
R> out_dglars <- dglars(status ~., family = "binomial",
+ data = breast, control = list(method = "dgLARS"))
```

The method function `print.dglars()` can be used to print the basic information contained in a `dglars` object, i.e., the call that produced the `dglars` object with a five-column table showing the names of predictors included or excluded from the active set, the sequence of γ values used to compute the dgLARS solution curve and the corresponding deviance and fraction of explained deviance, respectively. The number of non-zero estimated coefficients is also reported. Next code shows a part of the output obtained calling the `dglars` object `out_dglasso`

```
R> out_dglasso

Call:  dglars(formula = status ~., family = "binomial",
     data = breast)

        Sequence         g       Dev
                   3.51858   71.3935   0.00000   1
          +SHGC4
                   2.75502   64.5074   0.09645   2
                   2.34736   60.3404   0.15482   2
                   2.17491   58.8377   0.17587   2
                   2.10563   58.2907   0.18353   2
                   2.07327   58.0462   0.18695   2
                   2.05755   57.9298   0.18858   2
                   2.04980   57.8730   0.18938   2
                   2.04595   57.8449   0.18977   2
```

```
                2.04211   57.8170   0.19016    2
      +WI.2389
                1.97513   57.1786   0.19911    3
                1.97408   57.1688   0.19924    3
         +COX2
                1.48084   52.2190   0.26857    4
                1.46357   52.0736   0.27061    4

                  ...       ...       ...     ...

                0.08483    0.9333   0.98693   31
        +PRKCZ
                0.05066    0.3308   0.99537   32
                0.05000    0.3221   0.99549   32

Algorithm pc ( method = dgLASSO ) with exit = 0
```

The output shows that at $\gamma^{(1)} = 3.51858$, i.e., the starting value of the Rao score test statistic, the predictor SHGC4 makes the smallest angle with the tangent residual vector then it is included in the active set. The predictor WI.2389 is included in the active set at $\gamma^{(2)} = 2.04211$, this means that for any $\gamma \in [\gamma^{(2)}; \gamma^{(1)})$ only the intercept and the coefficient associated to SHGC4 are different from zero; consequently the number of non-zero estimates is equal to 2. The third predictor is included at $\gamma^{(3)} = 1.97408$, which means that for any $\gamma \in [\gamma^{(3)}; \gamma^{(2)})$ the number of non-zero estimates is equal to 3. This process goes on until the parameter γ is equal to the control argument g0 which is fixed to 0.05. The estimated coefficient path can be extracted from the dglars object using the method function coef.dglars(). More informations about the estimated sequence of models can be obtained using the method function summary.dglars()

```
summary(object, k = c("BIC", "AIC"),
     complexity = c("df", "gdf"),
     digits = max(3, getOption("digits") - 3), ...)
```

where object is a fitted dglars object. To choose the best solution point, the R function summary.dglars() computes the measure of goodness-of-fit

$$\textit{residual deviance} + k \times \textit{complexity}, \qquad (1)$$

where k is a non-negative value used to weight the complexity of the fitted model. Using the argument complexity the user can choose between two different definitions of complexity of a fitted model, i.e., the well-known number of estimated non-zero coefficients (complexity = "df") and the notion of generalized degrees-of-freedom [3] (complexity = "gdf"). Setting k = "BIC" and complexity = "df", which are the default values, for definition (1), the function summary.dglars reports the bayesian information criterion (BIC) [7]. The akaike information criterion (AIC) [1] can be easily computed setting `k = "AIC"` and `complexity = "df"`. The user can also define own measures of goodness-of-fit setting `k` as any non-negative value. The following `R` code shows that the output printed by `summary.dglars()` is divided into two different sections.

```
R> summary(out_dglasso, k = "BIC", complexity = "df")

Call:  dglars(formula = status ~ ., family = "binomial",
     data = breast)

          Sequence         g       Dev  df     BIC    Rank
                     3.51858   71.3935   1   75.34     22
            +SHGC4
                     2.75502   64.5074   2   72.41     19
                     2.34736   60.3404   2   68.24     12
                     2.17491   58.8377   2   66.74      7
                     2.10563   58.2907   2   66.19      6
                     2.07327   58.0462   2   65.95      5
                     2.05755   57.9298   2   65.83      4
                     2.04980   57.8730   2   65.78      3
                     2.04595   57.8449   2   65.75      2
                     2.04211   57.8170   2   65.72      1 <-
         +WI.2389
                     1.97513   57.1786   3   69.03     14
                     1.97408   57.1688   3   69.02     13
             +COX2
                     1.48084   52.2190   4   68.02     11
                     1.46357   52.0736   4   67.88     10

                        ...       ...  ...     ...    ...

                     0.08483    0.9333  31  123.42     91
            +PRKCZ
                     0.05066    0.3308  32  126.77    135
                     0.05000    0.3221  32  126.76    134

===============================================

Best model identified by BIC criterion
( k = 3.951244 and complexity = df ):

 y ~ SHGC4

Coefficients:

   Int.    SHGC4
 0.2619  -3.2184
```

```
BIC :  65.72

===

Algorithm pc ( method = dgLASSO ) with exit = 0
```

The first part of the output completes the information printed out by `print.default()` showing the BIC. The ranking of the estimated models obtained by this measure of goodness-of-fit is also shown and the corresponding best model is identified by an arrow on the right. The second section shows the formula of the identified best model and the corresponding estimated coefficients. From the previous output we can see that the best model identified by the BIC criterion is that one with only the predictor `SHGC4`.

The user can plot the output from the `dglars()` function using the method function

```
plot.dglars(x, k = c("BIC", "AIC"),
    complexity = c("df", "gdf"), g.gof = NULL, ...)
```

where `x` is a fitted `dglars` object while the arguments `k` and `complexity` are equal to the arguments of the `summary.dglars()` function. With the following R code:

```
R> out_dglasso <- dglars(status ~., family = "binomial",
+ data = breast, control = list(dg_max = 0.1))
R> par(mfrow = c(1, 3))
R> plot(out_dglasso, k = "BIC", complexity = "df")
```

we first reduce the step size setting the control parameter `dg_max = 0.1` and then we plot the output from the `dglars.fit()` function. As we have done for the `summary.dglars()` function, we use the BIC criterion to select the best model. As shown in Fig. 1, when we fit the dgLARS solution curve using the PC algorithm, the `plot.dglars()` function produces three different plots, namely the plots showing the sequence of the BIC as a function of γ and the plots showing the paths

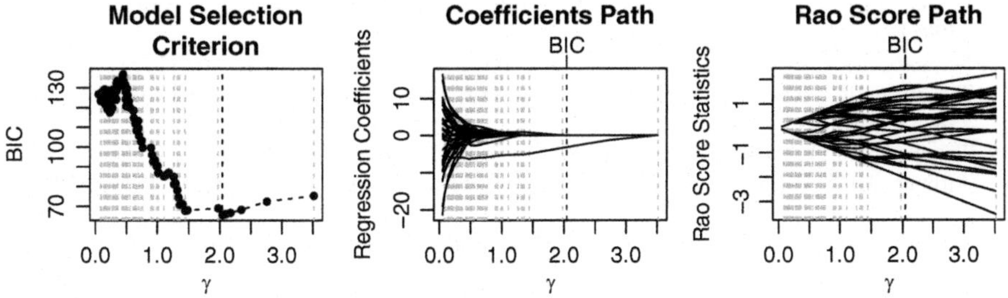

Fig. 1 Plot of the path of the BIC values computed for the estimated logistic regression model, the coefficient path and the path of the Rao score test statistics

of the coefficients and of the Rao score test statistics. The last plot is not available when the dgLARS solution curve is fitted using the ccd algorithm. The values of the tuning parameter corresponding to a change in the active set are identified by vertical dashed gray lines, while the optimal value of the tuning parameter γ, according to the BIC, is identified by a black dashed line.

4 Conclusions

In this paper we have described the `R` package `dglars`. This package implements the differential geometric extension of the method proposed in [3]. The core of the package are two functions implementing a PC algorithm and a ccd algorithm to compute the dgLARS solution curve. In order to implement these two algorithms in an efficient way, the main code is written in `Fortran 90`. The use of the main functions of the proposed package is shown by means of a logistic regression model. The output of the functions is presented in a way that is easy to interpret for people familiar with standard `lm()` or `glm()` output.

References

1. Akaike, H.: Information Theory as an extension of the maximum likelihood principle. In: Petrov, B.N., Czaki, F. (eds.) Second International Symposium on Information Theory, pp. 267–281. Akademiai Kiado, Budapest (1973)
2. Augugliaro, L., Mineo, A.M., Wit, E.C.: Differential geometric LARS via cyclic coordinate descent method. In: Proceedings of COMPSTAT 2012, pp. 67–79. Limassol, Cyprus (2012)
3. Augugliaro, L., Mineo, A.M., Wit, E.C.: Differential Geometric least angle regression: a differential geometric approach to sparse generalized linear models. J. R. Stat. Soc. B **75**(3), 471–498 (2013)
4. Efron, B., Hastie, T., Johnstone, I., Tibshirani, R.: Least angle regression. Ann. Stat. **32**(2), 407–499 (2004)
5. Fan, J., Li R.: Variable selection via nonconcave penalized likelihood and its oracle properties. J. Am. Stat. Assoc. **96**(456), 1348–1360 (2001)
6. `R` Development Core Team: `R`: A Language and Environment for Statistical Computing. `R` Foundation for Statistical Computing, Vienna (2012). http://www.R-project.org/ [ISBN 3-900051-07-0]
7. Schwarz, G.: Estimating the dimension of a model. Ann. Stat. **6**(2), 461–464 (1978)
8. Tibshirani, R.: Regression shrinkage and selection via the lasso. J. R. Stat. Soc. B **58**(1), 267–288 (1996)
9. Troyanskaya, O., Cantor, M., Sherlock, G., Brown, P., Hastie, T., Tibshirani, R., Bolstein, D., Altman, R.B.: Missing value estimation methods for DNA. Bioinformatics **17**(6), 520–525 (2001)
10. Wit, E.C., McClure, J.D.: Statistics for Microarrays: Design, Analysis and Inference. Wiley, Chichester (2004)
11. Zhang, C.H.: Nearly unbiased variable selection under minimax concave penalty. Ann. Stat. **38**(2), 894–942 (2010)

A Depth Function for Geostatistical Functional Data

Antonio Balzanella and Romano Elvira

Abstract In this paper we introduce a depth measure for geostatistical functional data. The aim is to provide a tool which allows to get a center-outward ordering of functional data recorded by sensors placed on a geographic area. Although the topic of ordering functional data has already been addressed in the literature, no proposal analyzes the case in which there is a spatial dependence among the curves. With this aim, we extend a well-known depth measure for functional data by introducing a new component in the measurement, which accounts for the spatial covariance. An application of the proposed method to a wide range of simulated cases shows its effectiveness in discovering a useful ordering of the spatially located curves.

Keywords Depth functions • Functional data ordering • Geostatistical functional data

1 Introduction

Recently, functional data analysis [4] was extended to the study of geostatistical functional datasets [1]. In this context, each curve is a sample of a continuous spatial functional process so that the dataset to analyze is made by units which include a spatial component (usually in $\Re^2$) and a functional component.

The interest in the analysis of geostatistical functional data is motivated by the a priori assumption that spatially near observations tend to be more similar than spatially far ones, so that there is a spatial dependence to be considered in the analysis. This assumption finds its application in real-world scenarios where sensors, located on spatial regions, monitor environmental variables such as temperature, humidity, and precipitation.

In this context this paper introduces a new depth function whose aim is to provide a center-outward ordering of curves taking into account the spatial dependence.

A. Balzanella (✉) • E. Romano
Department of Political Science "Jean Monnet", Second University of Naples, Caseate, Italy
e-mail: antonio.balzanella@unina2.it; elvira.romano@unina2.it

I. Morlini et al. (eds.), *Advances in Statistical Models for Data Analysis*,
Studies in Classification, Data Analysis, and Knowledge Organization,
DOI 10.1007/978-3-319-17377-1_2

Depth functions are a widely used tool for providing a center-outward ordering in multivariate data. Their role is to define multivariate analogues of univariate rank and order statistics via depth-induced "contours" [6].

The definition of the center-outward ordering is carried out by assigning a real nonnegative and bounded value (depth) to each multivariate data point according to the rule that the highest depth value is attributed to the most central data point while the lowest depth is assigned to the least central one. Due to this criterion, the depth functions are useful tools for identifying a median in the data cloud as well as outliers, which correspond, respectively, to the data point having the highest depth and to the data points having the lowest depth value. They are still useful tools for computing quantile-based statistics such as skewness and kurtosis.

The concept of ordering has been introduced for univariate and multivariate functional data by several authors. The definitions of depth are mainly based on two different notions: the first is a generalization of the classical depth, defined on integrals of univariate depths [2]; the second is a graphical approach, based on the graphical representation of functions [3]. Following the second definition, based on the graphic representation, the *depth* becomes the *band depth* definition.

The *band depth* provides a value of depth according to the graphical inclusion of a curve in the sample of curves. The inclusion of the whole curve inside several possible bands graphically obtained by the curves is evaluated. The induced order statistics starts from the most central sample curve (the median) which has the highest depth and moves outward according to decreasing depth values [2, 3].

This definition of *band depth* takes into account the whole graph of a curve. However, it could happen that the graph of a curve is in a band for a proportion of time. In order to overcome this problem, the *modified band depth* has been introduced. In this case, the depth value associated to each curve is obtained by evaluating the ratio between the portion of curve included in the band and the temporal interval.

Our proposal is to introduce in the evaluation of the curve centrality the spatial covariance among the curves. If the analyzed curves are generated by a spatially stationary and isotropic process, our depth function tends to assign a higher depth value to curves which are in the center of the spatial region and lower values to the curves on the boundary. This result finds its main application in contexts where it is necessary to find a center or the outliers taking into account the spatial location of the sensor which records the curve.

The paper is organized as follows: Sect. 2 provides a formal introduction to geostatistical functional data; Sect. 3 introduces the details of the proposed method; Sect. 4 evaluates the proposed method on a wide range of simulated datasets; Sect. 5 gives conclusions and perspectives.

2 Geostatistical Functional Data

Geostatistical functional data may be defined as curves generated by the spatial functional process $\{\chi_s : s \in D \subseteq \mathbb{R}^d\}$, where s is a generic data location in a fixed d-dimensional Euclidean space $D \subseteq \mathbb{R}^d$ with positive volume. We assume to observe a set of functions at n locations $(\chi_{s_1}(t), \ldots, \chi_{s_i}(t), \ldots, \chi_{s_n}(t))$ for $t \in T = [a, b] \subseteq \mathbb{R}$ and $s_i \in D$, for $i = 1, \ldots, n$ defining the set of functional observations. Each function is assumed to belong to a Hilbert space. For each t, we assume that the process is a second-order stationary functional random process, which formally means that the expected value $\mathbb{E}(\chi_s(\cdot))$ and the variance $\mathbb{V}(\chi_s(\cdot))$ do not depend on the spatial location, that is,

- $\mathbb{E}(\chi_s(t)) = m(t), \forall\, t \in T,\ s \in D$
- $\mathbb{V}(\chi_s(t)) = \sigma^2(t), \forall\, t \in T,\ s \in D$

In addition, we have that

- $\mathbb{C}ov(\chi_{s_i}(t), \chi_{s_j}(t)) = \mathbb{C}(h, t)$, with $h = \|s_i - s_j\|, \forall\, t \in T, \forall\, s_i, s_j \in D$
- $\frac{1}{2}\mathbb{V}(\chi_{s_i}(t) - \chi_{s_j}(t)) = \gamma(h, t) = \gamma_{s_i s_j}(t)$, with $h = \|s_i - s_j\|, \forall\, t \in T, \forall\, s_i, s_j \in D$

3 A Band Depth for Geostatistical Functional Data

Let $(\chi_{s_1}(t), \ldots, \chi_{s_i}(t), \ldots, \chi_{s_n}(t))$ with $t \in T = [a, b] \subseteq \mathbf{R}$ be a set of geostatistical functional observations.

We define the notion of depth and modified band depth for geostatistical functional data by generalizing the band depth for functional data. The spatial dependence among the curves is thus formalized by the spatial covariance function which can be considered as a weight on geostatistical functional data.

For all $s \in D$ the graph of a function χ_{s_i} is the subset of the plane $G(\chi_{s_i}) = \{(t, \chi_{s_i}(t)) : t \in T\}$.

The band in $\Re^2$ obtained by the curves $(\chi_{s_{i_1}}, \ldots, \chi_{s_{i_k}})$ is

$$B\left(\chi_{s_{i_1}}, \ldots, \chi_{s_{i_k}}\right) = \left\{(t, y_s) : t \in T \min_{r=1,\ldots,k} \chi_{s_{ir}}(t) \le y_s \le \max_{r=1,\ldots,k} \chi_{s_{ir}}(t)\right\} \qquad (1)$$

We define the fraction of bands as the proportion of bands inside several possible bands that simultaneously are graphically included and spatially correlated with a curve χ_{s_i}.

It can be defined as:

Definition 1 (Fraction of Bands) Let J be the number of a set of geostatistical functional data determining a band, where $J \geq 2$ is a fixed value and $B\left(\chi_{s_{i_1}}, \ldots, \chi_{s_{i_j}}\right)$ a band delimited by j geostatistical functional observations containing the whole graph of the geostatistical curve χ_{s_i}. The fraction of bands is

$$SB_n^j(\chi_{s_i}) = \binom{n}{j}^{-1} \sum_{1 \le s_{i_1} \cdots \le s_{i_j} \le n} \sum_{s_{i_j}}^{J} \frac{1}{J} C(h_{s_{i_j}}).I\{G(\chi_{s_i}) \subset B\left(\chi_{s_{i_1}}, \ldots, \chi_{s_{i_j}}\right)\} \tag{2}$$

where:

- $\sum_{s_{i_j}}^{J} \frac{1}{J} C(h_{s_{i_j}})$ is the spatial weight (spatial component)
- $I\{A\}$ is the indicator function (functional component)

The spatial component $\sum_{s_{i_j}}^{J} \frac{1}{J} C(h_{s_{i_j}})$ is the average of the spatial covariances $C(h_{s_{i_j}}) = 1 - \gamma(h_{s_{i_j}})$ computed on $h_{i_j} = s_i - s_{i_j}$ distances for $j = 2, \ldots, J$. $C(h_{s_{i_j}})$ is normalized respect to their maximum.

The functional component is defined according to the band depth or modified band depth for functional data , thus we can have:

- $I\{G(\chi_{s_i}) \subset B\left(\chi_{s_{i_1}}, \ldots, \chi_{s_{i_j}}\right)\}$ the indicator function for the band depth.
- $\lambda\{A\left(\chi_{s_{i_1}}, \ldots, \chi_{s_{i_j}}\right)\}$ a function of proportion.

Following Definition 1 the depth function for geostatistical functional data can be defined as:

Definition 2 (Depth Function for Geostatistical Functional) Let J be the number of a set of geostatistical functional data determining a band, where J is a fixed value with $0 \le J \le 2$ and $B\left(\chi_{s_{i_1}}, \ldots, \chi_{s_{i_j}}\right)$ a band delimited by j geostatistical functional observations containing the whole graph of the geostatistical curve χ_{s_i}. Let $SB_n^j(\chi_{s_i})$ be the fraction of bands containing the curve χ_{s_i}. The depth for geostatistical functional data is

$$SB_{n,j}(\chi_{s_i}) = \sum_{j=2}^{J} SB_n^j \tag{3}$$

It can be seen as a spatially weighted banddepth for geostatistical functional data since it considers as a weight the spatial component in the proportion of band definition. The main characteristics of this function are:

- It provides a measure of the centrality of an observation with respect to a given geostatistical functional dataset.
- The curves with minor depth are outliers.
- The median function $m_{s_{nj}}$ is a curve with highest spatially weighted depth. It corresponds to the curve simultaneously have highest spatial covariance and which is maximally included in the band. Thus it is such to satisfy:

$$m_{s_{nj}} = argmax_{(\chi_{s_1}(t), \ldots, \chi_{s_i}(t), \ldots, \chi_{s_n}(t))} SB_{n,j}(\chi_{s_i}) \tag{4}$$

It is easy to demonstrate that some of the properties valid for the band depth for functional data are still valid for band depth for geostatistical functional data. These are:

- Proposition 1 (Monotonicity). *SB* is monotone with respect to the center of symmetry under distributions with symmetric marginal.
- Proposition 2 (Maximality at center). The median curve $m_{s_{nj}}$ is the unique function which maximizes the function 3.
- Proposition 3 (Vanishing at infinity). The band depth for geostatistical functional data presents good continuity properties.
- Proposition 4. *SB* is upper-semicontinuous.

4 Evaluations on Simulated Datasets

In order to evaluate the capability of the proposed depth measure, we have performed an extensive test on simulated data. The test is focused on comparing the results of our proposal to the outputs provided by the modified band depth in [3].

With this aim, we have generated 18 datasets of spatially dependent curves according to the setup in [5]. Each dataset is made by 196 curves located at $s_1, \ldots, s_{196}$ which correspond to sites on a grid of size 14×14 in the unit square. Each curve is formed by 50 equally spaced time points in $[0, 1]$.

Given a set of curves $(\chi_{s_1}(t), \ldots, \chi_{s_i}(t), \ldots, \chi_{s_n}(t))$ located at s_i in $\mathbb{R}^2$, for $i = 1, \ldots, n$ and $t \in T$, we assume that a generic curve $\chi_{s_i}(t)$ is generated by the general model: $\chi_{s_i}(t) = \mu_{s_i}(t) + \epsilon_{s_i}(t)\ t \in T$, with mean $\mu_{s_i}(t)$ and $\epsilon_{s_i}(t)$ be a Gaussian random field with zero mean and a spatial functional covariance expressed by $C(h, u) = cov\left\{\chi_{s_i}(t_1), \chi_{s_j}(t_2)\right\}$, for any couple of locations s_i, s_j separated by a spatial distance $h = s_i - s_j$ and a temporal distance $u = t_1 - t_2$.

We have considered three covariance functions in the simulation process:

- Stationary purely spatial covariance function:

$$Cov_s(h) = (1 - \nu)\exp(-c\,|h|) + \nu I\{h = 0\} \tag{5}$$

where $c > 0$ controls the spatial correlation intensity and $\nu \in (0, 1]$ is the nugget effect.

- Separable spatiotemporal covariance function:

$$Cov_{SEP}(h, u) = cov\left\{\chi_{s_i}(t_1), \chi_{s_j}(t_2)\right\} = Cov_s(h)\left(1 + a\,|u|^{2\alpha}\right)^{-1} \tag{6}$$

where $u = |t_1 - t_2|$ is the time span, $a > 0$ is the scale parameter in time, which here is fixed to $a = 1$ for convenience, and $\alpha \in [0, 1]$ controls the strength of the functional variability.

- Symmetric but non-separable spatiotemporal covariance function:

$$Cov\,(h,u) = \frac{1-\nu}{1+au^{2\alpha}}\left[\exp\left\{-\frac{c\,\|h\|}{(1+a|u|^{2\alpha})^{\frac{\beta}{2}}}\right\} + \frac{\nu}{1-\nu}I\,\{h=0\}\right] \quad (7)$$

where the parameter $0 \le \beta \le 1$ controls the degree of non-separability.

Each spatiotemporal covariance function has been used for generating six datasets, however for space reasons we report and discuss only the results for three datasets (one for each covariance function).

The first dataset has been generated using the stationary purely spatial covariance function with a covariance value $c = 0.9$ and no nugget effect ($\nu = 0$). In Fig. 1, we report the results of the two compared methods.

For each one of the two methods, we have the functional boxplot introduced in [5], which, similarly to the classical boxplot, provides a graphical representation of the median, interquartile range, maximum and minimum, as a result of the ordering induced by the used depth function. We still have a double view of the depth value assigned to each curve in the dataset corresponding to a spatial location on the squared grid (the two plots represent a different graphical view of the same data).

By looking at the plots in Fig. 1, we can derive that the proposed method tends to set higher values of depth at the center of the spatial region while curves located at the boundary have lower values of depth. This means that on this squared spatial grid, the depth values assume a dome shape. This effect of the spatial covariance is higher when the value of the parameter c is higher. We can still observe that our

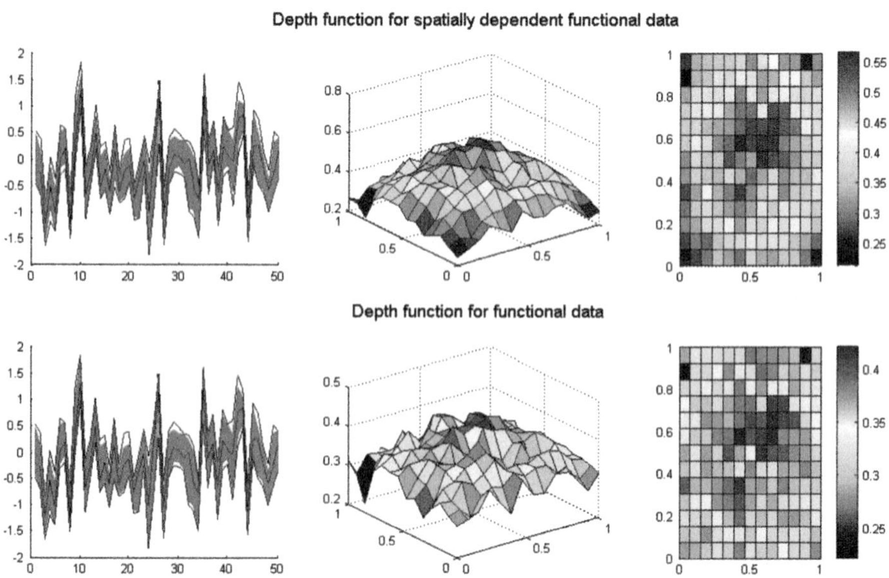

Fig. 1 Main results for the Dataset 1 using a stationary purely spatial covariance function

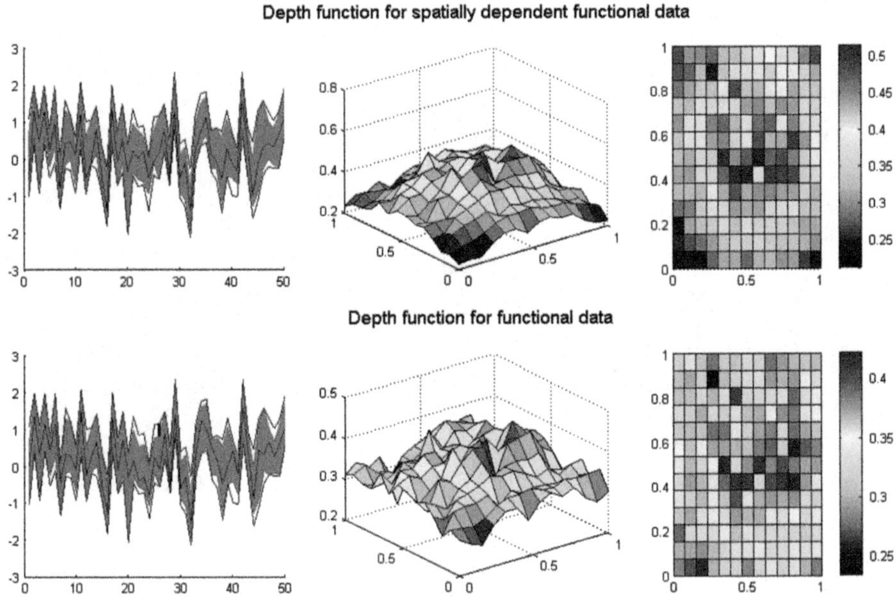

Fig. 2 Main results for the Dataset 2 using a separable spatiotemporal covariance function

method tends to extend the range of the depth values; in fact, in Fig. 1, the ranges are [0–57–005] and [0.42–004], respectively, for the spatially dependent depth for functional data and for the depth for functional data. This is a consequence of the convex shape of the covariance function.

In Fig. 2, we reports the results for the second dataset which has been generated using the separable spatiotemporal covariance function with parameters $c = 1.5$ and $\alpha = 0.4$.

As before, the proposed depth function provides higher values of depth at the center of the spatial region and lower values on the boundary; however, the effect of the weighting scheme produces here a higher impact due to the value of spatial dependence c. The range is wider, as in the previous case.

The third dataset, whose results are available in Fig. 3, uses a symmetric but non-separable spatiotemporal covariance function with parameters $c = 0.4$, $\beta = 0.5$, and $\alpha = 0.4$. It still confirms the effectiveness of the proposal in giving a higher depth to spatially central curves; however, due to the curve shape and, so, to a covariance function which is not convex along the whole domain, the range is not wider as before.

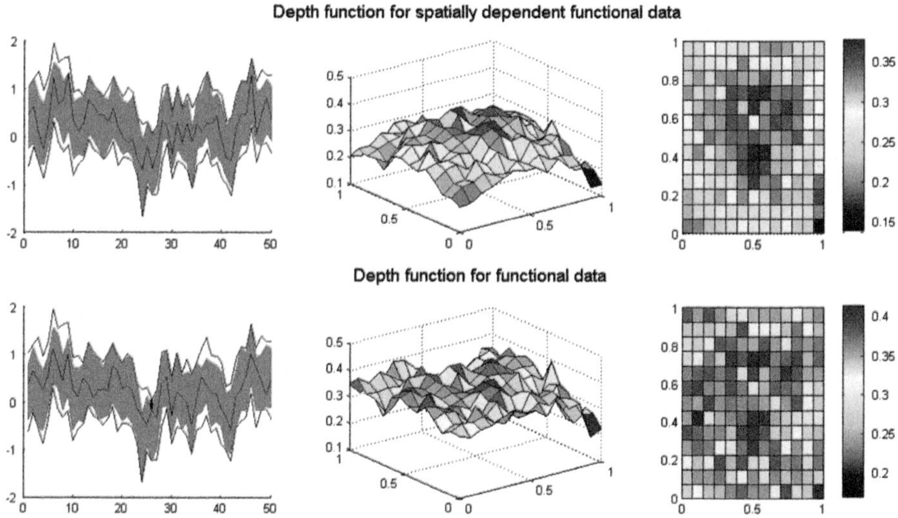

Fig. 3 Main results for the Dataset 3 using a symmetric but non-separable spatiotemporal covariance function

5 Conclusions

In this paper, we have shown how to incorporate the spatial information in the curves ordering and in the definition of a median curve. From the application on simulated data we have highlighted that in the central region of the geographic space the depth is higher than on the boundary. This is an interesting feature from a geostatistical point of view. Future developments will be the introduction of directional variograms in order to deal with the covariance structures which change according to the spatial direction.

References

1. Delicado, P., Giraldo, R., Comas, C. Mateu, J.: Statistics for spatial functional data: some recent contributions. Environmetrics **21**, 224–239 (2010)
2. Fraiman, R., Muniz, G.: Trimmed means for functional data. Test **10**, 419–440 (2001)
3. Lopez-Pintado, S., Romo, J.: On the concept of depth for functional data. J. Am. Stat. Assoc. **104**, 718–734 (2009)
4. Ramsay, J.O., Silverman, B.W.: Functional Data Analysis, 2nd edn. Springer, New York (2005)
5. Sun, Y., Genton, M.G.: Functional boxplots. J. Comput. Graph. Stat. **20**, 316–334 (2011)
6. Tukey, J.: Mathematics and picturing data. In: Proceedings of the 1975 International Congress of Mathematics, vol. 2, pp. 523–531 (1975)

Robust Clustering of EU Banking Data

Jessica Cariboni, Andrea Pagano, Domenico Perrotta, and Francesca Torti

Abstract In this paper we present an application of robust clustering to the European union (EU) banking system. Banks may differ in several aspects, such as size, business activities and geographical location. After the latest financial crisis, it has become of paramount importance for European regulators to identify common features and issues in the EU banking system and address them in all Member States (or at least those of the Euro area) in a harmonized manner. A key issue is to identify using publicly available information those banks more involved in risky activities, in particular trading, which may need to be restructured to improve the stability of the whole EU banking sector. In this paper we show how robust clustering can help in achieving this purpose. In particular we look for a sound method able to clearly cut the two-dimensional space of trading volumes and their shares over total assets into two subsets, one containing *safe* banks and the other the *risky* ones. The dataset, built using banks' balance sheets, includes 245 banks from all EU27 countries, but Estonia, plus a Norwegian bank. With appropriate parameters, the `TCLUST` routine could provide better insight of the data and suggest proper thresholds for regulators.

1 Introduction

The latest financial crisis has driven, and it keeps driving, the scientific community to develop tools able to address some of the issues which have a clear impact on financial stability. In this paper we face the issue of classifying bank's riskiness in terms of trading activities given that *modern* banking business models heavily mix trading and retail businesses (e.g. universal banking model). This relates to the fact

J. Cariboni (✉) • A. Pagano
European Commission, Joint Research Centre, Institute for the Protection and Security of the Citizen, Financial and Economic Analysis Unit, Ispra site, Italy
e-mail: jessica.cariboni@jrc.ec.europa.eu; andrea.pagano@jrc.ec.europa.eu

D. Perrotta • F. Torti
European Commission, Joint Research Centre, Institute for the Protection and Security of the Citizen, Global Security and Crisis Management Unit, Ispra site, Italy
e-mail: domenico.perrotta@ec.europa.eu; francesca.torti@jrc.ec.europa.eu

I. Morlini et al. (eds.), *Advances in Statistical Models for Data Analysis*, Studies in Classification, Data Analysis, and Knowledge Organization, DOI 10.1007/978-3-319-17377-1_3

that lately the European Commission, as well as individual countries such as United States, United Kingdom, France, Germany, have been working on setting up specific regulations to separate certain trading activities from retail ones (see [3]), in order to enhance the stability of the whole banking sectors.

Separation should concern only the *riskiest* banks, which are heavily engaged in the *riskiest* trading activities. This contribution focuses on two specific measures of trading activities which can be derived from banks' balance sheets, specifically the amount of total trading activities (that we will refer to as *TradAct*) and their relative share compared to the total assets (TA) portfolio: $ShareTradAct = TradAct/TA$.

In this context, we have been called to develop an approach to help policy makers in setting suitable thresholds for dividing the (*TradAct*, *ShareTradAct*) two-dimensional space into two separate zones, one including banks which could be considered for possible structural separation and the second with the other banks. In this paper, we propose an approach based on robust clustering methods (see [6–8]) and a recently proposed use of the bayesian information criterion (BIC). The monitoring of the BIC gives a solid ground to the choice of the right number of clusters as well as the most suitable cluster shape. The approach is applied to a dataset of banks' balance sheet extracted from the commercial bank data provider SNL Financial (http://www.snl.com/). The database covers 245 European Union banks for the years 2006–2011, for which consolidated data have been considered.

Different definitions to estimate the amount of trading activities based on balance sheet data have been proposed (see [4]). These definitions try to distinguish different types of trading, e.g. proprietary trading versus market making. The proposed robust clustering approach is applicable to all definitions.

2 Robust Approach for Clustering SNL Data

A clustering problem like ours is traditionally addressed in the model-based clustering framework, i.e. with a finite mixture of distributions, where each mixture component corresponds to a group in the data. A common reference model for the mixture components is the multivariate Gaussian distribution, estimated using the EM algorithm in the popular MCLUST [5]. Then, each observation is assigned to the group to which it is most likely to belong. The determination of the right number of groups is still today an outstanding unsolved problem, usually approached with the BIC or AIC criteria.

For our data such models are insufficient, because they do not account for the presence of outliers, which may occur as noise-like structures or as a small tight group of observations in specific areas of the space. In both cases, the presence of outliers can considerably bias the estimation of the centroids and shape (covariance structure) of the groups and seriously affect the final clustering. For this reason, we opted for a robust counterpart of the normal mixture modeling known in the literature as *Robust Trimmed Clustering* or TCLUST [6, 7]. The robustness capacity of TCLUST comes from the trimming approach, i.e. the possibility to leave a proportion α of observations, hopefully the most outlying ones, unassigned.

The TCLUST approach is defined through the search of k centers $m_1, \dots, m_k$ and k shape matrices $U_1, \dots, U_k$ solving the double minimization problem:

$$\arg\min_{\mathbf{Y}} \min_{\substack{m_1, \dots, m_k \\ U_1, \dots, U_k}} \sum_{j=1,\dots,k} \sum_{x_i \in \mathbf{Y}} (x_i - m_j)' U_j^{-1} (x_i - m_j) \quad (1)$$

where $\mathbf{Y} \subset \{x_1, \dots, x_n\} : |Y| = [n(1-\alpha)]$, i.e. $\mathbf{Y}$ ranges on the class of subsets of size $[n(1-\alpha)]$ within the sample $\{x_1, \dots, x_n\}$ (being $[\cdot]$ the integer part operator).

The shape matrices U_j are covariance matrices of the different groups, which can handle elliptically contoured clusters and that are properly constrained to restrict the relative variability among the groups and avoid spurious solutions. For example, to control the relative group sizes and also the deviation from spherical structures, it is sufficient to constrain the ratio between the maximum and minimum eigenvalues to be smaller or equal than a fixed constant. On the other hand, a constraint on the ratio between the maximum and minimum covariance determinants limits the relative volumes of the ellipsoids.

An important parameter of any clustering approach is the number of groups k, which is crucial in our problem. Note that in TCLUST there is also an implicit extra group to consider, which is the group of the $n \cdot \alpha$ trimmed observations. In other words, the choices of k and α in TCLUST are related and should be addressed simultaneously. With our data we have tested different approaches.

1. The traditional approach used in TCLUST is the so-called classification trimmed likelihood curves plot [8]. The plot monitors the classification trimmed likelihoods for different applications of TCLUST with varying values of k and α. The idea is to identify, by visual inspection of the plot, combinations of k and α that determine an increase of the likelihoods. Unfortunately, in our case, the curves obtained were difficult to interpret, even after restricting the search to a single trimming level fixed at $\alpha = 0.04$, chosen on the basis of prior information on the problem.
2. The second approach that we tried is based on the Forward Search of [2]. Originally introduced for detecting masked outliers, the Forward Search is not yet applicable as a fully automatic clustering tool. However, it can be used to infer k by repeating searches from many different randomly chosen subsets. The repeated process should reveal the presence of multiple populations as separated peaks in plots monitoring the trajectory of the values of the minimum Mahalanobis distance of observations from the data centroids [1]. With our data, the Forward Search random start plot clearly highlights the presence of nonhomogeneous data, but distinct peaks corresponding to different groups were not clearly identifiable.
3. A third natural approach is based on the idea of monitoring the AIC or BIC criteria for different k values. The same approach can be used to decide on α and

the restriction factor. In the present work we have fixed $\alpha = 3.5\,\%$ and we have monitored BIC values as a function of k and the restriction factor *restrfact*.

The AIC and BIC are, respectively, given by $2\log L + 2m$ and $2\log L + m\log(n)$, where $\log L$ is the negative of the maximized log-likelihood, n is the number of observations and m is the number of estimated parameters. If we neglect the constraints on the covariance matrices, it can be easily seen that the TCLUST parameters are $m = k \cdot p + k \cdot p \cdot (p+1)/2$, where p is the number of variables. Recently, Gordaliza et al. [9] have derived the effective number of parameters to be used for penalizing the likelihoods $\log L$ when restrictions on the covariance matrices are adopted in the TCLUST model. In this work we used successfully this newly introduced approach.

To run the method on our data we used a MATLAB implementation developed in the framework of the FSDA project[1] [10]. The original implementation by the TCLUST authors,[2] available in R, has been also applied with almost identical results.

3 Interpretation and Use of the Clusters

In the following, we present results obtained using the `tclust` routine. In particular, we use `tclust` to

1. Detect the existence of multiple groups in the data
2. Select the optimal number of clusters as well as the most suitable value for the restriction factor via BIC monitoring
3. Identify precisely the clusters

Since the variables are defined on different scales (one is expressed in billion euros while the others are percentages) to improve the stability of the statistical analysis we have standardized the data to have mean 0 and standard deviation 1 (z-scores). By looking at the scatter plot of standardized variables (*TradAct*, *ShareTradAct*) (Fig. 1), one can clearly see the existence of two well-separated groups. From the scatter plots it is quite evident that the main problem is identifying clusters in the left-bottom corner whose original values for the *ShareTradAct* are very small. Clearly the data are far from the model assumptions and it is natural to consider some data transformations, e.g. a log-transformation. The appropriateness of this choice is confirmed by the Forward Search random start plot in Fig. 2. In fact, the plot shows that, independently from the starting subset that initializes the search, the trajectories of the minimum Mahalanobis distance

[1]The toolbox can be downloaded at these web addresses: http://www.riani.it and https://fsda.jrc.ec.europa.eu.

[2]http://cran.r-project.org/web/packages/tclust/index.html, CRAN R-package for TCLUST.

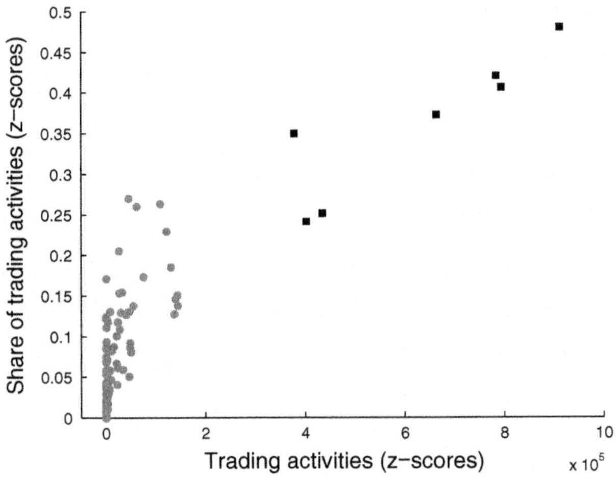

Fig. 1 SNL data: few very large banks are easily identifiable

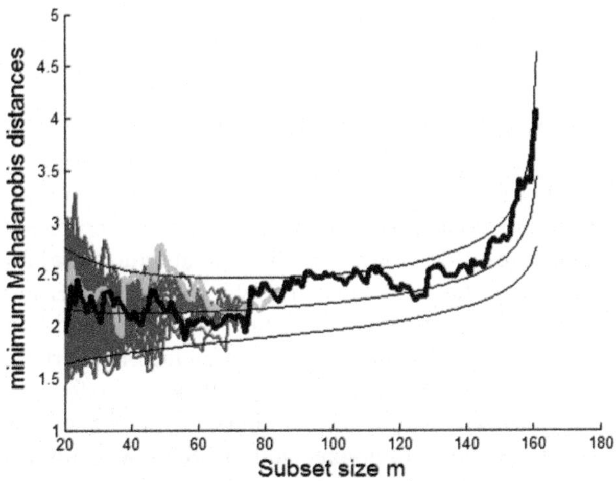

Fig. 2 Forward Search random start plot on log-transformed data

are within the confidence bands derived for such statistic under null hypothesis. However, the systematic tendency in the central part of the curve, between steps 80 and 120, indicates the presence of some structure in the data, such as two overlapping groups.

The monitoring of the BIC criterion, in the form adapted to TCLUST by [9], to the log-transformed data confirms this point. The right panel of Fig. 3 shows the monitoring of the BIC value (y-axis) for different number of groups (x-axis) and various values of the TCLUST restriction factor. For a number of restriction factors

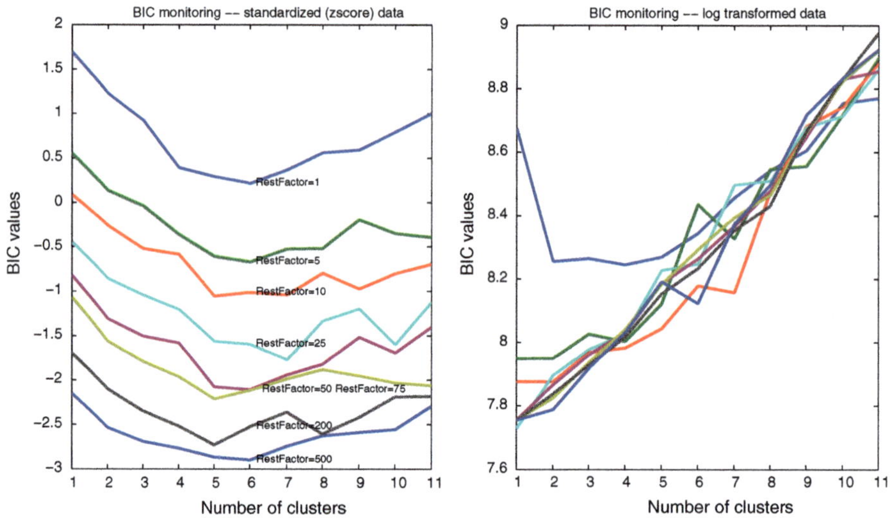

Fig. 3 BIC scores for original data (*left panel*) and log-transformed data (*right panel*). The scores have been monitored for different restriction factor (RF) values

we observe a step when the number of clusters is between 1 and 2. Then, the curves all increase. This is indication of two groups in the log-transformed data.

On the original data (Fig. 3, left panel), the BIC monitoring suggests a number of clusters between 5 and 7. In addition, the lower level curves corresponding to larger restriction factors suggest that elongated elliptical structures are more appropriate than those closer to the spherical *k*-means, which corresponds to the restriction factor 1 at the top of the plot. This is not the case of the log-transformed data, for which the BIC curves for the different restriction factors overlap considerably.

As expected at this point, TCLUST applied to the log-transformed data produces two well-defined clusters and only few outliers, as shown in the left panel of Fig. 4. The `tclust` function as implemented in FSDA was run using a trimming percentage of 3.5 % and a default restriction factor equal to 50.

From the right panel of Fig. 4, one can see the seven clusters and the clear outlying banks, found in the original data by `tcluts`. This clustering, from an operational point of view, is a suitable categorization of the banks with respect to the shares of trading activities.

The trimming level, set to 3.5 %, has the effect to separate the largest banks from the rest of the population. This value was triggered by the problem, to address a limited number of European banks that are so large compared to the others to be subject to supervision independently by their activities, for the potential impact they have on the whole system. The obvious alternative to the trimming parameter is to remove a priori these outlying banks from the analysis. We prefer the trimming parameter because it can flexibly address unforeseen changes in the banking sector and, therefore, avoid its periodical manual revision.

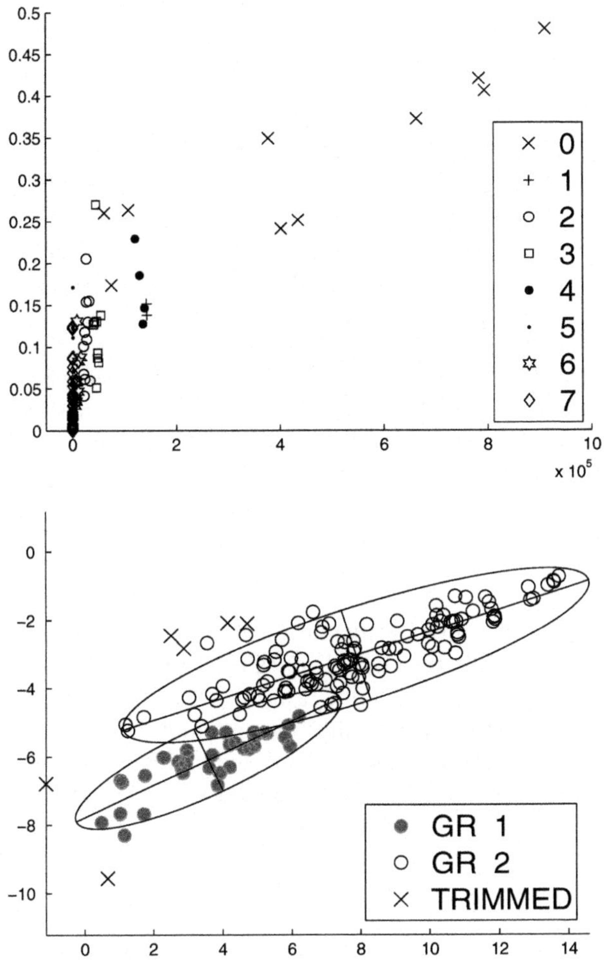

Fig. 4 `tclust` clusters in original data (*left panel*) and log-transformed data (*right panel*)

The choice of the restriction factor value is less crucial in our case. Considering the clusters as ellipsoids, the parameter imposes an upper bound on the ratio between the major and the minor diameters. By setting it to 50, we allow lot of flexibility compared to the k-means spherical clusters (that can be obtained with a restriction factor equal to 1), but, at the same time, we avoid to detect too elongated spurious clusters.

4 Conclusions

The aim of the work is to find an efficient and robust procedure to separate *safe* banks form *risky* ones with respect to the volumes and the shares of their trading activities. Because of the need to only use publicly available data, as in banks' balance sheet, and because of the heavily mixed business model banks usually use, it is not easy to set thresholds which can clearly divide the two-dimensional space, defined by trading volumes and their shares over total assets. The heterogeneity of the banking network makes the classical clustering algorithms, such as *k*-means, unusable.

Instead, we have successfully applied TCLUST and other robust tools available through the FSDA Matlab toolbox and the TCLUST R-package in CRAN. With these tools we have been able to give a clear indication on where to draw the separating line thresholds. More specifically, we have used a semi-automatic tuning of the parameters in the `tcluts` routine via a BIC monitoring recently adapted to TCLUST.[3] In this way we were able to properly determine the right number of clusters and the appropriate value for the restriction factor which gives raise to a clusters' classification in line with regulators desiderata.

The BIC monitoring, which uses the classification likelihood based on parameters estimated using the mixture likelihood, is the key ingredient to choose the number of the groups. The peculiarity of TCLUST is that the effective number of estimated parameters depends on the restriction factor, which reduces the space of the model parameter values. The implementation of the exact derivation of this dependency, documented in [9] and experimented in this work, will be available in the `tclust.m` function of the FSDA MATLAB toolbox (footnote 1).

References

1. Atkinson, A.C., Riani M.: Exploratory tools for clustering multivariate data. Comput. Stat. Data Anal. **52**, 272–285 (2007)
2. Atkinson, A.C., Riani, M. Cerioli, A.: Exploring Multivariate Data with the Forward Search. Springer, New York (2004)
3. Blundell-Wignall, A., Atkinson, P., Roulet, C.: Bank business models and the separation issue. OECD J. Financ. Mark. Trends **2013**(2), 69–91 (2013)
4. European Commission Directorate General Internal Market and Services: Reforming the structure of the EU banking sector. Consultation paper, Annex 2 (2013)
5. Fraley, C., Raftery, A.E.: Model-based clustering, discriminant analysis, and density estimation. J. Am. Stat. Assoc. **97**, 611–631 (2002)
6. Garcia-Escudero, L.A., Gordaliza, A., Matran, C., Mayo-Iscar, A.: A general trimming approach to robust cluster analysis. Ann. Stat. **36**, 1324–1345 (2008)

[3]The authors thank Marco Riani and Agustìn Mayo Iscar for the fruitful discussions on the approach.

7. Garcia-Escudero, L.A., Gordaliza, A., Matran, C., Mayo-Iscar A.: A review of robust clustering methods. In: Advanced in Data Analysis and Classification, pp. 89–109. Springer, New York (2010)
8. Garcia-Escudero, L.A., Gordaliza, A., Matran, C., Mayo-Iscar, A.: Exploring the number of groups in robust model based clustering. Stat. Comput. **21**(4), 585–599 (2011)
9. Garcia-Escudero, L.A., Gordaliza, A., Mayo-Iscar A.: A constrained robust proposal for mixture modeling avoiding spurious solutions. Adv. Data Anal. Classif. **8**(1), 27–43 (2014)
10. Riani, M., Perrotta, D., Torti, F. FSDA: a MATLAB toolbox for robust analysis and interactive data exploration. Chemometr. Intell. Lab. Syst. **116**, 17–32 (2012)

Sovereign Risk and Contagion Effects in the Eurozone: A Bayesian Stochastic Correlation Model

Roberto Casarin, Marco Tronzano, and Domenico Sartore

Abstract This research proposes a Bayesian multivariate stochastic volatility (MSV) model to analyze the dynamics of sovereign risk in eurozone CDS markets during the recent financial crisis. We follow an MCMC approach to parameters and latent variable estimation and provide evidence of significant volatility shifts in asset returns, strong simultaneous increases in cross-market correlations, as well as sharp declines in correlations patterns. Overall, these findings are highly consistent with various empirical characterizations of contagion put forward in the literature, allowing us to conclude that the recent financial crisis generated severe contagion effects in sovereign debt markets of eurozone countries.

Keywords Bayesian methods • Contagion • Credit default swap • Multivariate stochastic volatility

1 Introduction

Modelling and forecasting contagion between financial markets are crucial issues for systemic risk analysis and the development of macro-prudential policies for crisis prevention. The aim of this paper is to investigate the contagion effects in the eurozone during the latest financial turmoil focusing on sovereign credit default swap (CDS) spread.

The CDS is a financial instrument ensuring protection against credit risk. The CDS spread represents the annual cost of this financial instrument, expressed in basis points with respect to the nominal value of the underlying corporate or sovereign bond. As widely recognized, CDS spreads accurately reflect market evaluation about credit risk, thus de facto replacing analogous evaluations obtained

---

R. Casarin (✉) • D. Sartore
University Ca' Foscari of Venice, Venezia, Italy
e-mail: r.casarin@unive.it; sartore@unive.it

M. Tronzano
University of Genova, 16146 Genova, Italy
e-mail: m.tronzano@mclink.it

I. Morlini et al. (eds.), *Advances in Statistical Models for Data Analysis*,
Studies in Classification, Data Analysis, and Knowledge Organization,
DOI 10.1007/978-3-319-17377-1_4

through rating criteria. Applied research on CDS spreads plays therefore a major role when evaluating the intensity of an international financial crisis, particularly as regards the detection of contagion across markets in different countries. The 5-year maturity for CDS spreads, employed in the present analysis, has been selected because it closely reflects the typical maturity of a sovereign bond. However, since CDS are negotiated in OTC markets, data on CDS spreads may also refer to alternative temporal horizons.

In order to capture both shifts in volatility and correlation we follow [7] and propose a Bayesian multivariate stochastic volatility (MSV) model (e.g., see [1, 6]) with Markov-switching stochastic volatility and correlation. The correlation and volatility regimes should identify contagion across markets following the standard definition of contagion present in the literature (see [5]).

The structure of the paper is as follows. Section 2 introduces the stochastic correlation model. Section 3 describes briefly the Bayesian inference approach used. Section 4 presents the results for eurozone sovereign risk. Section 5 concludes.

2 A Markov-Switching Stochastic Correlation Model

Let $\mathbf{y}_t = (y_{1t}, \ldots, y_{mt})' \in \mathbb{R}^m$ be a vector-valued time series, representing the log-differences in the CDS rate, $\mathbf{h}_t = (h_{1t}, \ldots, h_{mt})' \in \mathbb{R}^m$ the log-volatility process, $\Sigma_t \in \mathbb{R}^m \times \mathbb{R}^m$ the time-varying covariance matrix, and $s_t \in \{0, 1\}$ a two-state Markov chain. We consider the following Markov-switching stochastic correlation model (*MSSC*):

$$\mathbf{y}_t = \Sigma_t^{1/2} \boldsymbol{\varepsilon}_t, \qquad \boldsymbol{\varepsilon}_t \sim \mathcal{N}_m(\mathbf{0}, I_m), \tag{1}$$

$$\mathbf{h}_t = \mathbf{b}_{s_t} + B_{s_t} \mathbf{h}_{t-1} + \boldsymbol{\eta}_t, \qquad \boldsymbol{\eta}_t \sim \mathcal{N}_m(\mathbf{0}, \Sigma_\eta), \tag{2}$$

with $\boldsymbol{\varepsilon}_t \perp \boldsymbol{\eta}_s \ \forall\, s, t$, and $\mathcal{N}_m(\boldsymbol{\mu}, \Sigma)$ the m-variate normal distribution, with mean $\boldsymbol{\mu}$ and covariance matrix Σ, and $\mathbf{b}_i$ and B_i, $i = 0, 1$, parameters to be estimated. The probability law governing s_t is $\mathbb{P}(s_t = j | s_{t-1} = i) = p_{ij}$ with ($p_{ij} \geq 0$ and $\sum_{j=0,1} p_{ij} = 1$, $i = 0, 1$. As regards the conditional covariance matrix Σ_t, we assume the decomposition (see [1, 2]):

$$\Sigma_t = \Lambda_t \Omega_t \Lambda_t, \tag{3}$$

with $\Lambda_t = \text{diag}\{\exp\{h_{1t}/2\}, \ldots, \exp\{h_{mt}/2\}\}$ a diagonal matrix with the log-volatilities on the main diagonal and $\Omega_t = \tilde{Q}_t^{-1} Q_t \tilde{Q}_t^{-1}$ the stochastic correlation matrix with $\tilde{Q}_t = (\text{diag}\{\text{vecd}\, Q_t\})^{1/2}$ and $Q_t^{-1} \sim \mathcal{W}_m(\nu, S_{t-1})$ where

$$S_{t-1} = \frac{1}{\nu} Q_{t-1}^{-d/2} \bar{Q}_t Q_{t-1}^{-d/2}, \quad \bar{Q}_t = \sum_{k=0,1} \mathbb{I}_{\{k\}}(s_{1,t}) \bar{D}_k,$$

and $\bar{D}_k$, $k \in \{0, 1\}$, is a sequence of positive definite matrices, which captures the long-term dependence structure between series in the different regimes and d is a scalar parameter.

3 Bayesian Inference

Define $\mathbf{y} = (\mathbf{y}_1', \ldots, \mathbf{y}_T')'$, $\mathbf{z} = (\mathbf{h}, \mathbf{q}, \mathbf{s})$, with $\mathbf{h} = (\mathbf{h}_0', \ldots, \mathbf{h}_T')'$, $\mathbf{s} = (s_0', \ldots, s_T')'$, and $\mathbf{q} = (\text{vech}(Q_0)', \ldots, \text{vech}(Q_T)')'$. The complete-data likelihood function of the MSSC model is

$$\mathcal{L}(\mathbf{y}, \mathbf{z}|\boldsymbol{\theta}) \propto \prod_{t=1}^{T} \frac{1}{\Sigma_t|^{1/2}} e^{-\frac{1}{2}\mathbf{y}_t'\Sigma_t^{-1}\mathbf{y}_t} \frac{1}{|\Sigma_\eta|^{1/2}} e^{-\frac{1}{2}\eta_t'\Sigma_\eta^{-1}\eta_t}$$
$$\cdot 2^{-\frac{m\nu}{2}} \Gamma_m(\nu/2)^{-1} |S_{t-1}|^{-\frac{\nu}{2}} e^{-\text{tr}\left(\frac{1}{2}S_{t-1}^{-1}Q_t^{-1}\right)} |Q_t^{-1}|^{\frac{\nu-m-1}{2}}$$
$$\cdot p_{00}^{(1-s_t)(1-s_{t-1})} (1 - p_{00})^{s_t(1-s_{t-1})} p_{01}^{s_t s_{t-1}} (1 - p_{01})^{(1-s_t)s_{t-1}},$$

where $\Gamma_m(\nu/2)$ is the m-variate gamma function and $\boldsymbol{\theta} = (\boldsymbol{\theta}_1', \boldsymbol{\theta}_2', \boldsymbol{\theta}_3')'$, with $\boldsymbol{\theta}$ in three sub-vectors: $\boldsymbol{\theta}_1 = (\boldsymbol{\phi}', \text{vech}(\Sigma_\eta)')'$, with $\boldsymbol{\phi} = \text{vec}(\Phi)$, where $\Phi = (\phi_1, \ldots, \phi_m)$ has in the columns the vectors $\phi_j = (b_{0,j}, b_{1,j}, (B_{0,j1}, \ldots, B_{0,jm}), (B_{1,j1}, \ldots, B_{1,jm}))'$, $j = 1, \ldots, m$; $\boldsymbol{\theta}_2 = (\nu, d, \text{vech}(\bar{D}_0), \text{vech}(\bar{D}_1))'$; $\boldsymbol{\theta}_3 = (p_{00}, p_{11})'$. We specify the following prior distributions:

$$\boldsymbol{\phi}|\Sigma_\eta \sim \mathcal{N}_{m(2m+2)}\left(\mathbf{0}, \Sigma_\eta \otimes 10I_{2m+2}\right), \ \Sigma_\eta^{-1} \sim \mathcal{W}_m\left(10, 4I_m\right),$$
$$d \sim \mathcal{U}_{(-1,1)}, \ \nu \sim \frac{1}{\Gamma(10)}(\nu - m)^{10-1} \exp\{-(\nu - m)\} \mathbb{I}_{(m,+\infty)}(\nu),$$
$$\bar{D}_i^{-1} \sim \mathcal{W}_m(10, 0.1I_m), \ p_{ii} \sim \mathcal{U}_{(0,1)}, \ i = 0, 1.$$

The posterior approximation is obtained by Gibbs sampling. Following [1], the sampler iterates over different blocks of parameters and latent variables. The iteration j, $j = 1, \ldots, J$, of the Gibbs sampler constitutes of the following steps.

Sample $\boldsymbol{\theta}_1^{(j)} \sim f(\boldsymbol{\theta}_1|\boldsymbol{\theta}_2^{(j-1)}, \boldsymbol{\theta}_3^{(j-1)}, \mathbf{y}, \mathbf{z}^{(j-1)})$, by drawing iteratively from the following Gaussian and Wishart distributions:

$$(\boldsymbol{\phi}|\Sigma_\eta, \boldsymbol{\theta}_2, \boldsymbol{\theta}_3, \mathbf{y}, \mathbf{z}) \sim \mathcal{N}_{m(2m+2)}\left(\bar{\boldsymbol{\mu}}_1, \Sigma_\eta \otimes \bar{\Upsilon}_1^{-1}\right) \tag{4}$$
$$(\Sigma_\eta^{-1}|\boldsymbol{\theta}_2, \boldsymbol{\theta}_3, \mathbf{y}, \mathbf{z}) \sim \mathcal{W}_m\left(\bar{\mu}_2, \bar{\Upsilon}_2\right), \tag{5}$$

with $\bar{\boldsymbol{\mu}}_1 = \hat{\boldsymbol{\phi}}$, $\bar{\Upsilon}_1 = (10I_{2m+2} + Z'Z)$, $\bar{\Upsilon}_2 = (1/4I_m + (H - Z\hat{\Phi})'(H - Z\hat{\Phi}))^{-1}$, and $\bar{\mu}_2 = 10 + T - (2m + 2)$, where $\hat{\Phi} = (Z'Z)^{-1}Z'H$ and $\hat{\boldsymbol{\phi}} = \text{vec}(\hat{\Phi})' = (I_m \otimes (Z'Z)^{-1}Z')$.

Sample $\boldsymbol{\theta}_2^{(j)} \sim f(\boldsymbol{\theta}_2|\boldsymbol{\theta}_1^{(j)}, \boldsymbol{\theta}_3^{(j-1)}, \mathbf{y}, \mathbf{z}^{(j-1)})$, by sampling iteratively from the following full conditional distributions. The full conditional of ν is

$$f(\nu|, d, \bar{D}_0, \bar{D}_1, \boldsymbol{\theta}_1, \boldsymbol{\theta}_3, \mathbf{y}, \mathbf{z})$$
$$\propto \exp\left\{\ln(\nu - m)^{3-1} + \frac{\nu T m}{2}\ln(\nu) - T\ln\Gamma_m(\nu/2) - \frac{\nu}{2}U_T\right\}\mathbb{I}_{(m,+\infty)}(\nu), \quad (6)$$

where $U_T = \sum_{t=1}^{T}\left[d\ln|Q_{t-1}^{-1}| - \ln|Q_t^{-1}| + \mathrm{tr}\left(Q_{t-1}^{-\frac{d}{2}}\bar{Q}_t^{-1}Q_{t-1}^{-\frac{d}{2}}Q_t^{-1}\right) + \ln(|\bar{Q}_t|)\right]$ $+mT\ln(2) + 2\cdot 10$. The full conditional of d is

$$f(d|\nu, \bar{D}_0, \bar{D}_1, \boldsymbol{\theta}_1, \boldsymbol{\theta}_3, \mathbf{y}, \mathbf{z}) \propto \exp\left\{-d\left(\frac{\nu}{2}\sum_{t=1}^{T}\ln|Q_{t-1}^{-1}|\right)\right.$$
$$\left.-\frac{1}{2}\mathrm{tr}\left(\sum_{t=1}^{T}Q_{t-1}^{\frac{d}{2}}\bar{Q}_t^{-1}Q_{t-1}^{\frac{d}{2}}Q_t^{-1}\right)\right\}\mathbb{I}_{(-1,1)}(d). \quad (7)$$

The full conditional distributions of the long-run components $\bar{D}_i$, $i = 0, 1$ of the correlation matrix are

$$f(\bar{D}_i^{-1}|\nu, d, \boldsymbol{\theta}_1, \boldsymbol{\theta}_3, \mathbf{y}, \mathbf{z}) \quad (8)$$
$$\propto |\bar{D}_i|^{-\frac{\bar{\mu}_{3+i}-m-1}{2}}\exp\left\{-\frac{1}{2}\mathrm{tr}\left[\bar{\Upsilon}_{3+i}^{-1}\bar{D}_i^{-1}\right]\right\}g(\bar{D}_i^{-1}), \quad i = 0, 1,$$

with $\bar{\mu}_{3+i} = 10 + \nu n_i$, $\bar{\Upsilon}_{3+i}^{-1} = 10 I_m + \nu\sum_{t=1}^{T}Q_{t-1}^{d/2}Q_t^{-1}Q_{t-1}^{d/2}\mathbb{I}_{\{i\}}(s_t)$, $n_i = \sum_{t=1}^{T}\mathbb{I}_{\{i\}}(s_t)$. Sampling from the four full conditional distributions given above is obtained by Metropolis-Hastings. The proposals are similar to those in [1].

Sample $\boldsymbol{\theta}_3^{(j)} \sim f(\boldsymbol{\theta}_3|\boldsymbol{\theta}_1^{(j)}, \boldsymbol{\theta}_2^{(j)}, \mathbf{y}, \mathbf{z}^{(j-1)})$, by simulating iteratively from the full conditional distributions

$$f(p_{ii}|\boldsymbol{\theta}_1, \boldsymbol{\theta}_2, \boldsymbol{\theta}_3, \mathbf{y}, \mathbf{z}) \propto g(s_0)p_{ii}^{n_{ii}}(1 - p_{ii})^{n_{i1-i}}, \quad (9)$$

$i = 0, 1$, where $g(s_0) = p_{00}^{s_0}p_{10}^{1-s_0}/(p_{00} + p_{10})$, $n_{ij} = \sum_{t=1}^{T}\mathbb{I}_{\{j\}}(s_t)\mathbb{I}_{\{i\}}(s_{t-1})$ $i, i \in \{0, 1\}$. In line with approach used in the previous step of the Gibbs sampler, we apply a M.-H. sampler.

Sample $\mathbf{h}^{(j)} \sim f(\mathbf{h}|\boldsymbol{\theta}^{(j)}, \mathbf{y}, \mathbf{q}^{(j-1)}, \mathbf{s}^{(j-1)})$. Due to the Markov property of the process for $\{\mathbf{h}_t\}_t$, the full conditional distribution of $\mathbf{h}_t$ is

$$f(\mathbf{h}_t|\boldsymbol{\theta}, \mathbf{y}, \mathbf{h}_{t+1}, \mathbf{h}_{t-1}, \mathbf{q}, \mathbf{s}) \propto$$
$$\propto \exp\left\{-\frac{1}{2}\mathrm{tr}\left[(\mathbf{h}_t - \boldsymbol{\mu}_{ht})'\Upsilon_{ht}^{-1}(\mathbf{h}_t - \boldsymbol{\mu}_{ht})\right]\right\}g(\mathbf{h}_t), \quad (10)$$

where $g(\mathbf{h}_t) = \exp\left\{-\frac{1}{2}\mathrm{tr}\left[\boldsymbol{\varepsilon}_t'\Sigma_t^{-1}\boldsymbol{\varepsilon}_t\right]\right\}$, $\mathbf{1} = (1,\ldots,1)'$, $\boldsymbol{\mu}_{ht} = \Upsilon_{ht}(\Sigma_\eta^{-1}(\mathbf{b}_{s_t} + B_{s_t}\mathbf{h}_{t-1}) + B_{s_{t+1}}'\Sigma_\eta^{-1}(\mathbf{h}_{t+1} - \mathbf{b}_{s_{t+1}}) - \frac{1}{2}\mathbf{1})$, and $\Upsilon_{ht} = (\Sigma_\eta^{-1} + B_{s_{t+1}}'\Sigma_\eta^{-1}B_{s_{t+1}})^{-1}$. We apply a M.-H. with proposal distribution $\mathbf{h}_t^{(*)} \sim \mathcal{N}_m(\boldsymbol{\mu}_{ht}, \Upsilon_{ht})$. We proceed in a similar way for the full conditional distribution of h_T. Sampling $\mathbf{q}^{(j)} \sim f(\mathbf{q}|\boldsymbol{\theta}^{(j)}, \mathbf{y}, \mathbf{h}^{(j)}, \mathbf{s}^{(j-1)})$ is obtained by single-move Gibbs sampler with full conditionals

$$
\begin{aligned}
&f(Q_t^{-1}|\boldsymbol{\theta}, \mathbf{y}, \mathbf{h}, Q_{t+1}, Q_{t-1}, \mathbf{s}) \\
&\propto |Q_t^{-1}|^{\frac{\nu+1-m-1}{2}} \exp\left\{-\frac{1}{2}\mathrm{tr}\left[S_{t-1}^{-1}Q_t^{-1}\right]\right\} g(Q_t^{-1}). \qquad (11)
\end{aligned}
$$

$t = 1, \ldots, T-1$, with

$$
g(Q_t^{-1}) = |Q_t^{-1}|^{-\frac{\nu d}{2}} |\tilde{Q}_t| \exp\left\{-\frac{1}{2}\mathrm{tr}(S_t^{-1}Q_{t+1}^{-1} + (\tilde{Q}_t\Lambda_t^{-1}\boldsymbol{\varepsilon}_t\boldsymbol{\varepsilon}_t'\Lambda_t^{-1}\tilde{Q}_t)Q_t^{-1})\right\}.
$$

We apply a M.-H. algorithm with proposal distribution $Q_t^{-1*} \sim \mathcal{W}_m(\mu_{Qt}, \Upsilon_{Qt})$, $\mu_{Qt} = \nu + 1$ and $\Upsilon_{Qt}^{-1} = S_{t-1}^{-1} + \tilde{Q}\Lambda_t^{-1}\boldsymbol{\varepsilon}_t\boldsymbol{\varepsilon}_t'\Lambda_t^{-1}\tilde{Q}$. We proceed in a similar way for the full conditional of Q_T^{-1}.

Sampling $\mathbf{s}^{(j)} \sim f(\mathbf{s}|\ \boldsymbol{\theta}^{(j)}, \mathbf{y}, \mathbf{h}^{(j)}, \mathbf{q}^{(j)})$ is obtained by generating sequentially from the full conditional distributions of s_t, $t = 1, \ldots, T$,

$$
\begin{aligned}
f(s_t|\boldsymbol{\theta}, \mathbf{y}, \mathbf{h}, \mathbf{q}, s_{t-1}, s_{t+1}) &\propto f(\mathbf{h}_t|\mathbf{h}_{t-1}, s_t, \boldsymbol{\theta}) f(Q_t|Q_{t-1}, s_t, \boldsymbol{\theta}) \\
&f(s_t|s_{t-1}, \boldsymbol{\theta}) f(s_{t+1}|s_t, \boldsymbol{\theta}). \qquad (12)
\end{aligned}
$$

We apply a global Metropolis-Hastings step with the transition as proposal distribution.

4 Contagion Effects in the Eurozone

We consider the closing spread for the 5 years CDS on the sovereign debt of the developed countries in the eurozone. The series are sampled at a daily frequency for the period 08 August 2008 to 20 June 2012 and for the following countries: United Kingdom (UK), Cyprus, France, Germany, Greece, Ireland, Italy, Portugal, Spain (source for Greece is Datastream, for the other countries Bloomberg). Thus the dataset has 916 observations and nine series.

We study the contagion between three areas. The first one is UK, the second one is France and Germany (denoted with EU1), and the third one is Cyprus, Greece, Ireland, Italy, Portugal, and Spain (denoted with EU2). In order to obtain aggregated CDS indexes we consider an equally weighted average of the country CDS indexes.

Unit-root test on the series leads us to conclude in favor the non-stationarity. Thus we consider first differences of the series, which are stationary.

One of the most common empirical regularity usually associated with the existence of contagion is represented by volatility spillovers, namely simultaneous increases in the volatility of asset returns across markets (see [3] for a comprehensive discussion of alternative definitions of contagion in the literature). A second well-known empirical characterization of contagion is proposed in [5], where contagion is defined as a significant increase in cross-market correlation between asset returns in the aftermath of an (exogenous) crisis event. More recently, Corsetti et al. [3, 4] have criticized the approach taken in [5] observing that, for given factor loadings, correlation between asset returns will rise only to the extent that the variance of the common factor is relatively large with respect to that of idiosyncratic asset noise. In this perspective, Corsetti et al. [4] argue that correlation between asset returns will not necessarily rise, if contagion exists, but might also significantly decrease.

We now interpret the empirical evidence obtained with our Bayesian MSSC model on the CDS dataset. Focusing on the most common definition of contagion (volatility spillovers) our results document various shifts from low to high volatility regimes, evenly distributed along the sample (see stepwise lines in Fig. 1). Since these volatility shifts are estimated through a MSV model, this evidence is clearly consistent with the existence of various contagion episodes in CDS markets for different eurozone areas.

Turning to the characterization of contagion proposed in [5] (significant increase in cross-market correlations), this empirical regularity is highly supported by our estimates. As shown in Fig. 1, we document frequent episodes of increases in cross-market correlations for all pairs of the eurozone areas considered. Most interestingly, focusing on Greek's sovereign debt crisis, we find strong evidence of contagion in the period from May 2010 to October 2011 (as witnessed by frequent and persistent increases in cross-area correlations and the shifts to the high-volatility regime, see gray vertical bars in Fig. 1). Actually, as regards this specific period, it is possible to establish close connections between various contagion episodes and the following important periods described in the ECB crisis timeline:

- From 23 April to 10 May 2010, the first financial support to Greece takes place. More specifically, Greece seeks financial support from Euro area countries and IMF (23 April) and S&P downgrades Greek rating to BB+ (27 April). The 2 May the first loan package for Greece is agreed and ECB assesses Greek government's adjustment programme. ECB supports Greece's effort for fiscal consolidation independently of the external rating on the sovereign debt (3 May). Finally from 9 May to 10 May EU aims to boost financial stability and the ECB introduces the security market programme and the 7 June the European Financial Stability Facility is established.
- From 28 November to 7 December 2010, the EU-IMF package for Ireland is agreed. More specifically the 28 November ECB assesses Ireland's economic

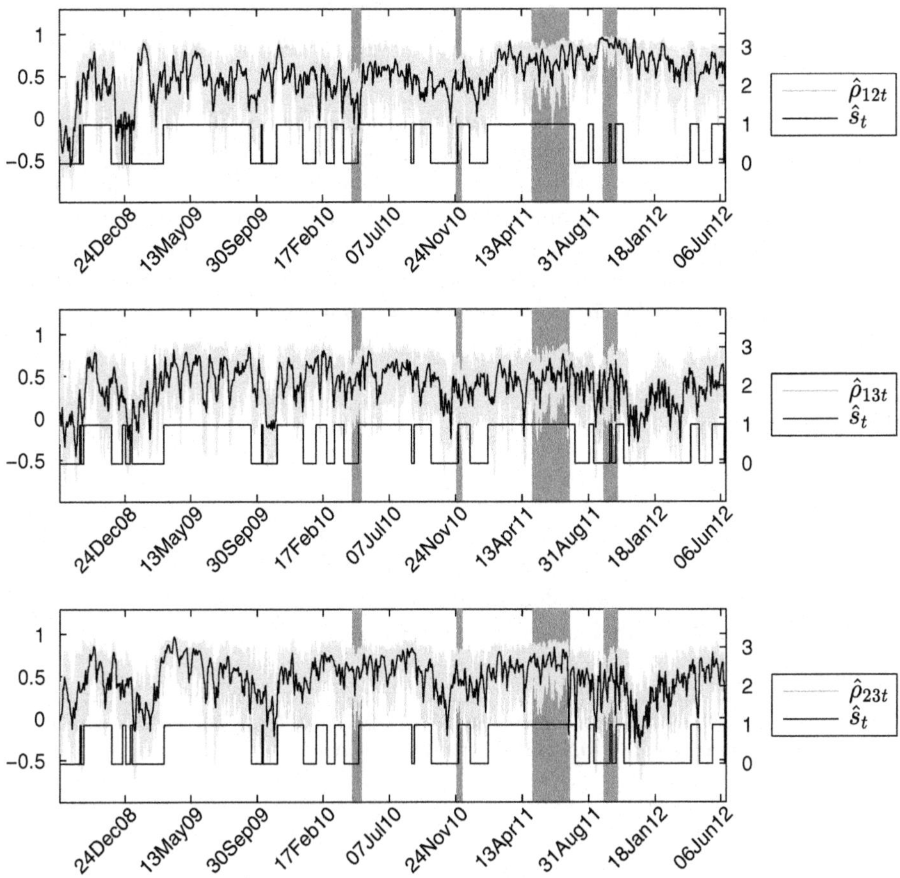

Fig. 1 Posterior means (*solid lines, left axes*) and 95 % credibility regions (*light gray areas, left axes*) of the correlation Ω_t. Each figure includes $\hat{s}_t$ (stepwise, *right axes*). *Vertical bars* represent some important periods from the ECB crisis timeline described in the main text

and financial adjustment programme and the 7 December the package for Ireland is agreed.

- From 5 May to 21 July 2011, Portugal and Greece receive financial aids. More specifically, the 5 May ECB welcomes Portugal's economic and financial adjustment programme and the 17 May EU Council approves aid to Portugal. Statement by EC, ECB, and IMF on Greece the 3 June and on Ireland the 14 July. The 21 July EU leaders announce in Brussels the second package of financial aid to Greece.
- October 2011 further measures are adopted to preserve financial stability. Among others, 6 October ECB announces the details of the refinancing operations and of the second covered bond purchase programme. Then, there is a statement by EC, ECB, and IMF on the fifth review mission to Greece (11 October) and the

Enhanced European Financial Stability Facility becomes fully operational (13 October). The 26 and 27 of October the EU leaders agree on additional measures including new financial aids for Greece.

Finally, turning to [4] critical remarks (strong decreases in cross-market correlations may as well point out the existence of contagion effects) our empirical evidence is also in line with the above point. Actually, focusing on time-varying correlation patterns in Fig. 1, many sharp declines are apparent along the sample period, for all pairs of eurozone CDS markets considered. To sum up, the MSSC model estimated in the present study provides very strong support for the existence of contagion effects in the eurozone sovereign CDS markets.

5 Conclusion

This research develops a Bayesian MSV model to analyze the dynamics of sovereign risk in eurozone CDS markets during the recent financial crisis. This model is applied to daily CDS spreads from August 2008 to June 2012 for nine eurozone countries. Empirical estimates document that the recent financial crisis generated severe contagion effects in the eurozone countries.

Acknowledgements Research supported by funding from the European Union, 7th Framework Programme FP7/2007-2013, grant agreement SYRTO-SSH-2012-320270, and by the Italian Ministry of Education, MIUR, PRIN 2010-11 grant MISURA.

References

1. Asai, M., McAleer, M.: The structure of dynamic correlations in multivariate stochastic volatility models. J. Econ. **150**, 182–192 (2009)
2. Bollerslev, T.: Modelling the coherence in short-run nominal exchange rates: a multivariate generalized arch approach. Rev. Econ. Stat. **72**, 498–505 (1990)
3. Corsetti, G., Pericoli, M., Sbracia, M.: Some contagion, some interdependence: more pitfalls in tests of financial contagion. J. Int. Money Financ. **24**, 1177–1199 (2005)
4. Corsetti, G., Pericoli, M., Sbracia, M.: Financial contagion: the viral threat to the wealth of nations. In: Kolb, R.W. (ed.) Financial Contagion, pp. 11–20. Wiley, New York (2011)
5. Forbes, K., Rigobon, R.: No contagion, only interdependence: measuring stock market comovements. J. Financ. **57**, 2223–2261 (2002)
6. Philipov, A., Glickman, M.: Multivariate stochatic volatility via wishart processes. J. Econ. Bus. Stat. **24**, 313–328 (2006)
7. Roberto C., Marco T., Domenico S.: Bayesian Markov Switching Stochastic Correlation Models, Working Papers No 2013:11, Department of Economics, University of Venice Ca' Foscari.

Female Labour Force Participation and Selection Effect: Southern vs Eastern European Countries

Rosalia Castellano, Gennaro Punzo, and Antonella Rocca

Abstract The aim of this paper is to explore the main determinants of women's job search propensity as well as the mechanism underlying the selection effect across the four European countries (Italy, Greece, Hungary and Poland) with the lowest female labour force participation. The potential bias due to the overlap in some unobserved characteristics is addressed via a bivariate probit model. Significant selection effects of opposite signs are found for the Greek and Polish labour markets.

Keywords Cross-country analysis • Female labour propensity • Heckman correction

1 Introduction

Getting through any types of gender gap in labour market appears to be a crucial matter in contributing to the social progress and economic growth of a country. In this field, some important aspects are well captured by the increasing trend of female labour force participation which has been characterizing most European countries during the last few years. Many factors contributed to the raise in female activity rates, from the increase in the women's educational attainment, the change in their social attitudes and labour market opportunities to the desire of keeping higher standards of living and the need of economic independence in response to the rise of the instability of couple relationship [4, 5].

Similarly, a great importance in explaining the increase in female activity rates is also due to some specific economic traits of labour market functioning [12]. Indeed, a high local unemployment and lower household incomes could produce the need to increase the economic resources for their members' sustenance, while a higher

R. Castellano (✉) • G. Punzo • A. Rocca
Department of Management and Quantitative Studies, University of Naples "Parthenope", 13, via Generale Parisi, Naples 80132, Italy
e-mail: lia.castellano@uniparthenope.it; gennaro.punzo@uniparthenope.it; antonella.rocca@uniparthenope.it

I. Morlini et al. (eds.), *Advances in Statistical Models for Data Analysis*, Studies in Classification, Data Analysis, and Knowledge Organization, DOI 10.1007/978-3-319-17377-1_5

degree of labour market rigidity could make difficult for women to reconcile their work with child and home care.

However, the different dynamics in labour market participation between the genders, which inevitably reflect social, cultural and economic norms and incentives, and the potential differences in behaviours between working and non-working women require to deal with important methodological issues. Indeed, the idea of this work was inspired by some empirical results coming from the analysis of sample selection effects on female employees based on women's wage equations tested over 26 European countries through the Heckman procedure. Lambda coefficients, consistently significant and negatively signed for each country (except for Norway), suggest a negative correlation between the error terms of the selection probit and the primary wage models. It means that unobserved factors, which make labour participation more likely, tend to be associated with lower potential returns.

In this field, the aim of the paper is twofold. First, it points to explore the mechanisms underlying the selection effect in women's job search process across the European countries with the lowest levels of female participation to labour market, i.e., two Southern European countries, Italy and Greece, whose economic dynamics are quite similar, *vs* two Eastern countries, Hungary and Poland, which reflect differences in economic characteristics during the years of post-transition. Second, after a close examination of national socio-economic background and market labour frameworks, the paper aims at exploring the main determinants of women's job search propensity and interpreting cross-country differentials in the behaviour of women who are actively looking for a job in the light of the main peculiarities of the potential sample selection effect into occupation.

2 A Socio-Economic Framework

Over the last years, the female employment rate has been rising throughout Europe, reaching on average 58.2 % in 2007, close to the Lisbon target (60 % in 2010), with an increase of 13.23 percentage points over the decade 1997–2007 (epp.eurostat.ec.europa.eu). However, although the female participation in labour market has been increasing and the male–female gap has been decreasing (at European level it has passed from 20 % in 1997 to 14.7 % in 2007), the female employment rates are still consistently lower than their male counterpart everywhere with large cross-national differentials. As they say, women are characterized by a different job-seeking behaviour [10] and, in general, they appear to be less likely than men to be employed or looking actively for a job. More precisely, the female labour participation rates still vary from the lowest values of Southern—i.e., Italy (50.7) and Greece (54.9)—and some Eastern countries—i.e., Hungary (55.1) and Poland (56.5)—to the highest ones for the well-developed economies of North Europe—i.e., Iceland (82.7), Sweden (76.8) and Denmark (76.4), against a EU-27 average of 63.2 per cent (epp.eurostat.ec.europa.eu).

Cross-country differences in the patterns of female labour force and their changes over time arise from a complex interaction among institutional, cultural and socio-economic dynamics [9]. Indeed, the regulations of national labour markets in terms of hiring and firing structure, working-time regulations, the tax and benefit systems, the more or less restrictive policies for balancing work and family life—which involve different activities related to paid work and unpaid caring as well as to social life, personal development and civic participation, the cost and availability of child and elder care services and human resource management practices of firms [2]—may also strongly affect the women's work choices and propensities.

However, although Italy, Greece, Hungary and Poland strongly differ in terms of labour market flexibility, level of tertiarization of the economy, women's participation in higher education programs and social policies, these countries share low rates of female labour participation. In particular, in Italy and Greece, where the decline of marriages and the increase of births outside marriage undermined the male breadwinner model, the transition from care force to workforce has still weak social supports for childcare. Although a number of interesting family-friendly schemes were introduced, measures to support women in balancing work and family responsibilities and in combining work flexibility with a series of rights and guarantees are not really effective. Indeed, in Greece, the labour flexibility is now at low levels if compared to the EU-average and the need for new working time arrangement is often perceived, while in Italy a greater attention has been paying to reconciliation issues. Nevertheless, attempts to increase flexibility (i.e., part-time, atypical works, job-sharing, innovative working time arrangements, telework, supplementary services) have not still reached the desired effects in terms of female labour force participation and quality of their work. Moreover, the higher levels of income inequalities and public debts may distract governments from adequate gender equality policies which are officially in force but not very actively pursued. As a consequence, in the Greek labour market, the gender gap in employment rates is the highest one (higher than 25 %) against values lower than 5 % in Finland and Sweden.

Just like Greece, also Poland and Hungary show the lowest rates of female part-time and high levels of poverty and unemployment. In these Eastern countries, the female participation in the labour market and gender pay gaps—which appeared on the surface like the Nordic countries during the socialist-type regime, whose policies strongly encouraged women to work—worsened for the period of transition and the work-life balance was not the main target for their governments. However, since 2005, Hungarian and Polish Governments, in cooperation with some non-governmental organizations, have been promoting the idea of the *family-friendly* workplace in order to favour the reconciliation of work and family life and some rules are now adopted on equal treatment and gender discrimination.

3 Data and Methodology

As widely documented [1, 11, 14], in countries where the female participation in labour market is still low, there could be problems of sample selection because working women could be unrepresentative of the entire female population. Indeed, beyond differences in male and female behaviours in labour market, women who do not work may differ in some important *unmeasured* ways (i.e., individual status, family-specific or socio-cultural background) from women who choose to belong to labour market and this may even lead to biased estimates of structural parameters relevant to the behaviour of working women. For example, in the classical wage equations [13], where the logarithm of earnings is modelled on a set of human capital variables (i.e., education and labour market experience), it is likely that women's earnings are biased because women who are working form a self-selected (and not a random) sub-sample. The two-stage Heckman procedure [8] is a way to correct for this selectivity: in the first step, the female labour propensities are estimated on a set of women's characteristics through a probit model which provides the correction term (λ), equal to the inverse of Mill's ratio, to include as additional predictor in the regression estimated in the second step.

Our analysis draws upon the 2007 EU-SILC data (European Union-Survey on Income and Living Conditions) and it is focused on all adult women aged 16–65. As anticipated above, empirical results of women's wage equations (Table 1), estimated over 26 European countries through the Heckman procedure, showed lambda coefficients significant and negatively signed for each country (except for Norway):

Table 1 *Lambda* coefficients over 26 European countries on the Mincerian wage equation. Year 2007

Country	λ coefficient	Country	λ coefficient
Austria	−0.29807**	Italy	−0.16126**
Belgium	−0.54024**	Latvia	−0.96101**
Cyprus	−0.39348**	Lithuania	−0.78201**
Czech Republic	−0.66854**	Luxembourg	−0.30978**
Denmark	−2.08190**	the Netherlands	−0.61911**
Estonia	−1.20022**	Norway	0.09560
Finland	−1.05783**	Poland	−0.59788**
France	−0.73739**	Portugal	−0.31405**
Germany	−0.22876**	Slovakia	−1.62603**
Greece	−0.25662**	Slovenia	−0.53048**
Hungary	−0.57680**	Spain	−0.30628**
Iceland	−0.76499**	Sweden	−1.35408**
Ireland	−0.65874**	the United Kingdom	−0.08448*

**Significant at 1 %; *significant at 5 %

This suggests a negative correlation between the error terms of selection probit and primary wage models. It means that unobserved factors, which make female labour force participation more likely, tend to be associated with lower potential returns.

Exploring 2007 EU-SILC data, the women's propensity to work (number of women actively looking for a job on the number of unemployed women) was higher than 70 % for Greek, Italian and Hungarian women, while just in Poland females' propensity was lower than their male counterpart. However, current levels of female participation in labour market strongly affect who is actively looking for a job; thus, in order to control for the potential overlap in unobserved characteristics influencing both the women's propensity to work and their propensity to look actively for a job, an ML (maximum likelihood) bivariate probit model is estimated [7].

The first probit model estimates the probability that a woman is *not occupied*:

$$y_i^{F*} = X_i^F \beta + \nu_i^F \tag{1}$$

with $\nu_i^F \sim N(0, \sigma_\nu^2)$ and where the latent variable y_i^{F*} drives the observed outcome of not working y_i through the following measurement equation:

$$y_i^F = \begin{cases} 1, & \text{if } y_i^{F*} > 0 \\ 0, & \text{otherwise} \end{cases} \tag{2}$$

Focusing on the subset of women who do not work, the probability of being *actively searching a job* is given by

$$S_i^{F*} = X_i^F \gamma + W_i^F \delta + \epsilon_i^F \tag{3}$$

with $\epsilon_i^F \sim N(0, \sigma_\epsilon^2)$, including additional covariates (W_i^F) concerning the equivalized household income and size, the individual health status and geographical area of residence. More specifically, W_i^F is a set of identifying restrictions that could be significant determinants of the endogeneous variable (to be actively searching a job), but also orthogonal to the residuals of the main equation [7], which is not significantly associated with the probability of being not occupied. S_i^{F*} drives the observed outcome of being actively searching a job through the following measurement equation:

$$S_i^F = \begin{cases} 1, & \text{if } S_i^{F*} > 0 \\ 0, & \text{otherwise} \end{cases} \tag{4}$$

In this way, the potential for unobserved heterogeneity that could produce a correlation between error terms of the two probit models is considered. Therefore, not only the true effects of searching a job but also the effect on professional condition of having these unobservable characteristics are captured [6]. If the error terms ν_i and ϵ_i, jointly distributed as bivariate normal with zero means and unit variances, are significantly and positively correlated ($\rho > 0$), unobserved factors increase both the probability of being a not-occupied female and looking for a job; for significantly negative ρ, the reverse is true, while not significant ρ shows the absence of selection effect and the equivalence of using the bivariate or two separate probit models.

4 Main Empirical Evidence

By justifying the bivariate probit model in the effort to limit the risk of selection bias, it allows to estimate the probability of the event to be actively searching for a job upon the condition to be unemployed.

Using Stata software, several explanatory variables are tested according to a stepwise procedure. A first set of covariates detects some socio-demographic characteristics at individual (i.e., marital status, educational attainment, age, health status) and household level (i.e., dependent children, household income and composition), while a second set includes location characteristics of each respondent (i.e., area of residence and urbanization degree) in order to explore the role of territorial perspective in the women's job search propensity.

Significant selection effects of opposite signs are found for Greece and Poland (Table 2), while in Italian and Hungarian labour markets the non-sample selection could derive from a lack of link between the mechanisms of job search and the status of unemployed. In this light, the significance of lambda coefficients for the Heckman correction in the women's wage equations could denote a sample selection which exclusively involves women who do not participate at all to labour force. The harsh Greek scenario and the difficulties to find a job drive both the propensity of being not occupied and negatively the propensity of actively seeking employment. In Poland, the unmeasured factors associated to a lower propensity to search a job act in the opposite direction. While in all countries a higher female propensity to work concerns families where more members are already occupied, just in Poland, the women's job search propensity appears to be not linked to financial household problems or marital status; anyway, Polish high-educated women are less likely to be looking for a job. Finally, although sub-national differences occur (women living in the North–West or Centre of Italy and in the North of Hungary are more likely to be actively searching), the presence of children discourages women to be active in the labour market everywhere.

As the classical human capital theory suggests [3, 13, 15] and consistent with other empirical studies [16, 17], our results emphasize the crucial role of education and age in determining both the propensity to work and the propensity to search a job. Indeed, a higher educational attainment significantly increases the job search propensity everywhere; job search is expected to pay more educated females off more than the less educated ones, while younger women are usually more active in search. Certainly, the latter is a negative effect which leads older women to decrease their search effort because of discouragement.

Table 2 Bivariate probit estimates of *not working* and *actively searching a job* for females

Variables	Italy	Greece	Hungary	Poland
Actively searching for a job				
Intercept	−1.2127***	−1.7942***	−1.4347***	−1.3072***
Age class (ref: 16–24 years)				
–*Younger* [25–40 years]	0.3672***	0.4762***	−0.3924*	0.4272***
–*Older* [41–65 years]	−0.4022***	−0.0109***	0.9193*	−0.2661***
Marital status (1 if *married*)	−0.3349***	−0.5128***	−0.2277**	0.0395
Children (1 if *with children*)	−0.3355***	−0.6085***	−0.2834**	−0.4220***
Urbanization degree (1 if *densely*)	−0.0576	−0.0773	0.0446	0.0786
Educational attainment (ref: *low*)				
–*Medium* (ISCED97: 3;4)	0.2387***	0.4134***	0.2798***	0.5373***
–*High* (ISCED97: 5)	0.6332***	1.1202***	0.6385***	−0.2661***
Ratio[a]	0.2999	1.8186***	1.0821**	0.4861*
Health (1 if *chronic*)	−0.0598	−0.0731	0.7179***	−0.3056***
Equivalized household income	−2.2E−5***	−2.5E−5***	−0.0002***	−4.19E−5
Equivalized household size	0.1588***	0.2206***	0.1319**	0.0952**
Geographical area (*NUTS*)[b]				
–*Area 1*	0.2062*	0.2018	−0.2331**	−0.0181
–*Area 2*	0.1085	0.0279	−0.1726**	−0.2322***
–*Area 3*	0.1532*	0.0658	–	−0.0186
–*Area 4*	−0.0891	–	–	−0.0837
–*Area 5*	–	–	–	0.0569
Not working				
Intercept	3.5105***	3.5638***	3.8445***	3.2892***
Age (years)	−0.0092***	−0.0115***	−0.0140***	−0.0016
Marital status (1 if *married*)	0.0962**	0.0677	−0.2143***	−0.3350***
Children (1 if *with children*)	0.5082***	0.4875***	0.4473***	0.2833***
Urbanization degree (1 if *densely*)	0.1256***	0.1002**	0.0352	0.0871***
Educational attainment (ref: *low*)				
–*Medium* (ISCED97: 3;4)	−04432***	−0.4274***	−0.5901***	−0.8561***
–*High* (ISCED97: 5)	−0.9438***	−0.9553***	−0.9686***	−1.4804***
Ratio[+]	−4.4691***	−4.3516***	−4.30451***	−3.5013***
Wald χ^2	429.33	317.55	188.26	313.94
Correlation (ρ)	−0.2390	0.6858**	0.1859	−0.3482**

***Significant at 1 %; **significant at 5 %; *significant at 10 %

[a](number of wage earners–1)/(number of household members)

[b]NUTS1 codes: Italy: 1 North-West, 2 North-East, 3 Centre, 4 South (ref.: Isles); Greece: 1 Voreia, 2 Kentriki, 3 Attiki (ref.: Nisia Aigaiou, Kriti); Hungary: 1 Central, 2 Transdanubia (ref.: Greath Plain and North); Poland: 1 Centralny, 2 Poludniowy, 3 Wschodni, 4 Polnocno-Zachodni, 5 Poludniowo-Zachodni (ref.: Polnocny)

5 Conclusions

The growth of female labour force participation is a feasible channel for increasing per capita GDP and, in turn, for narrowing the gender gaps. This is of great importance, mainly in recent years characterized by a reduced economic growth even for the most developed European countries. The emphasis of EU institutions on economic and social policies devised to support gender equality and innovative forms of work organization and legislation produced a further increase in the female labour force participation, driving national governments on the definition of various measures for reconciling work and family life.

However, substantial cross-national differences in the levels of female participation in the labour market still persist. In this light, in Europe, the well-known contraposition between the most developed Northern economies, on the one side, and the Southern and Eastern countries, on the other one, whose economic growth is obstructed by socio-economic problems, is too much simplistic and lacking of significance in explaining these differentials.

In this paper, as regards countries with the lowest female participation rates, an in-depth analysis of determinants of women's job search activity has been carried out taking into account the specific economic framework and the influence of household composition. Indeed, institutions surely play a crucial role in stimulating the women's participation in the labour market through initiatives increasing flexibility or different kinds of employment and labour tax policies, although the decision to be active in the labour market is also strongly affected by the choices in education and fertility.

The analysis has shown a significant selection effect only for Greece and Poland. In particular, in Greece, the negative sign of selection effect could highlight a strong influence of financial problems and high levels of unemployment in female propensity of actively looking for a job; on the other side, Polish women seem to be driven in their decisions in finding a job by opposite factors. In Italy and Hungary, higher levels of unemployment, the weak diffusion of part-time and the persistence of the male breadwinner model don't let emerge any predominant aspect; indeed, in these countries, even if some programs for reconciling motherhood with professional career are in force, they are not actively pursued and the burden of childbearing is still often left to the family.

Furthermore, having children seems to be a problem even now because it still negatively affects the propensity to search a job, probably due to unsuccessful mix of conciliation policies. Some other common factors across countries are also identified, such as the direct relationship between educational level and propensity to work. However, although women's work propensity should be higher in countries where effective social policies aimed primarily to reconciling motherhood with professional life are in force, the macroeconomic scenario and the strictness of labour market institutions may negatively affect their participation.

References

1. Albrecht, J., Van Vuuren, A., Vroman, S.: Counterfactual distributions with sample selection adjustments: econometric theory and an application to the Netherlands. Labour Econ. **16**(4), 383–396 (2009)
2. Anxol, D., Fagan, C., Cebrian, I., Moreno, G.: Patterns of labour market integration in Europe. A life course perspective on time policies. Soc. Econ. Rev. **5**(2), 233–260 (2007)
3. Becker, G.S.: Human Capital. Columbia University Press, New York (1964)
4. Blau, F., Ferber, M., Winkler, A.: The Economics of Women, Men and Work. Prentice Hall, New Jersey (2010)
5. Castellano, R., Punzo, G., Rocca, A.: Intergenerational mobility and gender gap: evidence from mediterranean countries. In: Proceedings of 46th Scientific Meeting of the Italian Statistical Society, Rome (2012)
6. Fleming, C.M., Kler, P.: I'm too clever for this job: a bivariate probit analysis on overeducation and job satisfaction in Australia. Appl. Econ. **40**(9), 1123–1138 (2011)
7. Green, W.H.: Econometric Analysis. Prentice Hall, New Jersey (1997)
8. Heckman, J.: Sample Selection Bias as a Specification Error. Econometrica **47**(1), 153–161 (1979)
9. Jaumotte, F.: Female Labour Force Participation: Past Trends and Main Determinants in OECD Countries. OECD Economics Department Working Papers, 376. OECD Publishing, Paris (2003)
10. Kahn, L.M., Low, S.A.: An empirical model of employed search, unemployed search and non-search. J. Hum. Resour. **XIX**(1), 104–117 (1984)
11. Maddala G.S.: Limited-Dependent and Qualitative Variables in Econometrics. Cambridge University Press, Cambridge (1983)
12. McConnell, C.R., Brue, S.S., Macpherson, D.A.: Contemporary Labor Economics, 9th edn. McGraw Hill Irwin, New York (2010)
13. Mincer J.: Investment in human capital and personal income distribution. J. Polit. Econ. **66**, 281–302 (1958)
14. Mulligan, C., Rubinstein, Y.: Selection, investment and women's relative wage over time. Q. J. Econ. **123**(3), 219–277 (2008)
15. Schultz, T.W.: Investment in human capital. Am. Econ. Rev. **51**(1), 1–17 (1961)
16. Smirnova, N.: Job search behavior of unemployed in Russia. Bank of Finland Discussion Papers 13, pp. 1–36 (2003)
17. Smith, S.: Labor Economics, Routledge, London (2003)

Asymptotics in Survey Sampling for High Entropy Sampling Designs

Pier Luigi Conti and Daniela Marella

Abstract The aim of the paper is to establish asymptotics in sampling finite populations. Asymptotic results are first established for an analogous of the empirical process based on the Hájek estimator of the population distribution function and then extended to Hadamard-differentiable functions. As an application, asymptotic normality of estimated quantiles is provided.

Keywords Empirical processes • Hájek estimator • Quantiles • Sampling design

1 Introductory Aspects

Asymptotic results in sampling finite populations are widely used in different contexts. All results are concerned with the asymptotic normality of (usually linear) statistics, under different conditions and sampling plans. Among several papers devoted to this subject, in [15] asymptotic properties of L-statistics are obtained under stratified two-stage sampling. In [8] asymptotic properties for the Horvitz–Thompson estimator under rejective sampling are obtained; extensions to different sampling designs are in [1, 19]. In [7] the estimation of the population distribution function and quantiles is studied in case of a stratified cluster sampling. Clusters are selected from each stratum without replacement and with equal inclusion probabilities. Similar results are obtained in [15]. In [10] estimators based of auxiliary variables are introduced and carefully studied. Estimators of the population distribution function based on calibration are in [3, 11].

All the above mentioned papers (and several others, as well) focus on the problem of estimating the population distribution function (p.f.d.) or quantiles at a single point, or at a finite number of points. In this paper the main interest consists in

P.L. Conti (✉)
Sapienza Università di Roma, P.le A. Moro 5, 00185 Roma, Italy
e-mail: pierluigi.conti@uniroma1.it

D. Marella
Università Roma Tre, Via del Castro Pretorio 20, 00185 Roma, Italy
e-mail: daniela.marella@uniroma3.it

I. Morlini et al. (eds.), *Advances in Statistical Models for Data Analysis*,
Studies in Classification, Data Analysis, and Knowledge Organization,
DOI 10.1007/978-3-319-17377-1_6

estimating the *whole* p.d.f. or quantile function. This requires the construction of an "infinite dimensional" asymptotic theory for sampling finite populations that parallels, as far as possible, the classical theory of nonparametric statistics. For the important class of "high entropy" sampling designs, similarities and differences between finite populations results and classical nonparametrics will be discussed. The paper is organized as follows. In Sect. 2 the main technical aspects and the basic assumptions on which the present paper relies are briefly introduced. In Sect. 3 asymptotic results for the estimation of population distribution function are first established and then extended to Hadamard-differentiable functions (4). As an application, in Sect. 5 the problem of quantile function estimation is dealt with.

2 Notations and Assumptions

Let $\mathscr{U}_N$ be a finite population of N units, labeled by integers $1, \ldots, N$. Let Y be the variable of interest and for each unit i, denote by y_i the value of Y ($i = 1, \ldots, N$). Let further $\boldsymbol{y}_N = (y_1, \ldots, y_N)$. For each real y, the *population distribution function* (p.d.f., for short) is defined as

$$F_N(y) = \frac{1}{N} \sum_{i=1}^{N} I_{(y_i \leq y)}, \quad y \in \mathbb{R}, \tag{1}$$

where

$$I_{(y_i \leq y)} = \begin{cases} 1 \text{ if } y_i \leq y \\ 0 \text{ if } y_i > y \end{cases}, \quad i = 1, \ldots, N.$$

Now, for each unit i in $\mathscr{U}_N$, define a Bernoulli random variable (r.v.) D_i, such that the unit i is included in the sample if and only if (iff) $D_i = 1$, and let $\boldsymbol{D}_N$ be the N-dimensional vector of components $D_1, \ldots, D_N$. A (unordered, without replacement) sampling design P is the probability distribution of $\boldsymbol{D}_N$. In particular, $\pi_i = E_P[D_i]$ is the inclusion probability of unit i. The suffix P denotes the sampling design used to select population units. The sample size is $n_s = D_1 + \cdots + D_N$. A sampling design is of *fixed* size n iff $n_s = n$ for each sample.

Let $p_1, \ldots, p_N$ be N real numbers, with $p_1 + \cdots + p_N = n$. The sampling design is a *Poisson design* with parameters $p_1, \ldots, p_N$ if the r.v.s D_is are independent with $Pr_{Po}(D_i = 1) = p_i$ for each unit i, the suffix Po denoting the Poisson design.

The *rejective sampling* or *normalized conditional Poisson sampling* [8, 17] corresponds to the probability distribution of the random vector $\boldsymbol{D}_N$, under Poisson design, conditionally on $n_s = n$.

The *Hellinger distance* between a sampling design P and the rejective design P_R is defined as

$$d_H(P, P_R) = \sum_{D_1, \dots, D_N} \left(\sqrt{Pr_P(D_N)} - \sqrt{Pr_{P_R}(D_N)} \right)^2 . \tag{2}$$

The class of sampling designs we focus on do have a fundamental property: they asymptotically behave as the rejective sampling. More precisely, such a property is ensured by the assumptions listed below.

A1. $(\mathcal{U}_N;\ N \geq 1)$ is a sequence of finite populations of increasing size N.
A2. For each N, $y_1, \dots, y_N$ are realizations of a superpopulation $(Y_1, \dots, Y_N)$ composed by *i.i.d.* r.v.s Y_i with common d.f. F. In the sequel, we will denote by $\mathbb{P}$ the probability distribution of r.v.s Y_is and by $\mathbb{E}$, $\mathbb{V}$ the corresponding operators of mean and variance, respectively.
A3. For each population $\mathcal{U}_N$, sample units are selected according to a fixed size sample design with inclusion probabilities $\pi_1, \dots, \pi_N$ and sample size $n = \pi_1 + \dots + \pi_N$. Furthermore, for $d > 0$, $0 < f < 1$,

$$d_N = \sum_{i=1}^{N} \pi_i(1 - \pi_i) \rightarrow \infty, \quad \frac{1}{N} d_N \rightarrow d, \quad \lim_{N \to \infty} \frac{n}{N} = f \quad \text{as } N \rightarrow \infty.$$

A4. For each population $(\mathcal{U}_N;\ N \geq 1)$, let P_R be the rejective sampling design with inclusion probabilities $\pi_1, \dots, \pi_N$, and let P be the actual sampling design (having the same inclusion probabilities). Then

$$d_H(P, P_R) \rightarrow 0 \quad \text{as } N \rightarrow \infty.$$

A5. There exist two positive real numbers A, B such that

$$\lim_{N \to \infty} \frac{1}{N} \sum_{i=1}^{N} \frac{1}{\pi_i} = A < \infty, \quad \lim_{N \to \infty} \sum_{i=1}^{N} \frac{1}{(i\pi_i)^2} = B < \infty.$$

3 Estimation of the Population Distribution Function

The estimation of the p.d.f. (1) is an important problem in sampling finite populations. The simplest estimator is the Horvitz–Thompson estimator, given by

$$\hat{F}_{HT}(y) = \frac{1}{N} \sum_{i=1}^{N} \frac{D_i I_{(y_i \leq y)}}{\pi_i}. \tag{3}$$

Since $F_{HT}(+\infty) \neq 1$ with positive probability, the estimator (3) is not necessarily a proper distribution function. For this reason, we consider here the Hájek estimator of F_N

$$\hat{F}_H(y) = \frac{\sum_{i=1}^{N} \frac{1}{\pi_i} D_i I_{(y_i \leq y)}}{\sum_{i=1}^{N} \frac{1}{\pi_i} D_i} = \frac{\hat{F}_{HT}(y)}{\hat{F}_{HT}(+\infty)}, \tag{4}$$

which is actually a proper distribution function. Note that if the sampling design is simple random sampling (srs, for short), both the Horvitz–Thompson and Hájek estimator reduce to the empirical distribution function $\hat{F}_n$ given by

$$\hat{F}_n = \frac{1}{n} \sum_i D_i I_{(y_i \leq y)}. \tag{5}$$

Our main goal is to study the asymptotic "global" behaviour of the random function $\hat{F}_H(\cdot)$. In order to accomplish this, we define the (sequence of) random function(s)

$$W_N^H(y) = \sqrt{n}(\hat{F}_H(y) - F_N(y)), \ y \in \mathbb{R}; \ N \geq 1. \tag{6}$$

The main result of the present section is Proposition 1. A proof is in [4].

Proposition 1 *If the sampling design P satisfies assumptions A1–A5, with $\mathbb{P}$-probability 1, conditionally on $\mathbf{y}_N$ the sequence $(W_N^H(\cdot);\ N \geq 1)$ converges weakly, in $D[-\infty, +\infty]$ equipped with the Skorokhod topology, to a Gaussian process $W^H(\cdot) = (W^H(y);\ y \in \mathbb{R})$ that can be represented as*

$$W^H(y) = \sqrt{f(A-1)} B(F(y)), \ y \in \mathbb{R}, \tag{7}$$

where $(B(t);\ 0 \leq t \leq 1)$ is a Brownian bridge.

In classical nonparametric statistics, the empirical process $\sqrt{n}(\hat{F}_n(y) - F(y))$ converges weakly to a Gaussian process of the form $B(F(y))$. This result is apparently similar to Proposition 1, with two differences:

1. The centering factor F instead of F_N.
2. The absence of the finite population correction term $f(A-1)$, since in classical nonparametric statistics the observations are (realizations of) *i.i.d.* r.v.s, and there is essentially no sampling design.

The results can be particularized to the case of srs of size n. As previously said, in this case the Hájek estimator reduces to the empirical distribution function (5) and the limiting process can be written as

$$\sqrt{1-f} B(F(y)), \ y \in \mathbb{R} \tag{8}$$

where the term $\sqrt{1-f}$ is the finite population correction.

The asymptotic result of Proposition 1 is obtained conditionally to the population values $\mathbf{y}_N$; the expression "with $\mathbb{P}$-probability 1" means that the set of sequences $(y_i;\ i \geq 1)$ for which Proposition 1 fails does have probability zero.

It is important to stress that the probability involved in Proposition 1 is *only* the sample design probability. In other words, the only source of variability is the sampling design. Proposition 1 refers to design-based inference, where the values y_is are considered as *fixed*. In other words, the role played by the superpopulation model of assumption A2 is of secondary importance.

4 Asymptotics for Hadamard Differentiable Functions

The goal of the present section is to extend the result of Proposition 1 to population parameters more general than the p.d.f. Let $\theta_N = \phi(F_N)$ be a general functional of the p.d.f. F_N. Using the ideas of Sect. 3, it is "natural" to refer the estimator $\hat{\theta}_H = \phi(\hat{F}_H)$. In order to study its asymptotic behaviour, let us first re-scale the estimator itself by considering the sequence

$$\sqrt{n}(\hat{\theta}_H - \theta_N) = \sqrt{n}(\phi(\hat{F}_H) - \phi(F_N)). \tag{9}$$

If the functional $\phi(\cdot)$ is "smooth enough", it is fairly natural to expect that the asymptotic behaviour of (9) can be obtained by Proposition 1. The smoothness condition on $\phi(\cdot)$ that proves useful in this case is its *Hadamard differentiability* (cf. [18]). A map $\phi(\cdot) : D[-\infty, +\infty] \to E$, E being an appropriate normed space, is Hadamard differentiable at the "point" G, with Hadamard derivative $\phi'_G(\cdot)$, iff there exists a continuous, linear map $\phi' :\ D[-\infty, +\infty] \to E$ such that

$$\left\| \frac{\phi(G + th_t) - \phi(G)}{t} - \phi'_G(h) \right\|_E \to 0$$

as $t \downarrow 0$, $h_t \to h$, $\| \ \|_E$ being the norm on the vector space E.

Using the well-known functional delta method (cf. [18]), the following result is obtained.

Proposition 2 *If $\phi(\cdot)$ is (continuously) Hadamard-differentiable at F, with Hadamard derivative $\phi'_F(\cdot)$, then the asymptotic law of $\sqrt{n}(\phi(\hat{F}_H) - \phi(F_N))$ coincides with the asymptotic law of $\phi'_F(W^H)$, as N increases.*

In particular, if ϕ is real-valued, since $\phi'_F(\cdot)$ is linear and W^H is a Gaussian process, the law of $\phi'_F(W^H)$ is normal with mean zero and variance

$$\sigma^2_\theta = \mathbb{E}\left[\phi'_F(W^H)^2\right] \approx \frac{\sqrt{n}}{N} \sum_{i=1}^{N} \frac{D_i}{\pi_i} \tau_F(y_i), \tag{10}$$

where $\tau_F(y_i) = \phi'_F(I_{[y_i,+\infty)}(\cdot) - F(\cdot))$ is the *influence function* of θ. If $\theta'_F(\cdot)$ is continuous in F, then $\tau_F(y_i)$ can be approximated by

$$\hat{\tau}_{\hat{F}_H}(y_i) = \theta'_{\hat{F}_H}(I_{[y_i,+\infty)}(\cdot) - \hat{F}_H(\cdot)),$$

so that the following estimator of the (asymptotic) variance (10) is obtained:

$$\hat{V}_{YG} = -\frac{n}{N^2}\sum_{i=1}^{N}\sum_{j=1}^{N}\Delta_{ij}\left(\hat{\tau}_{\hat{F}_H}(y_i) - \hat{\tau}_{\hat{F}_H}(y_j)\right)^2 \frac{D_i D_j}{\pi_{ij}}. \tag{11}$$

As an application of Proposition 2 the asymptotic normality of estimated quantiles will be proved in the next section.

5 Quantile Estimation

In survey sampling the estimation of the population distribution function is an important problem, for several reasons. Even more important is the estimation of the population quantiles, because of the increasing demand of statistical data regarding poverty and inequality. Poverty and inequality measures are generally functions of (possibly cumulated) quantile estimates of the income or expenditure distribution.

The aim of this section is to study the problem of estimating the quantile function Q_N. The knowledge of the p.d.f. is essentially equivalent to the knowledge of the *population quantile function* (pqf, for short).

For each $0 < p < 1$, the *pth population quantile* $Q_N(p)$, say, is the left-continuous inverse of F_N computed at point p. In symbols:

$$Q_N(p) = F_N^{-1}(p) = \inf\{y : F_N(y) \geq p\}, \quad 0 < p < 1. \tag{12}$$

From (1) and (12) there is clearly a one-to-one map between F_N and Q_N, so that the estimation of Q_N is strictly related to the estimation of F_N.

The estimation of the p.d.f. $F_N(y)$ and the p.q.f. $Q_N(p)$ has been considered in several papers. In [20] confidence intervals for quantiles based on inverting confidence intervals for the p.d.f. are considered. In [12, 13], confidence intervals are studied for different sampling designs of interest (e.g. simple random sampling and stratified sampling). An interesting simulation study showing the (good) properties of Woodruff's confidence intervals is in [16]. In [9] the problem of estimating the p.d.f. F_N and the population median $Q_N(1/2)$ for general sampling designs with unequal first-order inclusion probabilities is dealt with.

The paper [7] is devoted to the estimation of the p.d.f. and quantiles in case of a stratified cluster sampling, where clusters are selected from each stratum without replacement and with equal inclusion probabilities. Asymptotic results are obtained under appropriate regularity conditions. Similar results are in [15].

Asymptotic properties of the sample quantiles under simple random sampling without replacement (srs, for short) are studied in [2].

In order to estimate the population quantile function the basic idea consists in inverting the Hájek estimator of the p.d.f. F_N. More specifically the pth quantile $Q_N(p)$ is estimated by

$$\hat{Q}_H(p) = \hat{F}_H^{-1}(p) = \inf\{y : \hat{F}_H(y) \geq p\} \tag{13}$$

with $0 < p < 1$.

If the sampling design is the simple random sampling (srs, for short), the estimator $\hat{Q}_H(p)$ reduces to the usual sample pth quantile. As previously stressed, for the p.d.f. F_N, our aim is to study the large sample behaviour of the whole estimated quantile function, that is the asymptotic behaviour of the estimator $\hat{Q}_H(p)$, as n, N increase. In order to accomplish this, we will study the large sample behaviour of the whole estimated quantile function $\hat{Q}_H(\cdot) = (\hat{Q}_H(p),\ \epsilon \leq p \leq 1-\epsilon)$, for positive ϵ. This is of course equivalent to study the behaviour of the random process $T_N^H(\cdot) = (T_N^H(p);\ \epsilon \leq p \leq 1-\epsilon)$, with

$$T_N^H(p) = \sqrt{n}(\hat{Q}_H(p) - Q_N(p)). \tag{14}$$

The random function $T_N^H(\cdot)$ is essentially the "finite population version" of the *quantile process*, which is of primary importance in nonparametric statistics (cf. [5, 6]). Its asymptotic behaviour can be obtained as a simple consequence of Proposition 2. In fact, if F is continuously differentiable with non-null derivative $\psi(y) = dF(y)/dy$, it is then easy to see (cf. Lemma 21.4 in [18]) that the quantile function is differentiable at F and that its Hadamard derivative is the map $h \mapsto -h/\psi(Q(\cdot))$. As a consequence, the following result holds.

Proposition 3 *Suppose that F is continuously differentiable with derivative $\psi(y) = dF(y)/dy$. Then, the sequence of random processes $\tilde{T}_N^H(\cdot) = (\sqrt{n}(\tilde{Q}_H(p) - \tilde{Q}_N(p));\ \epsilon \leq p \leq 1-\epsilon)$ converges weakly, in $D[\epsilon,\ 1-\epsilon]$ equipped with the Skorokhod topology, to a Gaussian process $T^H(\cdot) = (T^H(p);\ \epsilon \leq p \leq 1-\epsilon)$ that can be represented as*

$$T^H(p) = \sqrt{f(A-1)}\frac{B(p)}{\psi(Q(p))},\quad \epsilon \leq p \leq 1-\epsilon, \tag{15}$$

where $B(p)$ is a Brownian bridge.

In particular, Proposition 3 shows that conditionally on $\mathbf{y}_N$ the estimator $\hat{Q}_H(p) = \hat{F}_H^{-1}(p)$ is asymptotically normal with mean $Q_N(p)$ and variance $n^{-1}f(A-1)p(1-p)/\psi(Q(p))^2$. In symbols:

$$Pr_P\left(\sqrt{n}(\hat{Q}_H(p) - Q_N(p)) \leq z \,\middle|\, \mathbf{y}_N\right) \rightarrow$$
$$\Phi\left(\sqrt{f(A-1)}\frac{\sqrt{p(1-p)}}{\psi(Q(p))}z\right) \text{ as } N,\, n \rightarrow \infty, \tag{16}$$

where Φ is the standard normal d.f. and $Pr_P(\cdot|\mathbf{y}_N)$ denotes the probability w.r.t. the sampling design, *given* the y_is values of the N population units. This result is similar to the case of *i.i.d.* observations (cf. [14]), apart from the presence of the term $f(A-1)$. Intuitively speaking, this term represents the effect of the sampling design, i.e. a sort of "design effect" w.r.t. the standard case of *i.i.d.* observations.

6 Conclusions

In this paper the problem of estimating the d.f. F_N and functionals of F_N of a finite population is dealt with. Attention is focused on point estimation and asymptotic results are obtained for the class of high entropy designs. As application the asymptotic normality of estimated quantiles has been proved.

Further developments could regard the analysis of how asymptotic results change when the relationship between the variable of interest and the design variables is explicitly taken into account.

References

1. Berger, Y.G.: Rate of convergence to normal distribution for the Horvitz–Thompson estimator. J. Stat. Plan. Inference **67**, 209–226 (1998)
2. Chatterjee, A.: Asymptotic properties of sample quantiles from a finite population. Ann. Inst. Stat. Math. **63**, 157–179 (2011)
3. Chen, J., Wu, C.: Estimation of distribution function and quantiles using the model-calibrated pseudo empirical likelihood method. Stat. Sin. **12**, 1223–1239 (2002)
4. Conti, P.L.: On the estimation of the distribution function of a finite population under high entropy sampling designs, with applications. Sankhya 76-B, 234–259 (2014)
5. Csörgő, M.: Quantile Processes with Statistical Applications. SIAM, Philadelphia (1983)
6. Csörgő, M., Csörgő, S., Horváth, L.: An Asymptotic Theory for Empirical Reliability and Concentration Processes. Springer, Berlin (1986)
7. Francisco, C.A., Fuller, W.A.: Quantile estimation with a complex survey design. Ann. Stat. **19**, 454–469 (1991)
8. Hájek, J.: Asymptotic theory of rejective sampling With varying probabilities from a finite population. Ann. Math. Stat. **35**, 1491–1523 (1964)

9. Kuk, A.Y.C.: Estimation of distribution functions and medians under sampling with unequal probabilities. Biometrika **75**, 97–103 (1988)
10. Rao, J.N.K., Kovar, J.G., Mantel, J.G.: On estimating distribution functions and quantiles from survey data using auxiliary information. Biometrika **77**, 365–375 (1990)
11. Rueda M., Martinez, S., Martinez, H., Arcos, A.: Estimation of the distribution function with calibration methods. J. Stat. Plan. Inference **137**, 435–448 (2007)
12. Sedransk, J., Meyer,J.: Confidence intervals for the quantiles of a finite population: simple random and stratified random sampling. J. R. Stat. Soc. B **40**, 239–252 (1978)
13. Sedransk, J., Smith, P.: Lower bounds for confidence coefficients for confidence intervals for finite population quantiles. Comput. Stat. Theory Methods **12**, 1329–1344 (1983)
14. Serfling, R.J.: Approximation Theorems of Mathematical Statistics. Wiley, New York (1980)
15. Shao, J.: L-statistics in complex survey problems. Ann. Stat. **22**, 946–967 (1994)
16. Sitter, R.R., Wu, C.: A note on Woodruff confidence intervals for quantiles. Stat. Probab. Lett. **52**, 353–358 (2001)
17. Tillé, Y.: Sampling Algorithms. Springer, New York (2006)
18. Van Der Vaart, A.: Asymptotic Statistics. Cambridge University Press, Cambridge (1998)
19. Víšek, J A.: Asymptotic distribution of simple estimate for rejective, Sampford and successive sampling. In: Jurečková, J. (ed.) Contributions to Statistics, pp. 263–275. Reidel Publishing Company, Dordrecht (1979)
20. Woodruff R.S.: Confidence intervals for medians and other position measures. J. Am. Stat. Assoc. **47**, 635–646 (1952)

A Note on the Use of Recursive Partitioning in Causal Inference

Claudio Conversano, Massimo Cannas, and Francesco Mola

Abstract A tree-based approach for identification of a balanced group of observations in causal inference studies is presented. The method uses an algorithm based on a multidimensional balance measure criterion applied to the values of the covariates to recursively split the data. Starting from an ad-hoc resampling scheme, observations are finally partitioned in subsets characterized by different degrees of homogeneity, and causal inference is carried out on the most homogeneous subgroups.

Keywords Average treatment effect • Balancing recursive partitioning • Regression trees • Resampling

1 Introduction

In experimental studies about the estimation of the effect of a treatment on a set of individuals the randomization of treatment assignment implies that the treated and control groups are balanced with respect to observed and unobserved covariates. Thus, unbiased estimators of causal effects can be obtained via simple comparison of the outcome variable across treated and control units. On the other hand, in observational studies the treated and control groups may have different distributions, so that the simple mean difference of the outcome between the two groups cannot be attributed solely to the treatment . In these situations a simple comparison may give a biased estimate. However, under suitable conditions [2, 4], unbiased estimates of treatment effects can be obtained after balancing the distribution of covariates across treated and control units. Matching methods are useful to identify similar

C. Conversano (✉)
Department of Mathematics and Informatics, Via Ospedale 72, 09124 Cagliari, Italy
e-mail: conversa@unica.it

M. Cannas • F. Mola
Dipartimento di Scienze Economiche ed Aziendali, Universitá di Cagliari, Viale S. Ignazio 17, 09123 Cagliari, Italy
e-mail: massimo.cannas@unica.it; mola@unica.it

I. Morlini et al. (eds.), *Advances in Statistical Models for Data Analysis*,
Studies in Classification, Data Analysis, and Knowledge Organization,
DOI 10.1007/978-3-319-17377-1_7

observations and for this reason they have been suggested for achieving balance. Matching requires that the covariate distributions of treated and control units share a common support of values otherwise (i.e., in case of lack of overlap) the comparison is not possible. The identification of the common support can be done together with matching (e.g., imposing a minimum distance for two units to be matched together) or before matching (e.g., by restricting the analysis to those units that are considered to belong to the common support). In the latter case the common support must be identified prior to matching, and various methods are available [3].

A hybrid approach to matching, and common support identification called random recursive partitioning (RRP) has been introduced in [3]. RRP defines, at each iteration, a random partition of the covariate space by growing regression trees with fictitious outcome $Z \sim U(0, 1)$. Each random partition yields a proximity matrix whose elements can be used to weight the difference of the outcome variable across treated and untreated units. By a suitable choice of a tuning parameter RRP can also be used to select a subset of observations belonging to the common support. Analyzing the data used in [1], RRP shows that both weighted estimators and estimators based on a selected subset of treated and control units provide reasonable results in comparison with traditional methods. In addition, it seems that the subsets selected through RRP have better covariate balance than the unselected ones.

In this paper a different approach based on recursive partitioning is proposed. It exploits a balancing property of tree-based methods by growing a sequence of trees on resampled versions of the original data. The aim of tree growing is to balance the set of covariates by means of a splitting criterion based on a multidimensional balancing measure. The resampling scheme uses a certain number of samples, and the tree obtained from each of these, to assign more sampling weight to the units belonging to the most homogeneous terminal nodes. As in RRP, the final proximity matrix can be used either to obtain causal estimates or to identify the common support. Since the proposed approach is mainly aimed at balancing covariates by using recursive partitioning, we name it *balancing recursive partitioning* (BaRPa).

The main features of BaRPa are described in Sect. 2, whereas Sects. 3 and 4 report the results of a simulation study and some concluding remarks.

2 Balancing Recursive Partitioning

Given an outcome variable Y, a set of covariates X_j $(j = 1, \ldots, p)$, and a treatment variable T observed on N cases, BaRPa can be applied in all the situations in which the assignment mechanism of the treatment variable T is such that the conditional distribution of $X_j|T = 0$ differs from the conditional distribution $X_j|T = 1$ for at least one j. In this framework, the basic idea supporting the implementation of BaRPa is that the balance of $X|T$ on the basis of the set of covariates X_j and the treatment variable T can be obtained in a recursive manner. Specifically, it is possible to exploit the capability of one of the $X_j|T$ in improving the balance of the whole set of covariates. BaRPa tries to obtain this balancing in a recursive way: data are partitioned by selecting a splitting covariate and its associated split point that minimizes the global imbalance of the set of covariates in at least one of the two resulting child nodes. As in [3], BaRPa allows us to estimate a proximity matrix which measures how close is a treated unit with all the untreated ones: this matrix is obtained by growing a tree on resampled subsets of the original data.

BaRPa can also be used to identify a subset of matched observations for which balance in covariate distribution holds. To this aim, BaRPa is orientated towards the search of a (possibly small) subset of observations whose covariate distribution is, on average, more balanced than the original distribution observed on the whole dataset. The balanced subset is detected in the first iteration (i.e., after growing the first tree). Resampling subsets of observations helps to assess if this detection was obtained by chance or if it really identifies a balanced subset.

2.1 Main Steps of BaRPa

1. Tree growing. A binary recursive partitioning of data is performed in order to identify subregions of the covariate space in which the distributions of each $X_j|T = 0$ and its corresponding $X_j|T = 1$ are more balanced. Node splitting is based on the idea that a node is split if either the right or the left child node is more balanced than the parent node. BaRPa uses the average standardized absolute mean difference (ASAM) to split a node: It searches the splitting covariate X_L^* and its associated split point x_L^* that minimize the average ASAM of all the covariates in the left child node, as well as the splitting covariate X_R^* and its associated split point x_R^* that minimize the average ASAM of all the covariates in the right child node. The split point (x_L^* or x_R^*) is the one providing the maximum decrease in imbalance (compared to the value of the same measure calculated in the parent node).
The tree growing process proceeds until all current terminal nodes cannot be split since they do not provide any improvement in the imbalance of any $X_j|T$. In addition, the user can specify, as stopping criterion, a minimum number ($n_{\min}$) of treated and control units to split a node.
2. Subset selection and weights updating. Once that a tree has been grown, BaRPa looks for the *best-balanced nodes*, i.e., the terminal nodes presenting the best balancing of each $X_j|T = 0$ with respect to $X_j|T = 1$. To avoid nodes with very few treated or untreated units, the final nodes must also have a ratio between the number of treated and untreated units between two user-specified values α and β. Denoting with N_T and N_C the number of treated and control units in the original data and with n_t and n_c the treated and control units belonging to the best-balanced nodes, BaRPa selects the subset $n_t + n_c$ and uses this subset to estimate the proximity matrix.
3. Estimation of the proximity matrix Π and of the average treatment effect. The proximity matrix Π is made up of N_T rows and N_C columns and is estimated after growing R binary trees on resampled versions of the original data. More precisely, in iteration r, a binary tree is grown to find a partition of the sampled

data. Then, using the subset of the best balanced nodes, a proximity measure $\pi_{ij}^{(r)}$ $(i = 1, \dots, N_T; j = 1, \dots, N_C)$ is derived as follows: $\pi_{ij}^{(r)}$ is set to 1 for treated and control units belonging to the selected subset $n_t + n_c$, whereas it is set to 0 for the remaining $N - (n_t + n_c)$ units. The final proximity matrix is the average of the $\pi_{ij}^{(r)}$s over the R samples:

$$\Pi = \left[\pi_{ij} = \frac{1}{R} \sum_{r=1}^{R} \pi_{ij}^{(r)} \right] \tag{1}$$

To estimate the average treatment effect, BaRPa uses the same estimators proposed in [3]: A weighted ATT estimator, based on weights $f_{ij} = \pi_{ij} / \sum_{j=1}^{N_C} \pi_{ij}$, is

$$\text{ATT}_W = \frac{1}{N_T} \sum_{i=1}^{N_T} \left[(Y|T=1)_i - \sum_{j=1}^{N_C} f_{ij} (Y|T=0)_j \right] \tag{2}$$

An ATT estimator based on normalized weights is also considered. Let $\pi_i^{max} = \max_{j \in (1,\dots,N_C)} \pi_{ij}$ indicate the maximum number of times a treated unit i has been matched to a control unit. Then the weights can be normalized by defining $q_i = \pi_i^{max} / \sum_{i=1}^{N_T} \pi_i^{max}$ in such a way that $\sum_{i=1}^{N_T} q_i = 1$. The normalized ATT estimator is

$$\text{ATT}_N = \frac{1}{N_T} \sum_{i=1}^{N_T} \left[(Y|T=1)_i - \sum_{j=1}^{N_C} f_{ij} (Y|T=0)_j \right] q_i \tag{3}$$

BaRPa also implements the *selected* ATT estimators, which are built solely on the treated units that have been matched at least $\lambda^*\%$ times with some other control units and so can be used to identify the common support. The value $\lambda^* \in (0, 1)$ is the maximum value of λ for which either more than $n_{\min}$ treated units or more than $n_{\min}$ control units can be selected. Alternatively, the selection threshold can be specified as $n_{\min} + k \cdot \sigma_\tau$ (τ is the number of units selected by the R trees and σ_τ is its standard deviation; k is a constant) in order to select larger subsets. Thus, selected estimators evaluate the treatment effect restricted to the portion of treated units that can be reliably matched. These estimators are denoted by $S.\text{ATT}_W$ and $S.\text{ATT}_N$. Of course, if the goal is estimating the average treatment effect on the controls (ATC), then ATC_W, ATC_N, $S.\text{ATC}_W$, and $S.\text{ATC}_N$ can be defined in a similar way.

4. Subsets Sampling and Stopping rule. As previously stated, after growing the first tree BaRPa selects a balanced subset $n_t + n_c$ of cases. In order to define weights for the estimation of the proximity matrix Π as well as to overcome the well-known instability problem characterizing recursive partitioning algorithms, BaRPa investigates, by growing additional trees on resampled data, if the selection of the first balanced subset is sensitive to small perturbation in the original data and if the same subset can be further refined. The sampling scheme depends on which average treatment effect is being estimated (ATT or ATC) and is motivated by the idea of finding the best (set of) counterpart(s) for each (treated or control) unit. In this respect, when estimating ATT resampling works by retaining, in each run, N_T units and by drawing $N_C = N_T$ cases from the original data with weights $\mathbf{w}_{N_C}$. Whereas, when estimating ATC, N_C cases are retained in each run and $N_T = N_C$ are drawn from the original data with weights $\mathbf{w}_{N_T}$. Weights $\mathbf{w}_{N_C}$ and $\mathbf{w}_{N_T}$ are updated, in each run, on the basis of the selected subset $n_t + n_c$. BaRPa stops the process of identification of the sequence of trees as soon as the relative change in value of one of the previously defined estimators is lower than a previously specified threshold.

3 Simulation Study

Design Factors The performance of BaRPa has been evaluated on simulated data. The design factors consider a dataset composed of $N = 1000$ cases on which a treatment effect is generated in order to obtain $P(T = 1) \approx 0.4$ as follows:

$$P(T = 1) = (1 + \exp(-(-0.22 + \sum_{i=1}^{4} B_i \cdot X_i + B_5\epsilon)))^{-1} \tag{4}$$

In (4), X_1, X_2 are numeric and derive from a mixture of two normal random variables: $X_1 \sim [0.5 \cdot N(0.1, 0.25) + 0.5 \cdot N(0.8, 0.375)]$; $X_2 \sim [0.5 \cdot N(0.3, 0.25) + 0.5 \cdot N(0.7, 0.375)]$. X_3 and X_4 are dichotomous and derive from $\tilde{X}_3$ and $\tilde{X}_4$, the latter being defined as mixtures of two uniform random variables: $\tilde{X}_3 \sim [0.5 \cdot U(0, 0.6) + 0.5 \cdot N(0.4, 1)]$ and $\tilde{X}_4 \sim [0.5 \cdot U(0, 0.6) + 0.5 \cdot N(0.6, 1)]$. In practice, X_3 (X_4) equals 1 if $\tilde{X}_3 > 0.5$ ($\tilde{X}_4 > 0.5$) and 0 otherwise. The error term is generated as: $\epsilon \sim N(0.5, 0.3)$. The B coefficients equals 0.10 for B_1, B_2, B_3, and B_5, and 0 for B_4, so that the treatment effect depends on X_1, X_2, and X_3 but not on X_4.

The outcome Y is generated as a linear combination of the four covariates and of the treatment T and is such that the *true treatment effect* G is 0.60:

$$Y = -1.50 + 0.60X_1 - 0.72X_2 - 0.73X_3 - 0.20X_4 + 0.60T + \epsilon \tag{5}$$

The simulated data described above are highly imbalanced: this imbalance can be observed from the empirical distribution of Y, X_1, and X_2, as well as from their conditional distributions shown in Fig. 1.

Results BaRPa has been applied on the simulated dataset in order to estimate G, which represents the Average Treatment effect on the Treated (ATT) via covariate balance adjustment. Thus, two natural and widely used performance metrics are the relative bias of the ATT estimator and the average imbalance of covariates in the selected subsets. We show the value of these metrics for several estimators obtained from BaRPa. In particular, the simulation study is aimed at the evaluation of the ability of the estimators defined in (2) and (3), as well as of the selected estimators

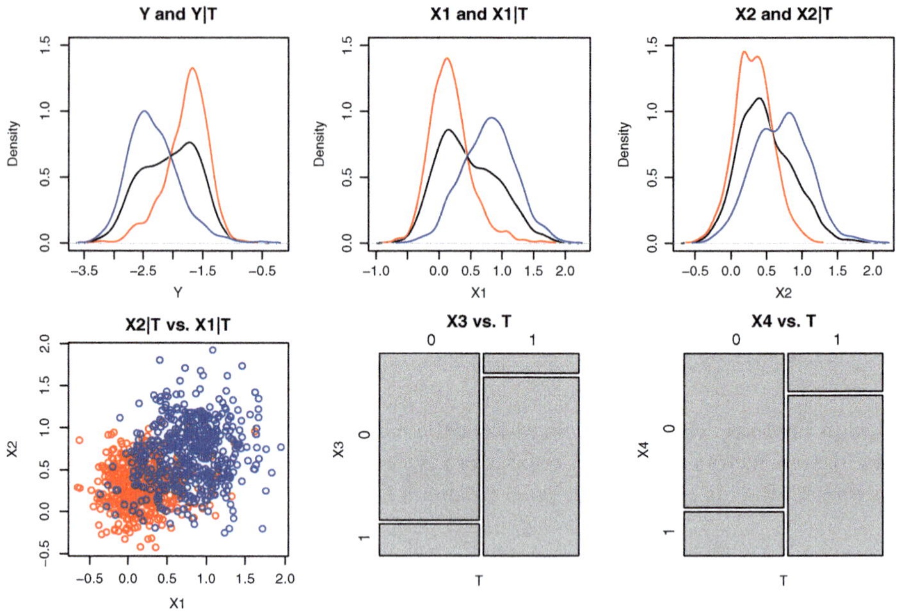

Fig. 1 *Top panel*. Imbalance in the distribution of the outcome Y (*left*) and of the continuous variables X_1 (*center*) and X_2 (*right*): *red lines* identify treated units, *blue lines* identify control ones. *Bottom panel*. Identification of the common support of X_1 and X_2 (*left*): *red points* identify treated units, *blue points* identify control ones Imbalance in the distribution of the categorical variables X_3 (*center*) and X_4 (*right*)

$S.\text{ATT}_W$ and $S.\text{ATT}_N$. BaRPa was applied by setting: $n_{\min} = \sqrt{N_T}$, $\alpha = 0.5$, $\beta = 2$ and by specifying a minimum relative change in the value of each considered estimators of 0.0001 from one iteration to the next one to stop the procedure. A previously performed pilot study has suggested that $n_{\min} = \sqrt{N_T}$ allows us to obtain a reasonable minimum size of a node to be split, while the values specified for α and β prevent from the situation in which a node with a few treated units and many control ones (or viceversa) is selected. To give insight into the information content of the output provided by BaRPa, the outcome of the first tree is summarized in Table 1. The original data are split according to the splitting and stopping criteria introduced in Sect. 2.1 and the final tree includes eight terminal nodes. Of these, only four are considered since they present a ratio between the number of treated units and that of the corresponding control units which is between $\alpha = 0.5$ and $\beta = 2$. As such, this tree selects 139 observations (58 treated units and 79 control ones) to be included in the selected subset $n_c + n_t$. This initial selection is updated in the following iterations by growing additional trees on selected samples of original data identified according to the resampling scheme defined in step 2 of Sect. 2.1.

Moving to selected estimators $S.\text{ATT}_W$ and $S.\text{ATT}_N$, their performance has been evaluated by setting $k = 0, 1, 2, 3$ for the "$n_{\min} + k\sigma_\tau$" rule specified in Sect. 2.1 in order to consider eight different selected estimators. Table 2 shows the performance

Table 1 Output of a tree grown by BaRPa on simulated data

$\mathcal{N}$	n	n_c	n_t	$\mu_{\text{ASAM}}(\mathcal{N})$	x_L^*	x_R^*	x_{ij}^*	X_j^*	Terminal node
1	784	392	392	1.99	0.12	0.78	0.78	X_1	
2	537	376	161	1.78	0.00	0.00	0.00	X_4	
3	247	16	231	0.01					*
4	254	245	9	1.76					*
5	283	131	152	1.40	0.24	0.54	0.54	X_1	
10	170	106	64	1.20	0.24	0.33	0.33	X_1	
11	113	25	88	0.03	0.28	0.46	0.46	X_2	
20	91	75	16	0.76	0.78	0.69	0.69	X_2	
21	**79**	**31**	**48**	**0.05**					*
22	39	18	21	0.12	0.67	0.68	0.67	X_1	
23	74	7	67	0.00					*
40	72	66	6	1.01					*
41	**19**	**9**	**10**	**0.20**					*
44	**16**	**7**	**9**	**0.41**					*
45	**23**	**11**	**12**	**0.13**					*

$\mathcal{N}$ indicates the node number. For each node: n is the number of observations; n_c and n_t are the number of control and treated units; $\mu_{\text{ASAM}}(\mathcal{N})$ is the average ASAM of all covariates for node N members; x_L^* (x_R^*) is the split point that minimizes the average ASAM of all the covariates in the left (right) child node; x_{ij}^* is the selected split point; "*" in the last column refers to tree nodes labelled as terminal. Rows with bold font indicated terminal nodes with a ratio between the number of treated units and that of the corresponding control units between $\alpha = 0.5$ and $\beta = 2$

Table 2 Results provided by BaRPa on simulated data (relative bias and average ASAM)

ATT	$\hat{G}$	$\|r_b(\hat{G})\|$	R	λ^*rule	λ^*	$\hat{n}_t(\hat{n}_c)$	μ_{ASAM}	$\Delta(\mu_{\text{ASAM}})$	μ_{ASAM999}
ATT_W	−0.746	2.24	311						
ATT_N	0.459	0.23	199						
$S.\text{ATT}_W$	0.612	0.02	19	$n_{\min} = \sqrt{N_T}$	0.48	20(20)	0.08	−0.96	1.96
$S.\text{ATT}_W$	0.501	0.16	79	$\sqrt{N_T} + \sigma_\tau$	0.27	54(54)	0.41	−0.79	1.97
$S.\text{ATT}_W$	0.427	0.29	48	$\sqrt{N_T} + 2\sigma_\tau$	0.11	86(86)	0.34	−0.83	1.97
$S.\text{ATT}_W$	0.418	0.30	103	$\sqrt{N_T} + 3\sigma_\tau$	0.03	137(122)	0.66	−0.67	1.97
$S.\text{ATT}_N$	0.463	0.23	816	$n_{\min} = \sqrt{N_T}$	0.53	20(21)	0.46	−0.77	1.96
$S.\text{ATT}_N$	0.479	0.20	202	$\sqrt{N_T} + \sigma_\tau$	0.28	54(54)	0.23	−0.88	1.96
$S.\text{ATT}_N$	0.503	0.16	300	$\sqrt{N_T} + 2\sigma_\tau$	0.10	93(86)	0.38	−0.81	1.96
$S.\text{ATT}_N$	0.482	0.20	182	$\sqrt{N_T} + 3\sigma_\tau$	0.04	123(119)	0.62	−0.69	1.96

The table reports, in each row, the results of the estimators defined in Sect. 2.1. $\hat{G}$ is the estimated value for the true treatment effect G and $|r_b(\hat{G})|$ is its relative bias; R denotes the number of samples; λ^*rule is the criterion used to define the empirical threshold λ^* for $S.\text{ATT}_W$ and $S.\text{ATT}_N$; $\hat{n}_t(\hat{n}_c)$ is the number of selected treated (control) units; μ_{ASAM} is the average ASAM obtained for the four covariates on the selected subset $\hat{n}_t(\hat{n}_c)$ and $\Delta(\mu_{\text{ASAM}})$ denotes the relative change of μ_{ASAM} with respect to the same measure computed on the original data; μ_{ASAM999} is the average μ_{ASAM} obtained on 999 independent samples composed of $\hat{n}_t$ treated and $\hat{n}_c$ control units of BaRPa on the simulated data. All the estimators, except ATT_W, present a reasonable, and in same cases extremely low, relative bias. As for balancing, the selected estimators consistently reduce the average ASAM of the original data (column $\Delta(\mu_{\text{ASAM}})$): in particular, $S.\text{ATT}_W$ with the $n_{\min}$ rule for the selection of λ^* provides the maximum reduction in imbalance by detecting a small subset composed of 20 treated and 20 control units. For comparison purposes, the same measure (μ_{ASAM}) has been computed on 999 bootstrap replications of the original data: reported values do not vary with respect to the same measure computed on the original data, thus confirming that original data are really imbalanced and enforcing the effectiveness of the results obtained for the different BaRPa's estimators.[1]

4 Concluding Remarks

A tree-based balancing algorithm named BaRPa which is invariant under monotonic transformation of data has been presented. BaRPa exploits a tree-based method balancing property by growing different trees on resampled versions of the original data based on a multidimensional balancing measure for node splitting. Next to tree growing, a proximity matrix that counts the proportion of times different units fall in the same terminal node of the tree is derived: it is the input for the derivation of classic causal estimators (e.g., ATT) which use all the data or a selected subset of matched observations. As such, BaRPa does not rely on a particular distance or on a specific model to be estimated. Results on simulated data provide evidence of the effectiveness of BaRPa in reducing imbalance in covariate distribution and selected versions of BaRPa estimators yield less biased estimators than the non-selected counterparts. The assessment of the properties of BaRPa's estimators and the definition of a stopping criteria that reduces computational complexity will both be addressed in future work.

References

1. Dehejia, R., Wahba, S.: Causal effects in nonexperimental studies: reevaluating the evaluation of training programs. J. Am. Stat. Assoc. **94**(448), 1053Ű1062 (1999)
2. Heckman, J.H., Ichimura, H., Smith, J., Todd, P.: Characterizing selection bias using experimental data. Econometrica **66**(5), 1017–1098 (1998)
3. Porro, G., Iacus, S.M.: Random recursive partitioning: a matching method for the estimation of average treatment effects. J. Appl. Econ. **24**, 163–185 (2009)
4. Rubin, D.B.: The use of matched sampling and regression adjustment to remove bias in observational studies. Biometrics **29**, 184–203 (1973)

[1]Results obtained by varying the number of observations ($N = 250$; $N = 500$), not reported here to save space, lead to the same conclusions.

Meta-Analysis of Poll Accuracy Measures: A Multilevel Approach

Rosario D'Agata and Venera Tomaselli

Abstract Following a meta-analysis approach as a special case of multilevel modelling, we identify potential sources of dissimilarities in accuracy measures of pre-election polls, carried out during Parliamentary elections in Italy from 2001 to 2008. The predictive accuracy measure, computed to compare the pre-electoral poll result to the actual result, is the dependent variable and the poll characteristics are the explanatory variables and are introduced in a hierarchical model. In the model each outcome is affected by a specific sampling error assumed to have a normal distribution and a known variance. The multilevel model approach decomposes variance components as well as meta-analysis random models. We propose a multilevel approach, in order to make the estimation procedure easier and more flexible than in a traditional meta-analysis approach.

Keywords Multilevel models • Pre-election polls

1 Meta-Analysis: A Case of Hierarchical Modelling

Envisaging to meta-analysis as a *special* case of multilevel modelling [2], we identify the potential sources of differences, by the estimation of an effect size over all the predictive accuracy measures of poll results. In this aim, we specify a random effects hierarchical model for a meta-analysis study of the measures, where each poll result is a level-1 unit and the poll is a level-2 unit [6], to check relationships between explanatory variables and dependent variables [3].

In order to compute an average effect size we can employ fixed or random effects models [1]. Specifying a fixed effects model, we assume that the true effect size is always the same across all studies. Formally [6]:

$$d_j = \delta_j + e_j \tag{1}$$

R. D'Agata (✉) • V. Tomaselli
University of Catania, Vitt. Emanuele II, 8 Catania, Italy
e-mail: rodagata@unict.it; tomavene@unict.it

I. Morlini et al. (eds.), *Advances in Statistical Models for Data Analysis*, Studies in Classification, Data Analysis, and Knowledge Organization, DOI 10.1007/978-3-319-17377-1_8

where:

- d_j is the outcome of study j ($j = 1, \ldots, J$)
- δ_j is the population value of the outcome of the j-th study
- e_j is the sampling error for this j-th study

As a consequence, the only error source is that produced by random sampling error or error e_j *within* studies, assumed to have normal distribution with known variance $\sigma^2_{e_j}$ [10].

Since we can suppose that the effect size across the studies will be similar but not identical, we can estimate different effect sizes across all studies. In this case, we specify a random effects model, in which the effect size, from each primary study, is estimated as a mean of a distribution of different true effect sizes δ_j across the studies, with an error term which is variance between studies. The observed effect d_j in (1) is sampled from a distribution of effects with true effect δ_j and variance $\sigma^2_{u_j}$ [7]. In turn, δ_j is a function of the mean of all true effects (γ_0) plus between studies error (u_j). Formally:

$$\delta_j = \gamma_0 + u_j \tag{2}$$

We can therefore rewrite (1) as:

$$d_j = \gamma_0 + u_j + e_j \tag{3}$$

- δ_j is the effect observed in the j-th study
- γ_0 is the estimate for the mean outcome across all the studies
- u_j is the between studies residual error term
- e_j is the within studies error term.

This is an *intercept only* or *empty* model, equivalent to the random effects model for meta-analysis [4], in which the variance of the residual errors $\sigma^2_{u_j}$ not equal to 0 and significant indicates that the outcomes across the studies are heterogeneous. In the model the effect size estimates are affected by two error sources: random sampling error or within studies error (e_j) and variance among the true effect sizes or between studies error (u_j).

In the multilevel approach to meta-analysis the dependent variable is the effect size of j-study. As data of primary level-1 units we use only summary statistics—i.e.: p-value, mean, correlation coefficient, odd-ratio, etc.—varying across studies as level-2 units [6].

Following the multilevel approach, in a random effects model, we can separate the variance of study outcomes into two components [5]:

- Within studies variance as sampling variance
- Between studies variance, due to the differences across the study results, computed in our application as predictive accuracy measures.

If the between variance is statistically significant, we can assess that the study outcomes are heterogeneous. To explain such a heterogeneity we include in a random effects model the study characteristics as explanatory predictors of the differences found in predictive accuracy measures across the studies. Estimating the effect size in the case of heterogeneous expected results by means of multilevel modelling is simpler than by traditional meta-analysis methods, because a multilevel approach is more flexible [6]. Furthermore, we can avoid the clustering of studies due to heterogeneous effect sizes across the studies. So, we do not need to identify any variable defining the membership of studies in a cluster.

Employing a multilevel approach (2) can be written as follows:

$$\delta_j = \gamma_0 + \gamma_1 Z_{1j} + \gamma_2 Z_{2j} + \cdots + \gamma_k Z_{kj} + u_j \tag{4}$$

where:

- δ_j is the effect size assumed as varying across the studies
- γ_0 is the mean of all true effects
- Z_{kj} are covariates as explanatory variables (study features)
- γ_k are the coefficients
- u_j is the error term, representing the differences across the studies, assumed to have normal distribution with known variance $\sigma^2_{u_j}$.

By substituting Eq. (4) into Eq. (1), the model can be written as:

$$d_j = \gamma_0 + \gamma_1 Z_{1j} + \gamma_2 Z_{2j} + \cdots + \gamma_k Z_{kj} + u_j + e_j. \tag{5}$$

So, the effect size estimate δ_j depends on study features Z_{kj}, on the error term u_j and on sampling error of each study e_j. The variance of uj ($\sigma^2_{u_j}$) could be considered a level-2 variance and indicates how much the outcomes vary across the studies.

In order to specify a multilevel model explaining the variance between studies, firstly we estimate an *empty* model (3) to check the homogeneity of outcomes, testing a null-hypothesis where the variance of the residual errors $\sigma^2_{u_j}$ is equal to 0 [8]. In meta-analysis with small sample and small variances, the Wald z-test is inaccurate for testing the variances [6]. Moreover, it is based on the assumption of normality and the variances have a χ^2-distribution with $df = j - k - 1$, where j is the number of studies and k is the number of covariates introduced into the model. So, we have to compute the deviance difference χ^2-test on the variances based on the sum of the squared residuals divided by their sampling variances or standard errors [6]. Formally:

$$\chi^2 = \sum_j \left[\frac{d_j - \hat{d}_j}{s.e.\,(d_j)} \right]^2. \tag{6}$$

If the null hypothesis is rejected, we have to estimate the proportion of variance due to the study characteristics or variance between the level-2 units. So, as in traditional multilevel modelling, we can compute the ρ intra-class correlation coefficient (ICC), as the ratio between the level-2 variance $\sigma^2_{u_j}$ and the total variance $(\sigma^2_{u_j} + \sigma^2_{e_j})$. In formula:

$$\rho = \frac{\sigma^2_{u_j}}{\sigma^2_{u_j} + \sigma^2_{e_j}}. \tag{7}$$

In multilevel meta-analysis we have only level-2 variables. We can therefore only calculate the level-2 variance. A reduction of ρ indicates how the considerable extent of the covariates (the features of the study) affects this variance [7].

2 The Accuracy Measure of Pre-Electoral Polls

Unlike other sample surveys, for pre-election polls, we can make comparisons between poll results and actual voting results. For pre-election poll data, we suppose that how poll respondents indicate they will vote and how they actually vote in a subsequent election will correspond. If a poll result reflects the same distribution of voting preferences as happens in the following election, we have an accurate predictor. The more a predictor of an election outcome is able to provide unbiased estimates of electoral preferences, the more accurate it is. In order to measure how accurate a poll outcome is, we choose to use a A_{ij} poll accuracy measure as a predictor of an election result. By transforming poll outcomes into accuracy measures, we standardize all results thus making them comparable to one another.

A_{ij} measure was used for the first time to assess the predictive ability of pre-election polls in the U.S. Presidential elections of 1948, 1996, 2000 and also in the 2002 election for the Offices of Governor and Senator [9]. A_{ij} measure[1] is computed as the ratio obtained by dividing two odds:

$$A_{ij} = \ln\left\{\left[s_{ij}/\left(1 - s_{ij}\right)\right] / \left[S_j/\left(1 - S_j\right)\right]\right\} \tag{8}$$

where:

- s_{ij} is the proportion of respondents favoring the s-competitor (party, coalition or candidate) in the i-th poll referring to the j-th population;
- $1 - s_{ij}$ is the proportion of respondents favoring all other competitors in the same i-th poll, for the same j-th population;
- S_j is the real proportion of votes polled by the same S-competitor in the same j-th population;

[1] A_{ij} measure is not affected by the size of the undecided voter category. Furthermore, it is standardized for the real election result. So, it is possible to study by means of a meta-analysis approach the origin of bias of the polls across different elections for race and time.

– $1 - S_j$ the actual proportion of votes polled by all other competitors in the same j-th population.

By dividing the poll odds by the election odds, we can obtain the value of odds ratio and, thus, the value of the A_{ij} measure as the natural logarithm of the odds ratio between the number of respondents who declare their intention to vote for the s_{ij}-competitor and those who intend to vote for all the others $(1 - s_{ij})$ in the i-th poll and for the j-th population, and the real number of votes for each of the two groups (S_j and $1 - S_j$) in the following election. The transformation of odds ratio by calculating its natural log is used to create a symmetric measure around 0 and to simplify the computation of the variance [9], taking into account the sampling error of the poll measure, assuming normal distribution and with a known variance.

Let s_{ij} and $1 - s_{ij}$ be random variables, where s_{ij} is the proportion of respondents preferring the s_{ij}-competitor and $1 - s_{ij}$ is the proportion of respondents who do not prefer the s_{ij}-competitor, in the i-th poll referring to the j-th population with sample size n_{ij}, with $[s_{ij} + (1 - s_{ij})] = 1$. Let $p(s_{ij})$ be the probability of preferring the s_{ij}-competitor and $[1 - p(s_{ij})]$ be the probability of not preferring the s_{ij}-competitor. The covariance matrix (Cov) of the vector $[s_{ij}, (1 - s_{ij})]$ is:

$$Cov\begin{bmatrix} s_{ij} \\ 1 - s_{ij} \end{bmatrix} = \frac{1}{n_{ij}}\begin{bmatrix} p(s_{ij})\left[1 - p(s_{ij})\right] & -p(s_{ij})\left[1 - p(s_{ij})\right] \\ -p(s_{ij})\left[1 - p(s_{ij})\right] & p(s_{ij})\left[1 - p(s_{ij})\right] \end{bmatrix} \tag{9}$$

so that the relative covariance matrix $(RelCov)$ is:

$$RelCov\begin{bmatrix} s_{ij} \\ 1 - s_{ij} \end{bmatrix} = \frac{1}{n_{ij}}\begin{bmatrix} \left[1 - p(s_{ij})\right]/p(s_{ij}) & -1 \\ -1 & p(s_{ij})/\left[1 - p(s_{ij})\right] \end{bmatrix}. \tag{10}$$

The variance of A_{ij} measure (8) for each i-poll is computed as:

$$Var\left(A_{ij}\right) = 1/\left[n_{ij}s_{ij}\left(1 - s_{ij}\right)\right]. \tag{11}$$

A_{ij} measure may take on positive, negative or null values. A positive value indicates an s_{ij} bias. If the A_{ij} measure value is negative, the poll is biased by an overestimated share of $(1 - s_{ij})$ compared to $(1 - S_j)$ election result. The A_{ij} measure is equal to 0, when the odds ratio is equal to 1. This last result occurs only if the poll result and the real voting result are exactly the same.

To explain the variance of the measure, we can use A_{ij} as y-dependent variable and the poll features as predictors. We can assess whether there are significant relationships between the ability of the poll to predict the election results and the characteristics of the poll [9], including referred territorial area, customer, sampling procedures, survey methods, time poll period, sample size, number of days from poll to election, vote gap between the two competitors, polling agency, type of election, etc.

3 Meta-Analysis of Pre-electoral Poll Accuracy

In this study, we propose a meta-analysis of poll predictive accuracy measures in order to analyse their variance in a dataset of 42 pre-election polls. These polls have been carried out before the fortnight press blackout, previous the National elections from 2001 to 2008, and published on the official website: www.sondaggielettorali.it.

In order to assess the accuracy of each poll, we compute the accuracy measure by employing the formula [8]. Figure 1 shows the distribution of accuracy measures in the 42 polls. On average, we note that the Centre-Right electoral outcome (*AccDx*) is basically underestimated (−0.0432), while the Centre-Left coalition performance (*AccSn*) is overestimated (0.0754). Over the period of the elections considered, an improvement occurs in the ability of polls to accurately forecast results. For the election in 2008, accuracy measures computed both for *Centre-Right* coalition and for *Centre-Left* coalition are very near to 0. This could be due to an improvement in the quality of the methods used to conduct the pre-election polls such as sampling techniques and survey methods.

In order to evaluate the relationship between poll characteristics and accuracy measure, a meta-analysis is conducted using a multilevel approach. As the first step, we specify an empty model with the aim of checking heterogeneity across the polls as level-2 units. As the dependent variable, we choose the accuracy measure for *Centre-Left* coalition computed as shown in (8). Formally the model is:

$$AccSn_j = \gamma_0 + u_j + e_j \tag{12}$$

where:

- $AccSn_j$ is the accuracy measure for *Centre-Left* coalition in the j-th poll;

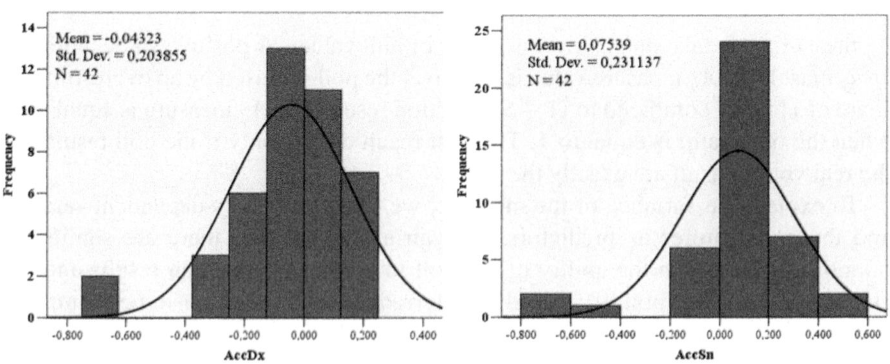

Fig. 1 Distribution of accuracy measures for the two coalitions

Table 1 Empty model: y = Accuracy of Centre-Left coalition ($AccSn_j$)

	β_{0j}	S.E.	Z	p-value
Fixed effects				
Intercept	0.078	0.035	2.229	<0.02
Random effects				
$\sigma^2_{u_j}$	0.047	0.011	4.273	<0.001
Deviance $= -4.839$				
Deviance difference test : $\chi^2 = 1089.7$; $df = 41$; p-value < 0.001				

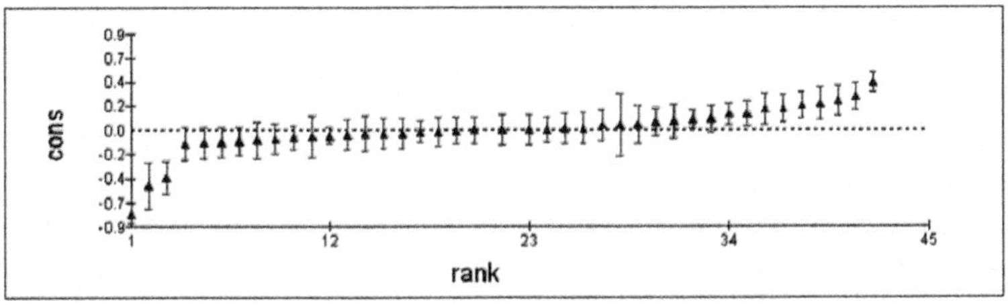

Fig. 2 Poll level residual plot (confidence intervals: 95 %) of empty model

- γ_0 is the estimate for the mean accuracy measure across all polls;
- u_j is the residual term for the j-th poll;
- e_j is the sampling error for the j-th poll computed by (11).

In the *empty* model[2] (Table 1), the value of the intercept (0.078) is significant (p-value < 0.02) and confirms the overestimation of the Centre-Left result previously observed (Fig. 1). The random component ($\sigma^2_{u_j}$) indicates how much the accuracy measures vary across the polls. It is estimated as 0.047. In order to test the homogeneity of accuracy measures across the polls, we compute the deviance difference χ^2-test on the residuals to check for a null hypothesis where $\sigma^2_{u_j}$ is equal to 0. The test produces a χ^2 equal to 1089.7 (p-value < 0.001). So, we have to reject the null hypothesis, because the p-value indicates the presence of heterogeneity in accuracy measures across the polls.

In the plot of poll level residuals (Fig. 2), analysing the confidence intervals computed for the 42 polls, we note a group of about 12 polls where the confidence intervals for their residuals do not overlap 0. We can, therefore, observe 12 polls that differ significantly from the mean accuracy measures at the 5 % confidence level. Furthermore, the proportion of systematic variance, computed by means of an ICC (7), is 0.90 and informs us that in 90 % of polls the difference in accuracy measures is due to the features of the polls. The presence of heterogeneity and the value of the ICC allow us to continue the analysis with the aim of explaining in a

[2]The models are estimated by employing RML algorithm implemented in *MlWin* software.

Table 2 Complete model: $y =$ Accuracy Centre-Left coalition ($AccSn_j$)

	β_{0i}	S.E.	Z	p-value
Fixed effects				
Intercept	0.675	0.224	3.013	<0.002
Customer: *agency* (Ref. Mass Media)	−0.179	0.056	−3.196	<0.001
Customer: *political organ* (Ref. Mass Media)	−0.280	0.110	−2.545	<0.006
Survey method: *CATI* e *CAWI* (Ref. *CATI*)	0.200	0.068	2.941	<0.002
Survey method: *CAWI* (Ref. *CATI*)	0.280	0.071	3.944	<0.001
Survey method: *CASI* (Ref. *CATI*)	0.369	0.14	2.636	<0.005
Sample size: $\ln(n_i/N_j)$	0.033	0.013	2.538	<0.006
Poll period (days)	0.055	0.021	2.619	<0.005
Days from poll to election	−0.019	0.006	−3.167	<0.001
Predicted gap	0.024	0.006	4.000	<0.001
Year: 2001 (Ref. 2008)	0.185	0.054	3.425	<0.001
Year: 2006 (Ref. 2008)	0.348	0.099	3.515	<0.001
Electoral winner: Centre-Left (Ref. Centre-Right)	−0.217	0.055	−3.945	<0.001
Random effects				
$\sigma^2_{u_i}$	0.007	0.002	3.500	<0.001
Deviance $= -68.327$				
Deviance difference test : $\chi^2 = 165.73$; $df = 29$; p-value < 0.001				

complete model the variance between polls in accuracy measures by means of the features of polls.

Table 2 shows that all predictors appear to be significant. Analysing the values of β_{0_j} coefficients, when the customer is a political organ or agency, the value of predictive accuracy measures tends to decrease compared to when the customer is one of the mass media (−0.179). The data collected by CAWI, either alone (0.280) or combined with *CATI* (0.200), appear to be linked to an increase in the value of accuracy. This is also true for *CASI* (0.369). All of the survey methods introduced in the model, compared to the *CATI* method, appear to have a positive relation to accuracy. The greater the sample size (0.033) and the longer the survey period (0.055), the more accurate the poll. The fewer days from poll to election (−0.019) and the greater the predicted gap (0.024) between the two coalitions, the more accurate the forecast is. Finally, if the winner is *Centre-Left* (−0.217), the accuracy measure decreases more than in elections won by the *Center-Right* coalition. Moreover, in the complete model the variance between the polls $\sigma^2_{u_j}$ is reduced from 0.047, observed in the empty model, to 0.007. The value of the deviance difference test is notably reduced from 1089.7 in empty model to 165.73, and it remains statistically significant, too. In addition, the proportion of systematic variance is reduced from 0.90 to 0.57. Comparing the poll residuals of the complete model, plotted in Fig. 3, with the poll residuals of the empty model (Fig. 2), we note that only one interval does not overlap 0.

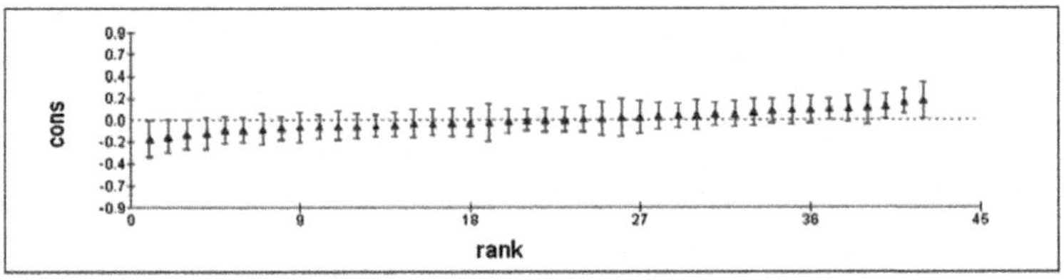

Fig. 3 Poll level residuals plot (confidence interval: 95 %) of complete model

4 Conclusions

The use of accuracy measures has made it possible to detect the predictive ability of each poll with a single value using a multilevel approach to meta-analysis. Specifying a random intercept model has allowed us to estimate the random component of the variance between the polls and to test the significance of heterogeneity among the results by means of the χ^2 residuals test. Thus, we were able to calculate the ICC in order to estimate the proportion of total variance, due either to the variance between the polls or the predictive accuracy measures across all the polls.

By means of a complete hierarchical model we obtained a reduction of the random component of the variance between the polls. So, the heterogeneity in the accuracy measures is explained by poll features as predictors in the estimated model. Nevertheless, a proportion of unexplained variance remains, as we note, both in the significance of the residuals test and the value of the ICC.

References

1. Borenstein, M., Hedges, L.V., Higgins, J.P.T., Rothstein, H.R.: Introduction to Meta-Analysis. Wiley, Chichester (2011)
2. Bryk, A., Raudenbush, S.W.: Hierarchical Linear Models: Applications and Data Analysis Methods. Sage, Newbury Park (1992)
3. Ellis, P.D.: The essential guide to effect sizes: statistical power, meta-analysis, and the interpretation of research results. Cambridge University Press, Cambridge (2010)
4. Hedges, L.V., Olkin, I., Statistiker, M., Olkin, I., Olkin, I.: Statistical Methods for Meta-Analysis. Academic, New York (1985)
5. Hox, J.: Multilevel Analysis: Techniques and Applications. Routledge, London (2010)
6. Hox, J., de Leeuw, E.: Multilevel models for meta-analysis. In: Reise, S.P., Duan, N. (eds.) Multilevel Modeling: Methodological Advances, Issues, and Applications, pp. 90–111. Lawrence Erlbaum Associates Publishers, Mahwah (2003)
7. Hunter, J.E., Schmidt, F.L.: Methods of meta-analysis: correcting error and bias in research findings. Sage, Thousand Oaks (2004)
8. Koetse, M.J., Florax, R.J., de Groot, H.L.: Consequences of effect size heterogeneity for meta-analysis: a Monte Carlo study. Stat. Methods Appl. **19**(2), 217–236 (2010)
9. Martin, E.A., Traugott, M.W., Kennedy, C.: A review and proposal for a new measure of poll accuracy. Public Opin. Q. **69**(3), 342–369 (2005)
10. Raudenbush, S.W., Bryk, A.: Hierarchical linear models: applications and data analysis methods, vol. 1. Sage, Thousand Oaks (2002)

Families of Parsimonious Finite Mixtures of Regression Models

Utkarsh J. Dang and Paul D. McNicholas

Abstract Finite mixtures of regression (FMR) models offer a flexible framework for investigating heterogeneity in data with functional dependencies. These models can be conveniently used for unsupervised learning on data with clear regression relationships. We extend such models by imposing an eigen-decomposition on the multivariate error covariance matrix. By constraining parts of this decomposition, we obtain families of parsimonious mixtures of regressions and mixtures of regressions with concomitant variables. These families of models account for correlations between multiple responses. An expectation-maximization algorithm is presented for parameter estimation and performance is illustrated on simulated and real data.

Keywords Concomitant variables • EM algorithm • Finite mixtures of regressions • Mixture models • Multivariate response

1 Introduction

Model-based clustering has become increasingly popular during the last decade. Parametric mixture models are used in model-based clustering; however, such models generally do not exploit covariates. Incorporating a regression structure can yield important insight when there is a regression relationship between some variables. Methodologies that deal with such data include finite mixtures of regression (FMR; [8, 15]) and finite mixtures of regressions with concomitant variables (FMRC; [24]), supported by the popular `flexmix` package [15]. Cluster-weighted models [11] are an alternative to FMR models and have recently been extended to deal with multivariate response [6]. However, these models are less

U.J. Dang (✉)
Department of Biology, McMaster University, Hamilton, ON, Canada
e-mail: udang@mcmaster.ca

P.D. McNicholas
Department of Mathematics & Statistics, McMaster University, Hamilton, ON, Canada
e-mail: mcnicholas@math.mcmaster.ca

I. Morlini et al. (eds.), *Advances in Statistical Models for Data Analysis*,
Studies in Classification, Data Analysis, and Knowledge Organization,
DOI 10.1007/978-3-319-17377-1_9

parsimonious by nature because they explicitly model the distribution of the covariates as well as the response given the covariates.

Multivariate correlated responses can be naturally integrated into FMR and FMRC models. However, `flexmix` currently does not account for correlated response variables for both FMR and FMRC. FMR models that deal with correlated response variables have recently been proposed [10, 21]. Experimental results using these models illustrated that ignoring this correlation can lead to estimated regression coefficients with larger mean square errors and may result in a worse fit to data [21]. However, these models do not decompose the covariance structure to gain parsimony, nor do they extend the FMRC model.

Here, FMR and FMRC are extended to deal with multiple correlated responses. Parsimonious versions of these models are developed by constraining the component covariance matrices using an eigen-decomposition (in Sect. 2.1). An expectation-maximization algorithm is described in Sect. 2.2. Performance is illustrated on simulated and real data and compared to popular existing methodologies like FMR and FMRC (Sect. 3) with some concluding remarks (Sect. 4).

2 Methodology

Let $\boldsymbol{X}_i$ and $\boldsymbol{Y}_i$ be random vectors defined on sample space $\boldsymbol{\Omega}$, for $i = 1, \ldots, N$, where $\boldsymbol{\Omega}_g$ can be partitioned into G disjoint groups. Here, the response vector $\boldsymbol{Y}_i$ has values in $\mathbb{R}^d$ and the explanatory vector $\boldsymbol{X}_i$ has values in $\mathbb{R}^p$. Then, the probability of the response given the covariates $p(\boldsymbol{y}_i|\boldsymbol{x}_i)$ can be decomposed as

$$p(\boldsymbol{y}_i|\boldsymbol{x}_i, \boldsymbol{\theta}) = \sum_{g=1}^{G} p(\boldsymbol{y}_i|\boldsymbol{x}_i, \boldsymbol{\Omega}_g)\pi_{ig}, \tag{1}$$

where $p(\boldsymbol{y}_i|\boldsymbol{x}_i, \boldsymbol{\Omega}_g)$ is the conditional density of $\boldsymbol{Y}_i$ given $\boldsymbol{x}_i$ and $\boldsymbol{\Omega}_g$, and π_{ig} are the mixing weights, where $\pi_{ig} > 0$ ($g = 1, \ldots, G$) and $\sum_{g=1}^{G} \pi_{ig} = 1$ for each i. The parameter $\boldsymbol{\theta}$ denotes the set of all parameters. $\boldsymbol{Y}|\boldsymbol{X}$ is assumed to be normally distributed with mean $\boldsymbol{\mu}_{y;g}$ and covariance matrix $\boldsymbol{\Sigma}_g$, for $g = 1, \ldots, G$. For the FMR model, $\pi_{ig} = \pi_g$, for $g = 1, \ldots, G$ and $i = 1, \ldots, N$. In addition to (1), the FMRC model assumes a concomitant variable multinomial logit model for the component mixing weights, i.e.,

$$\boldsymbol{\pi}_{ig}(\boldsymbol{x}_i, \boldsymbol{\alpha}_g) = \frac{\exp(\boldsymbol{\alpha}_g' \boldsymbol{x}_i)}{\sum_{h=1}^{G} \exp(\boldsymbol{\alpha}_h' \boldsymbol{x}_i)}, \tag{2}$$

with the first component as baseline. In other words, FMR only models the distribution of the $\boldsymbol{Y}|\boldsymbol{X}$, while FMRC models both the distribution of $\boldsymbol{Y}|\boldsymbol{X}$ and a logistic model of the concomitant variables (which may include the covariates).

This implies that for an FMRC model, the classification (dependent on the posterior probability) of an observation into a particular component is dependent on the covariates both through the mixing weights and $\boldsymbol{Y}|\boldsymbol{X}$. Note that for the purposes of this paper, the concomitant variables (usually denoted by $\boldsymbol{W}_i$) are the same as the explanatory variables $\boldsymbol{X}_i$.

2.1 Eigen Decomposition of $\boldsymbol{\Sigma}_{y|x}$

There are $d(d+1)/2$ free parameters in each component covariance matrix for a d-variate Gaussian mixture (cf. (1)). That this number increases quadratically with d is undesirable for all but very low dimensional data sets. To overcome this problem, $\boldsymbol{\Sigma}_g$ can be eigen-decomposed [1] and constraints can be imposed to give a family of mixture models [5], i.e., the gth component covariance matrix can be written as

$$\boldsymbol{\Sigma}_g = \lambda_g \boldsymbol{D}_g \boldsymbol{A}_g \boldsymbol{D}_g', \tag{3}$$

where λ_g is a constant, $\boldsymbol{D}_g$ is the orthogonal matrix of eigenvectors of $\boldsymbol{\Sigma}_g$, and $\boldsymbol{A}_g$ is a diagonal matrix with entries proportional to the eigenvalues of $\boldsymbol{\Sigma}_g$ with the constraint $|\boldsymbol{A}_g| = 1$. Geometrically, λ_g controls the volume, $\boldsymbol{D}_g$ the orientation, and $\boldsymbol{A}_g$ the shape of the gth component (Table 1).

Constraining the component covariance in (1) leads to two families (eFMR and eFMRC, respectively) of 14 models capable of modelling the correlation between responses. This is the first time that FMR and FMRC models have been used with eigen-decomposed covariance structures, i.e., the first parsimonious families of such models.

2.2 Parameter Estimation

Parameter estimation is described here for the most unconstrained model (VVV) from the eFMR and eFMRC families. Let $(\boldsymbol{x}_1, \boldsymbol{y}_1), \ldots, (\boldsymbol{x}_N, \boldsymbol{y}_N)$ be a sample of N independent observations. The observed likelihood function under Gaussian distributional assumptions is

$$L_0(\boldsymbol{\theta}|\boldsymbol{X}, \boldsymbol{Y}) = \prod_{i=1}^{N} p(\boldsymbol{y}_i|\boldsymbol{x}_i, \boldsymbol{\theta}) = \prod_{i=1}^{N} \left[\sum_{g=1}^{G} \phi_d(\boldsymbol{y}_i|\boldsymbol{x}_i, \boldsymbol{\chi}_g)\pi_{ig} \right]. \tag{4}$$

Here, ϕ_d denotes the probability density function for a d-dimensional multivariate Gaussian distribution, $\boldsymbol{\chi}_g = (\boldsymbol{B}_g, \boldsymbol{\Sigma}_g)$ refers to the parameters of the conditional distribution $p(\boldsymbol{y}|\boldsymbol{x})$. Here, the covariates are supplemented by a vector of ones such that $\boldsymbol{B}_g$ is a $(p+1) \times d$ matrix of regression intercepts and coefficients. Hence, the

Table 1 Geometric interpretation of the eigen decomposition of a covariance matrix

Name	Covariance	Volume	Shape	Orientation	Parameters
EII	λI	Equal	Spherical	–	1
VII	$\lambda_g I$	Variable	Spherical	–	G
EEI	$\lambda \boldsymbol{A}$	Equal	Equal	Axis-aligned	d
VEI	$\lambda_g \boldsymbol{A}$	Variable	Equal	Axis-aligned	$d+G-1$
EVI	$\lambda \boldsymbol{A}_g$	Equal	Variable	Axis-aligned	$dG-G+1$
VVI	$\lambda_g \boldsymbol{A}_g$	Variable	Variable	Axis-aligned	dG
EEE	$\lambda \boldsymbol{DAD}'$	Equal	Equal	Equal	$d(d+1)/2$
VEE	$\lambda_g \boldsymbol{DAD}'$	Variable	Equal	Equal	$d(d+1)/2+G-1$
EVE	$\lambda \boldsymbol{DA}_g\boldsymbol{D}'$	Equal	Variable	Equal	$(G-1)(d-1)+d(d+1)/2$
VVE	$\lambda_g \boldsymbol{DA}_g\boldsymbol{D}'$	Variable	Variable	Equal	$(G-1)d+d(d+1)/2$
EEV	$\lambda \boldsymbol{D}_g\boldsymbol{AD}_g'$	Equal	Equal	Variable	$Gd(d+1)/2-(G-1)d$
VEV	$\lambda_g \boldsymbol{D}_g\boldsymbol{AD}_g'$	Variable	Equal	Variable	$Gd(d+1)/2-(G-1)(d-1)$
EVV	$\lambda \boldsymbol{D}_g\boldsymbol{A}_g\boldsymbol{D}_g'$	Equal	Variable	Variable	$Gd(d+1)/2-(G-1)$
VVV	$\lambda_g \boldsymbol{D}_g\boldsymbol{A}_g\boldsymbol{D}_g'$	Variable	Variable	Variable	$Gd(d+1)/2$

$(p+1,\, d)$th element of $\boldsymbol{B}_g$ denotes the regression coefficient of the pth predictor on the dth response.

In (4), $(\boldsymbol{x}_1,\ldots,\boldsymbol{x}_N,\, \boldsymbol{y}_1,\ldots,\boldsymbol{y}_N)$ are considered incomplete in the context of the EM algorithm. The complete-data are $(\boldsymbol{x}_1,\ldots,\boldsymbol{x}_N,\, \boldsymbol{y}_1,\ldots,\boldsymbol{y}_N,\, z_1,\ldots,z_N)$, where z_{ig} is a component label indicator such that $z_{ig} = 1$ if $(\boldsymbol{x}_i',\, \boldsymbol{y}_i')'$ comes from the gth population and $z_{ig} = 0$ otherwise. Therefore, the complete-data log-likelihood is

$$\mathscr{L}_c(\boldsymbol{\theta}\,|\boldsymbol{x},\boldsymbol{y}) = \sum_{i=1}^{N}\sum_{g=1}^{G} z_{ig}\left[\log\phi_d(\boldsymbol{y}_i|\boldsymbol{x}_i,\,\boldsymbol{\chi}_g) + \log\pi_{ig}\right].$$

The E-step involves calculating the expected complete-data log-likelihood

$$Q\left(\boldsymbol{\theta},\boldsymbol{\theta}^{(k)}\right) = \mathbb{E}_{\boldsymbol{\theta}^{(k)}}\{\mathscr{L}_c(\boldsymbol{\theta}\,|\boldsymbol{x},\boldsymbol{y})\} = \sum_{i=1}^{N}\sum_{g=1}^{G}\tau_{ig}^{(k)}\left[Q_1\left(\boldsymbol{\chi}_g|\boldsymbol{\theta}^{(k)}\right) + \log\pi_{ig}^{(k)}\right],$$

where

$$Q_1\left(\boldsymbol{\chi}_g|\boldsymbol{\theta}^{(k)}\right) = \frac{1}{2}\left[-d\log 2\pi - \log|\boldsymbol{\Sigma}_g^{(k)}| - \left(\boldsymbol{y}_i - \boldsymbol{B}_g^{'(k)}\boldsymbol{x}_i\right)'\boldsymbol{\Sigma}_g^{(k)^{-1}}\left(\boldsymbol{y}_i - \boldsymbol{B}_g^{'(k)}\boldsymbol{x}_i\right)\right]$$

and

$$\tau_{ig}^{(k)} := \mathbb{E}_{\boldsymbol{\theta}^{(k)}}\{Z_{ig}|\boldsymbol{x}_i,\boldsymbol{y}_i\} = \frac{\pi_{ig}^{(k)}\phi_d\left(\boldsymbol{y}_i|\boldsymbol{x}_i,\boldsymbol{B}_g^{(k)},\boldsymbol{\Sigma}_g^{(k)}\right)}{\sum_{j=1}^{G}\pi_{ij}^{(k)}\phi_d\left(\boldsymbol{y}_i|\boldsymbol{x}_i,\boldsymbol{B}_j^{(k)},\boldsymbol{\Sigma}_j^{(k)}\right)}. \tag{5}$$

The M-step on the $(k+1)$th iteration of the EM algorithm involves the maximization of the conditional expectation of the complete-data log-likelihood with respect to $\boldsymbol{\theta}$. The updates for $\boldsymbol{B}_g^{(k+1)}$ and $\boldsymbol{\Sigma}_g^{(k+1)}$ are

$$\hat{\boldsymbol{B}}_g^{\prime(k+1)} = \sum_{i=1}^{N}\tau_{ig}^{(k)}\boldsymbol{y}_i\boldsymbol{x}_i'\left(\sum_{i=1}^{N}\tau_{ig}^{(k)}\boldsymbol{x}_i\boldsymbol{x}_i'\right)^{-1}, \tag{6}$$

$$\hat{\boldsymbol{\Sigma}}_g^{(k+1)} = \frac{\sum_{i=1}^{N}\tau_{ig}^{(k)}\left(\boldsymbol{y}_i-\hat{\boldsymbol{B}}_g'\boldsymbol{x}_i\right)\left(\boldsymbol{y}_i-\hat{\boldsymbol{B}}_g'\boldsymbol{x}_i\right)'}{\sum_{i=1}^{N}\tau_{ig}^{(k)}}, \tag{7}$$

respectively.

Now, for the VVV FMRC model, the algorithm consists of updating $\hat{\boldsymbol{\pi}}_{ig}$, τ_{ig}, $\hat{\boldsymbol{B}}_g$, and $\hat{\boldsymbol{\Sigma}}_g$ via (2), (5), (6), and (7), respectively. Parameter estimates for the concomitant parameters in (2) are estimated using function `multinom` from the `nnet` package [23] for R [18] with the dependent variables given by the *a posteriori* probability estimates τ_{ig}. On the other hand, for the VVV FMR model, the update for $\pi_{ig} = \pi_g$, for $g = 1,\ldots,G$ and $i = 1,\ldots,N$, is

$$\hat{\pi}_g^{(k+1)} = \frac{1}{N}\sum_{i=1}^{N}\tau_{ig}^{(k)},$$

and the updates for τ_{ig}, $\hat{\boldsymbol{B}}_g$, and $\hat{\boldsymbol{\Sigma}}_g$ are updated via (5), (6), and (7), respectively. For the other eFMR and eFMRC models, the M-step updates vary only with respect to the component covariance matrix $\boldsymbol{\Sigma}_g$ and are similar to those in [5].

2.3 *Model Selection and Initialization*

For choosing a "best" fitted model among a family of models, a model selection criterion like the BIC [20] is typically used [7]:

$$\text{BIC} = 2l(\hat{\boldsymbol{\theta}}) - m\log N,$$

where $l(\hat{\boldsymbol{\theta}})$ is the maximized log-likelihood and m is the number of free parameters. Even though mixture models generally do not satisfy the regularity conditions for the asymptotic approximation used in the development of the BIC [14], it has performed quite well in practice and has been used extensively (e.g., [9]).

Note that the EM algorithm can be heavily dependent on starting values. Singularities and convergence to local maxima are also well documented [22]. Initializing the EM algorithm multiple times using k-means [12] or random initializations can alleviate some of these issues. Specifically, our EM algorithms are each initialized from five starting values, where the first four are random and the other uses k-means clustering.

2.4 Convergence Criterion and Performance Assessment

An Aitken acceleration-based stopping criterion is used to determine the convergence of our EM algorithms. This criterion is at least as strict as lack of progress in likelihood in the neighbourhood of a maximum [17]. The Aitken acceleration at iteration k is

$$a^{(k)} = \frac{l^{(k+1)} - l^{(k)}}{l^{(k)} - l^{(k-1)}},$$

where $l^{(k)}$is the log-likelihood value at iteration k. An asymptotic estimate of the log-likelihood at iteration $k + 1$ is given by [2] as

$$l_A^{(k+1)} = l^{(k)} + \frac{l^{(k+1)} - l^{(k)}}{1 - a^{(k)}},$$

and the EM is algorithm is stopped when $l_A^{(k+1)} - l^k < \epsilon$, provided that this difference is positive [17]; this is similar to the criterion used by [16].

The adjusted Rand index (ARI; [13]) is used to compare predicted and true classifications when the true labels are known. The ARI calculates the agreement between true and estimated classification by correcting the Rand index [19] to account for chance. An ARI of one corresponds to perfect clustering, whereas the expected value of the ARI under random classification is zero.

3 Results

Performance of the proposed models is illustrated on simulated and real data. To facilitate comparison of the performance of the algorithms, the `flexmix` FMR and FMRC models are initialized with the same set of values as the eFMR and eFMRC models (Sect. 2.3). We used the `mixture` package [3] for the M-step updates for the 14 covariance structures.

3.1 Simulated Data

Data were generated from a two-component model with 275 observations in total (Simulation 1). A binomial model with $\pi_1 = 0.45$ was used to determine the component sizes. Here, the three-dimensional response ($d = 3$) was generated using an EEE covariance structure. Three covariates were generated ($p = 3$). For the first component, one came from a uniform distribution with support $[0, 3]$ and the others from a two-dimensional Gaussian distribution with mean $\boldsymbol{\mu}_{x1} = (0, 1)$. Covariates for the second group were generated from a uniform distribution with support $[-1, 5]$ and a two-dimensional Gaussian distribution with mean $\boldsymbol{\mu}_{x2} = (-3, 3)$. The covariance matrices of the normally distributed covariates for the two groups were

$$\begin{pmatrix} 1 & 0.8 \\ 0.8 & 1.2 \end{pmatrix} \text{ and } \begin{pmatrix} 1.2 & 0.4 \\ 0.4 & 1 \end{pmatrix},$$

respectively. The regression coefficient matrices used for the two groups were

$$\boldsymbol{B}_1 = \begin{pmatrix} -1.9 & 0.4 & -1.2 & -3 \\ 0 & -0.4 & 0.8 & -2 \\ -1 & 0.7 & 0.3 & 1 \end{pmatrix}' \text{ and } \boldsymbol{B}_2 = \begin{pmatrix} 2.5 & -0.5 & 1 & -4 \\ 2.3 & -1.3 & 1.9 & 2 \\ 1 & -2.7 & -2.3 & -1.3 \end{pmatrix}',$$

respectively. Lastly, the error for the two groups was simulated using an EEE covariance structure with mean $\mathbf{0}$ and

$$\boldsymbol{\Sigma}_1 = \boldsymbol{\Sigma}_2 = \begin{pmatrix} 1.31 & 0.77 & 0.68 \\ 0.77 & 1.70 & 1.06 \\ 0.68 & 1.06 & 1.90 \end{pmatrix}.$$

This corresponds to $\lambda_1 = \lambda_2 = 1.25$,

$$\boldsymbol{D}_1 = \boldsymbol{D}_2 = \begin{pmatrix} -0.45 & 0.72 & 0.53 \\ -0.62 & 0.18 & -0.76 \\ -0.65 & -0.67 & 0.36 \end{pmatrix},$$

and $\boldsymbol{A}_1 = \boldsymbol{A}_2$ (diagonal matrices) with entries $(2.7, 0.7, 1/(2.7 \times 0.7))$. An example data set from Simulation 1 is shown in Fig. 1.

A total of 50 samples were generated in R and run for $G = 1, \ldots, 4$. The parameter estimates for the selected model using the eFMR and eFMRC families were quite close to the generating values (results not shown). Summary statistics for the selected models are given in Table 2. Note that the range of the number of parameters fitted for the FMR and FMRC models is quite wide, implying that these models are overestimating the number of components. Specifically, the FMR and FMRC models overestimate the number of components 40 and 35 times, respectively. On the other hand, the selected eFMRC models always fitted the

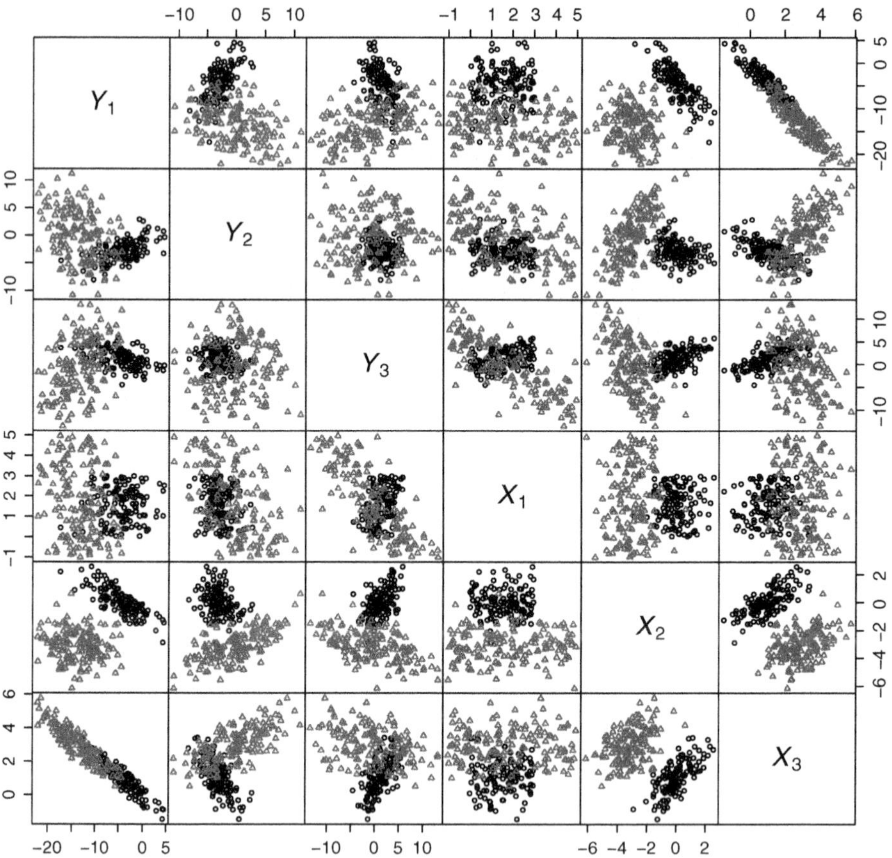

Fig. 1 Scatter plots depicting an example data set with three response variables and three covariates from Simulation 1

right number of components. The selected eFMR models fitted the right number of components 49 out of 50 times.

The above simulation was repeated with the same values for all generating parameters except for $\boldsymbol{A}_g$, i.e., the error matrices were simulated using an EVE covariance structure (Simulation 2). Here, $\boldsymbol{A}_1$ and $\boldsymbol{A}_2$ were diagonal matrices with entries $(2.7, 0.7, 1/(2.7 \times 0.7))$ and $(2.7, 1.7, 1/(2.7 \times 1.7))$, respectively. Summary statistics for the selected models are given in Table 3. While the eFMR and eFMRC families fit the right number of components all 50 times, the FMR and FMRC models overestimate the number of components 36 and 35 times, respectively.

Lastly, simulation 1 was repeated with the same values for all generating parameters except for $\boldsymbol{A}_g$ and λ_g, i.e., using a VVE covariance structure (Simulation 3). Here, $\boldsymbol{A}_1$ and $\boldsymbol{A}_2$ were diagonal matrices with entries $(2.7, 0.7, 1/(2.7 \times 0.7))$ and $(2.7, 1.7, 1/(2.7 \times 1.7))$, respectively, and $\lambda_1 = 1.25$ and $\lambda_2 = 2$. Summary statistics for the selected models are given in Table 4. While the eFMR and eFMRC

Table 2 Summary of different approaches using the 50 simulated data sets of Simulation 1

Statistic	FMR	FMRC	eFMR	eFMRC
ARI	0.64 (0.43, 1.00)	0.70 (0.49, 1.00)	0.96 (0.86, 1.00)	1.00 (0.96, 1.00)
$\mathscr{L}_0$	−1481 (−1538, −1389)	−1300 (−1376, −1201)	−1425 (−1476, −1381)	−1253 (−1293, −1209)
BIC	−3220 (−3332, −3130)	−2894 (−2995, 2779)	−3029 (−3127, −2937)	−2696 (−2778, −2609)
df	47 (31, 63)	53 (34, 72)	31 (31, 46)	34 (34, 35)

Values denote the medians (rounded to 2 decimals) with the ranges of the estimated statistics in parentheses. Here, $\mathscr{L}_0$ refers to the maximized log-likelihood value

Table 3 Summary of different approaches using the 50 simulated data sets of Simulation 2

Statistic	FMR	FMRC	eFMR	eFMRC
ARI	0.67 (0.45, 0.96)	0.74 (0.51, 1.00)	0.94 (0.86, 1.00)	1.00 (0.99, 1.00)
$\mathscr{L}_0$	−1519 (−1593, −1426)	−1342 (−1413, 1264)	−1425 (−1471, −1387)	−1244 (−1290, −1203)
BIC	−3283 (−3411, −3180)	−2956 (−3054, −2833)	−3036 (−3128, −2959)	−2690 (−2783, −2690)
df	47 (31, 63)	53 (34, 72)	33 (33, 34)	36 (36, 37)

Values denote the medians (rounded to 2 decimals) with the ranges of the estimated statistics in parentheses. Here, $\mathscr{L}_0$ refers to the maximized log-likelihood value

Table 4 Summary of different approaches using the 50 simulated data sets of Simulation 3

Statistic	FMR	FMRC	eFMR	eFMRC
ARI	0.67 (0.44, 0.99)	0.77 (0.50, 1.00)	0.94 (0.87, 1.00)	1.00 (0.99, 1.00)
$\mathscr{L}_0$	−1625 (−1718, −1551)	−1452 (−1552, 1369)	−1528 (−1572, −1491)	−1350 (−1392, 1310)
BIC	−3499 (−3611, −3384)	−3171 (−3296, −3041)	−3248 (−3336, −3173)	−2912 (−2992, −2827)
df	47 (31, 63)	53 (34, 72)	34 (34, 34)	37 (36, 37)

Values denote the medians (rounded to 2 decimals) with the ranges of the estimated statistics in parentheses. Here, $\mathscr{L}_0$ refers to the maximized log-likelihood value

families fit the right number of components all 50 times, the FMR and FMRC models overestimate the number of components 36 and 32 times, respectively.

Note that for all simulations, the parameter estimates for the selected model using the eFMR and eFMRC families were quite close to the generating values (results not shown). In all simulations, the eFMR and eFMRC families clearly perform much better than the FMR and FMRC models. Specifically, the models selected from both the eFMR and eFMRC families yielded higher average ARI and log-likelihood values. Furthermore, these models also yielded superior BIC values and estimated fewer parameters on average. Moreover, in contrast to the FMR and FMRC models, the models selected from the eFMR and eFMRC families also fit the right number of components. The eFMR and eFMRC families perform better because in contrast to the `flexmix` FMR and FMRC models, the proposed parsimonious models deal with correlations between the response variables.

3.2 Crabs Data

The crabs data set contains five morphological measurements on 200 crabs, split evenly between both sexes and two colours (blue and orange) of the species *Leptograpsus variegatus*. These data were originally introduced in [4] and are available as part of the `MASS` package [23] in R. The data are famous for having highly correlated measurements on width of frontal region just anterior to frontal tebercles (FL), width of posterior region (RW), carapace length (CL), carapace width (CW), and body depth (BD). The variables CW, FL, and RW reflect width measurements and were taken to be the response variables, with CL and BD as the predictor variables. Based on the two binary variables, sex and colour, there are four known groups in these data. Our algorithms were run for $G = 1, \ldots, 9$ (Table 5).

The selected eFMR model is a two-component VVI model with an ARI of 0.40. Because the VVI model assumes independence between the response variables, that is equivalent to the `flexmix` FMR model and unsurprisingly, FMR chooses a two-component model with an ARI of 0.40 (Table 6). Note that the estimated classification from the selected two-component eFMR model leads to good separation between sexes. If the class membership agreement is estimated based on only the sexes of the crabs, an ARI of 0.81 is achieved. FMRC did well, picking a four-component model (Table 6). However, the selected eFMRC model (VEE) also has four components with a higher ARI (0.84), while also being more parsimonious than the `flexmix` FMRC model.

Table 5 Model performance comparison for crabs data

Algorithm	Model	G	BIC	ARI	Parameters
FMR		2	−1178.45	0.40	25
FMRC		4	−1104.96	0.81	57
eFMR	VVI	2	−1178.38	0.40	25
eFMRC	VEE	4	−1069.36	0.84	54

Table 6 Cross-tabulations of true versus predicted group memberships for the crabs data

	FMRC				eFMRC				FMR		eFMR	
	1	2	3	4	1	2	3	4	1	2	1	2
BM	38	12			40	10			46	4	46	4
BF		48		2		49		1	4	46	4	46
OM			50				50		50		50	
OF			2	48			2	48	2	48	2	48

"B", "O", "M", "F" refer to blue, orange, male and female, respectively

4 Discussion

Families of parsimonious multivariate response FMR and FMRC models that can handle correlated response variables were proposed and illustrated. In a model-based clustering context, we showed that both eFMR and eFMRC families perform as well as or better than the `flexmix` FMR and FMRC models. Computationally, the algorithms were quite stable. However, to prevent fitting issues, the component sizes were computed before each M-step and a preset minimum size of the clusters was used (cf. [15]). For heavier tailed data, more robust distributions like the multivariate student-t distribution may be employed. Because the number of regression intercepts and coefficients estimated, i.e., $Gd(p + 1)$, can also increase quickly, more parsimonious models can be achieved using variable selection.

Acknowledgements This work is supported by a Alexander Graham Bell Canada Graduate Scholarship (CGS-D; Dang) as well as a Discovery Grant from the Natural Sciences and Engineering Research Council of Canada (McNicholas).

References

1. Banfield, J.D., Raftery, A.E.: Model-based Gaussian and non-Gaussian clustering. Biometrics **49**(3), 803–821 (1993)
2. Böhning, D., Dietz, E., Schaub, R., Schlattmann, P., Lindsay, B.G.: The distribution of the likelihood ratio for mixtures of densities from the one-parameter exponential family. Ann. Inst. Stat. Math. **46**(2), 373–388 (1994)
3. Browne, R.P., McNicholas, P.D.: 'mixture': Mixture Models for Clustering and Classification. R package version 1.0. (2013)
4. Campbell, N.A., Mahon, R.J.: A multivariate study of variation in two species of rock crab of the genus leptograpsus. Aust. J. Zool. **22**(3), 417–425 (1974)
5. Celeux, G., Govaert, G.: Gaussian parsimonious clustering models. Pattern Recogn. **28**(5), 781–793 (1995)
6. Dang, U.J., Punzo, A., McNicholas, P.D., Ingrassia, S., Browne, R.P.: Multivariate response and parsimony for Gaussian cluster-weighted models [arXiv preprint arXiv:1411.0560] (2014)
7. Dasgupta, A., Raftery, A.E.: Detecting features in spatial point processes with clutter via model-based clustering. J. Am. Stat. Assoc. **93**(441), 294–302 (1998)
8. DeSarbo, W.S., Cron, W.L.: A maximum likelihood methodology for clusterwise linear regression. J. Classif. **5**(2), 249–282 (1988)
9. Fraley, C., Raftery, A.E.: Model-based clustering, discriminant analysis, and density estimation. J. Am. Stat. Assoc. **97**(458), 611–631 (2002)
10. Galimberti, G., Soffritti, G.: A multivariate linear regression analysis using finite mixtures of t distributions. Comput. Stat. Data Anal. **71**, 138–150 (2014)
11. Gershenfeld, N.: Nonlinear inference and cluster-weighted modeling. Ann. N. Y. Acad. Sci. **808**(1), 18–24 (1997)
12. Hartigan, J.A., Wong, M.A.: A k-means clustering algorithm. J. R. Stat. Soc. C App. **28**(1), 100–108 (1979)
13. Hubert, L., Arabie, P.: Comparing partitions. J. Classif. **2**(1), 193–218 (1985)
14. Keribin, C.: Consistent estimation of the order of mixture models. Sankhya Ser. A **62**, 49–66 (2000)

15. Leisch, F.: FlexMix: a general framework for finite mixture models and latent class regression in R. J. Stat. Softw. **11**(8), 1–18 (2004)
16. Lindsay, B.G.: Mixture models: theory, geometry and applications. In: NSF-CBMS Regional Conference Series in Probability and Statistics, vol. 5 (1995)
17. McNicholas, P.D., Murphy, T.B., McDaid, A.F., Frost, D.: Serial and parallel implementations of model-based clustering via parsimonious Gaussian mixture models. Comput. Stat. Data Anal. **54**(3), 711–723 (2010)
18. R Core Team: R: A Language and Environment for Statistical Computing. R Foundation for Statistical Computing. Vienna, Austria (2014)
19. Rand, W.M.: Objective criteria for the evaluation of clustering methods. J. Am. Stat. Assoc. **66**(336), 846–850 (1971)
20. Schwarz, G.: Estimating the dimension of a model. Ann. Stat. **6**, 461–464 (1978)
21. Soffritti, G., Galimberti, G.: Multivariate linear regression with non-normal errors: a solution based on mixture models. Stat. Comput. **21**(4), 523–536 (2011)
22. Titterington, D.M., Smith, A.F.M., Makov, U,E.: Statistical Analysis of Finite Mixture Distributions, vol. 7. Wiley, New York (1985)
23. Venables, W.N., Ripley, B.D.: Modern Applied Statistics with S, 4th edn. Springer, New York (2002)
24. Wedel, M.: Concomitant variables in finite mixture models. Statistica Neerlandica **56**(3), 362–375 (2002)

Quantile Regression for Clustering and Modeling Data

Cristina Davino and Domenico Vistocco

Abstract This paper aims to propose an innovative approach to identify a typology in a quantile regression model. Quantile regression is a regression technique that allows to focus on the effects that a set of explanatory variables has on the entire conditional distribution of a dependent variable. The proposal concerns the use of multivariate techniques to simultaneously cluster and model data and it is illustrated using an empirical analysis. This analysis regards the impact of student features on the university outcome, measured by the degree mark. The analysis is based on the idea that the dependence structure could be different for units belonging to different groups.

Keywords Cluster analysis • Quantile regression • Unsupervised learning

1 Introduction

In many regression problems, the estimation of a single set of coefficients provides a misrepresentation of the true dependence structure if units belong to different groups. The solution to this issue becomes more difficult to achieve when group membership is not a priori known. A simplistic solution would consist in clustering units and later estimating different models for each group. This solution however would not permit to identify the impact of the groups on the dependent variable and it would require tools for comparing of the models estimated on different samples.

The aim of this paper is to propose an innovative approach for simultaneously clustering and modeling data. It is based on the conjoint use of multivariate methods and quantile regression to identify a typology in a dependence model.

C. Davino (✉)
Department of Political Sciences, University of Macerata, Communications and Intern. Relations, 62100 Macerata, Italy
e-mail: cristina.davino@unimc.it

D. Vistocco
Department of Economics and Law, University of Cassino, 03043 Cassino (FR), Italy
e-mail: vistocco@unicas.it

I. Morlini et al. (eds.), *Advances in Statistical Models for Data Analysis*, Studies in Classification, Data Analysis, and Knowledge Organization, DOI 10.1007/978-3-319-17377-1_10

Quantile regression, as introduced by Koenker and Basset [14], is an extension of the classical estimation of the conditional mean to the estimation of a set of conditional quantiles. It offers a complete view of a response variable providing a method for modeling the rates of changes at multiple points (conditional quantiles) of its conditional distribution [2, 13].

The rest of the paper is an extension of the supervised approach proposed by the authors Davino and Vistocco [3, 4]. The use of an unsupervised approach to classify units in the dependence model characterizes this proposal, whereas the former proposals exploit a priori defined groups.

In literature, a quite widespread approach to simultaneously identify a partition of the data and the related model is represented by clusterwise linear regression. The method is based on the hypothesis that there exist a finite number of unknown classes and each class is characterized by a different linear regression model. The literature concerning clusterwise linear regression is quite wide, from the starting works of Spath [19, 20] to the maximum likelihood approach of DeSarbo and Cron [5], until recent proposals [8, 11, 24]. Sharing the main goals of clusterwise linear regression, the method proposed in this paper exploits quantile regression and hierarchical clustering to model and partition data identifying a different dependence structure for each detected group. To pursue such aims, the method assigns a separate quantile model best representing each group. However, the different models are estimated on the total sample, making easier the comparisons among the group coefficients. The use of quantile regression allows us to study the dependence exploring the whole conditional distribution of the dependent variable unlike the clusterwise linear regression that focuses on the conditional mean. Furthermore, it offers well-known advantages with respect to robustness issues. Moreover, our approach attempts to overcome the main drawback of the original clusterwise linear regression, the a priori setting of the number of groups, while hierarchical clustering provides a data driven criterion to partition the sample. A proper comparison with clusterwise linear regression would require a wide simulation study that takes into account also the sensitivity of clusterwise linear regression solutions to the tuning parameters (e.g., the initial partition and the number of required groups) and it will be therefore subject of a specific paper.

The paper is organized as follows. Section 2 presents the dataset used to apply the proposed approach: it concerns the evaluation of the effectiveness of the university educational process. In Sect. 3 the methodology is described together with results deriving from the empirical analysis: students are grouped according to the relationship between the degree mark and their features. Some concluding remarks and future work directions are reported in Sect. 4.

2 A Dataset on Student University Outcome

The proposed approach is described in the following sections through an empirical analysis aiming at evaluating if and how the student features (socio-demographic and university experience attributes) affect the outcome of the university career, measured through the degree mark. As stated above, the underlying idea is that this effect can be very different for students belonging to different groups. Such groups are detected according to the relationship between the degree mark and the student features. The typology identification is embedded in a quantile regression model, and it is thus able to exploit the whole conditioned distribution of the degree mark.

The analysis is carried out on a random sample of 685 students who graduated from the University of Macerata [3], which is located in the Italian region of Marche. The survey was completed in 2007 and includes students who graduated between 2002 and 2005. The degree mark is measured on a discrete scale ranging between 66 and 110, with the "cum laude" mark coded as 110. The explicative variables included in the model pertain to the student profile. In particular, the following regressors have been considered: gender, place of residence during university education (Macerata and its province, Marche region, outside Marche), course attendance (no attendance, regular), foreign experience (yes, no), working condition (full-time student, working student), number of years to obtain a degree, diploma mark.

The density plot of the response variable (Fig. 1) shows the presence of a strong right skewness, further supporting the recourse to the dependence analysis outside the classical regression framework.

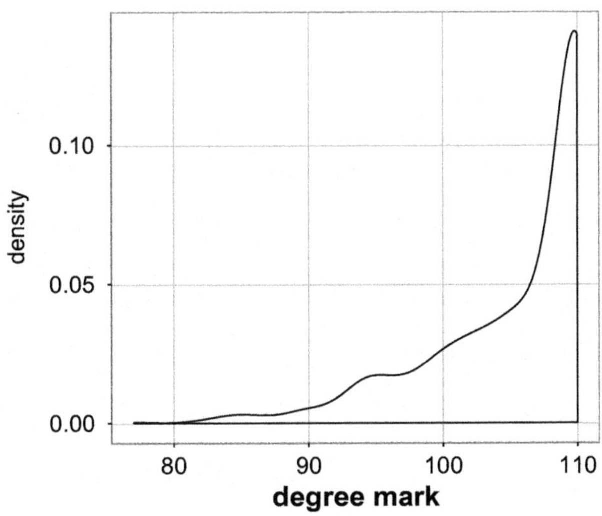

Fig. 1 Degree mark density

3 The Proposed Approach: Methodology and Results

The proposed unsupervised learning procedure is based on the joint use of hierarchical clustering and quantile regression. The approach is structured in the following four steps

1. Estimation of the global dependence structure
2. Identification of the best model for each unit
3. Identification of a typology
4. Estimation of the group dependence structure

Following sections detail the meaning of each step showing them in action on the student university outcome study.

3.1 *Estimation of the Global Dependence Structure*

In the first step, a quantile regression (QR) model is estimated on the whole sample:

$$Q_\theta(\hat{\mathbf{y}}|\mathbf{X}) = \mathbf{X}\hat{\boldsymbol{\beta}}(\theta) \tag{1}$$

where $0 < \theta < 1$ denotes the θth conditional quantile, $Q_\theta(.|.)$ is the corresponding conditional quantile function, $\mathbf{y}_{[n]}$ is the dependent variable (degree mark in the application) and $\mathbf{X}_{[n \times p]}$ is the matrix of the explanatory variables (students features in the application), n denoting the number of units and p the number of regressors.

Using a grid of k conditional quantiles, $\boldsymbol{\theta} = (\theta_1, \ldots, \theta_k)$, the model provides a coefficient matrix $\hat{\Theta}_{[p \times k]}$ with a generic element that can be interpreted as the rate of change in the θth quantile of the conditional distribution of the dependent variable per unit change in the value of the jth regressor. The value of k is therefore the number of estimated conditional quantiles. A fairly accurate approximation of the whole quantile process [15] can be obtained using a dense grid of equally spaced quantiles in the unit interval (0; 1) [2].

In Fig. 2, QR coefficients, obtained using a selected grid of quantiles ($\theta = [0.1, 0.25, 0.5, 0.75, 0.9]$), are graphically represented for the different features of the student profile. The horizontal axis displays the different quantiles, while the effect of each feature holding the others constant is represented on the vertical axis. QR confidence bands (in grey) are obtained through the bootstrap method for $\alpha = 0.1$ [17]. The solid lines parallel to the horizontal axis correspond to OLS coefficients, and the related confidence intervals are represented using dashed lines for $\alpha = 0.1$.

The graphical representation allows to visually catch the different effect of the student characteristics on the degree mark. Gender and residence during university education have a great influence on the lowest quantiles of the distribution: males and residents outside the Marche region show negative coefficients. A foreign

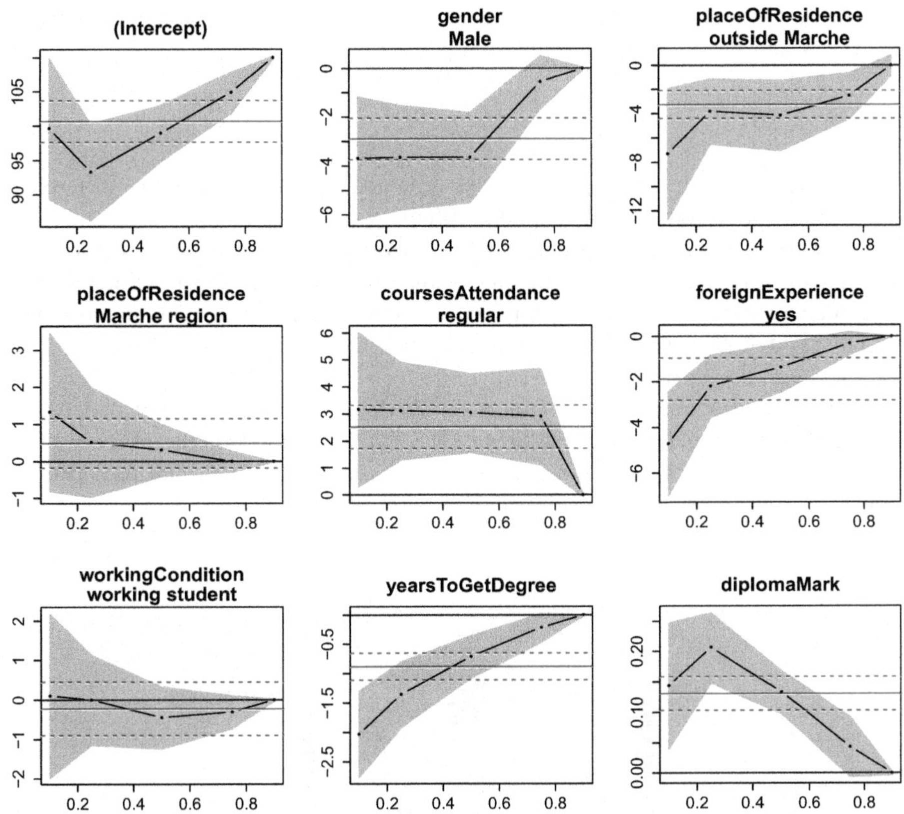

Fig. 2 OLS and QR coefficients and related confidence intervals

experience negatively influences the degree mark. This effect becomes null in the higher part of the distribution pointing out that very good students are not influenced by their university experiences abroad. Working students are less likely to get high degree marks but the QR results show how their impact is almost negligible. All the coefficients of the variable numbers of years to get a degree are negative, particularly for the lowest quantiles. Finally, the diploma mark always has a positive effect, but its value is very low for successful students.

3.2 *Identification of the Best Model for Each Unit*

In the second step, the coefficient matrix $\hat{\Theta}_{[p \times k]}$ and the regressor data matrix $\mathbf{X}$ are used to estimate the conditional distribution matrix of the response variable: $\hat{\mathbf{Y}} = \mathbf{X}\hat{\Theta}$. The generic element of the $\hat{\mathbf{Y}}_{[n \times k]}$ matrix is the estimate of the response variable in correspondence of the ith units according to the θth quantile.

The best model for each unit i is identified by the quantile able to better estimate the response variable and it is denoted as the *best quantile*:

$$\hat{\theta}_i^{\text{best}} : \text{argmin}_{\theta=1,\ldots,k} |y_i - \hat{y}_i(\theta)| \qquad (i = 1, \ldots, n) \tag{2}$$

The best quantile, $\hat{\theta}_i^{\text{best}}$, is therefore obtained by minimizing the difference between the observed and the estimated values.

From the $\hat{\mathbf{Y}}$ matrix it is then possible to extract the best estimated vector, $\hat{\mathbf{y}}_\theta^{\text{best}}$, identifying for each unit the estimated value corresponding to the assigned best quantile. Such vector provides both an accurate approximation of the response variable and embeds information on the dependence structure relating the response variable with the regressors.

Figure 3 reproduces the histograms of the dependent variable (left panel) and the estimated dependent variable using OLS (middle panel) or the proposed QR approach (right panel). For some considerations on the added value provided by considering $\hat{\mathbf{y}}_\theta^{\text{best}}$ instead of the classical OLS predicted values, the interest reader is referred to Davino and Vistocco [4].

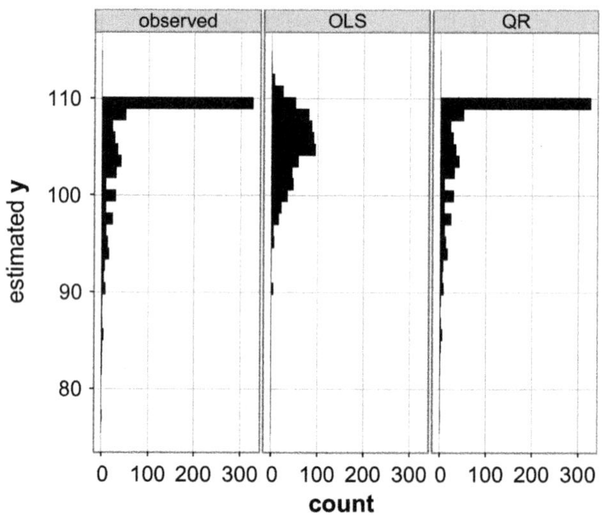

Fig. 3 Distribution of the dependent variable (*left panel*) and of the estimated dependent variable using OLS (*middle panel*) or the proposed QR approach (*right panel*)

3.3 Identification of a Typology

The third step of the proposed strategy aims to identify a typology on the basis of the QR results obtained in the previous step.

Units are grouped according to the best quantile they have been assigned because it can be considered as an indicator of a similar dependence structure. Working on the $\hat{\boldsymbol{\theta}}^{\text{best}}$ vector, it is possible to group units in clusters. The simplest criterion is based on its categorization. Albeit automatic methods (e.g. [21] or [18] rules) are available in literature, a certain degree of subjectivity remains. In the present paper a multivariate approach is proposed performing a hierarchical clustering [6, 10] on the estimated $\hat{\mathbf{Y}}$ matrix in order to classify units sharing similar patterns for the predicted values for all the considered quantiles.

Several criteria have been proposed in the literature to select the "best" partition by optimizing some cluster validity indexes (see, e.g., [12, 16, 23]). The seminal work of Milligan and Cooper [16] describes above 30 internal criterion measures coming from a wide variety of fields. More recently, other proposals combine the use of a cluster validity index with a searching strategy for exploring the extended hierarchy housed in a dendrogram [9] or exploit permutation tests in order to automatically detect a partition [1]. A competing method (GAP), proposed by Tibshirani et al. [22], permits to estimate the number of clusters starting from the output of any clustering algorithm. It is based on a Monte Carlo approach to derive the reference distribution of a test statistic and it requires as input the different partitions among which the optimal one has to be selected. Using the GAP statistics, the best partition is obtained by cutting the dendrogram in four groups (Fig. 4, left-hand side) with 318, 144, 154, and 64 observations, respectively.

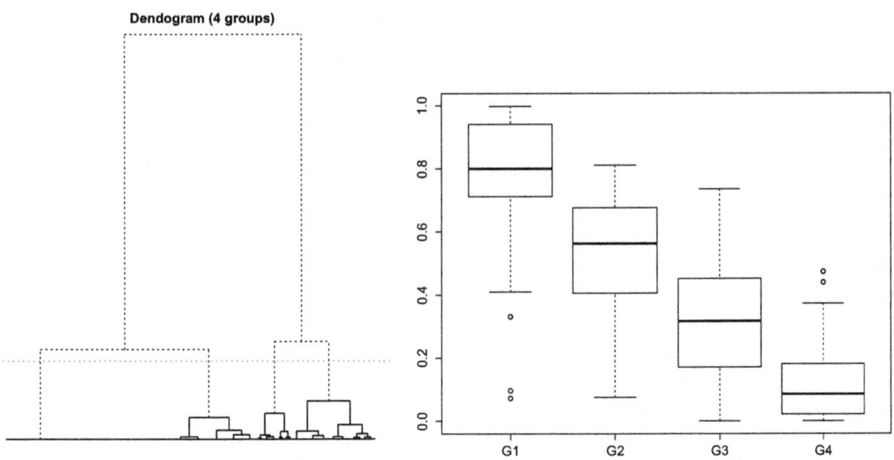

Fig. 4 Dendrogram and best partition (*left-hand side*) and distribution of the best quantiles in the identified groups (*right-hand side*)

In order to tailor a proper dependence structures as derived in the next step, a reference quantile is then associated to each group. The choice of the reference quantile is based on a synthesis measure of the distribution of the group best quantiles. Figure 4 (right-hand side) reveals a certain degree of skewness in the distribution of the best quantiles for each group. The $\hat{\boldsymbol{\theta}}^{\text{best}}$ median value of each group can then be considered as a robust reference quantile: $G1 = 0.800$, $G2 = 0.562$, $G3 = 0.315$, $G4 = 0.085$. The obtained reference quantiles are clearly distinct signaling an impact on different locations of the degree mark distribution played by the features of students belonging to each group, as it will be shown in the next section.

3.4 Estimation of the Group Dependence Structure

The four reference quantiles previously defined are used to estimate the group dependence structure. In particular, a QR model is carried out on the whole sample estimating the following four θ values: 0.800 ($G1$), 0.562 ($G2$), 0.315 ($G3$), 0.085 ($G4$). Estimating a model on the whole sample allows us to easily compare the group dependence structure through the evaluation of the statistical significance of the differences among the coefficients. In the QR framework such a comparison is based on the classical tools to test interquantile differences [7].

The QR results for the different groups are shown in Table 1 reporting the covariates on the rows and the groups on the columns; significant coefficients at $\alpha = 0.10$ are shown in bold. Reading the table by columns details information on the features mainly affecting each group, while a comparison of a specific coefficient among the different groups is provided by a row-wise inspection of the table. The reference quantiles play a crucial role in interpreting the results. For example, the effect on the degree mark of living outside Marche is negative for all the groups, but it is stronger for students belonging to group $G4$. On the other hand, as group 4

Table 1 Group effects estimates (in bold significant coefficients at $\alpha = 0.10$)

Variable	G1	G2	G3	G4
	$\theta = 0.800$	$\theta = 0.563$	$\theta = 0.315$	$\theta = 0.085$
Intercept	**109.00**	**98.87**	**97.17**	**96.51**
Gender (male)	0.00	**−2.87**	**−3.55**	**−3.23**
Place of residence (outside Marche)	**−2.00**	**−2.62**	**−4.61**	**−5.89**
Place of residence (Marche region)	0.00	0.12	1.11	1.27
Courses attendance (regular)	1.00	**3.12**	**3.28**	**3.14**
Foreign experience (yes)	0.00	−0.85	−2.00	**−5.46**
Working student	0.00	−0.75	0.00	0.26
Years to get a degree	0.00	**−0.50**	**−1.28**	**−1.82**
Diploma mark	0.00	**0.12**	**0.17**	**0.16**

is characterized by a reference quantile equal to 0.085, the negative effect of living outside Marche reduces the degree mark of 6.58 marks for student with a low degree mark, i.e. for the 8.7 % of students with the lowest marks. Moving toward the center of the conditional distribution, the effect is still negative but with a substantial numerical decrease. A foreign experience negatively influences the degree mark. This effect becomes null in group $G1$ pointing out that very good students are less influenced by their university experiences abroad.

It is worth to mention the peculiarities of $G1$ describing the effect of the covariates on the best performer students ($\theta = 0.800$). Most of the regressors do not play any effect on the 80th conditional percentile of the degree mark, which is a sign that the highest performances are related to other student features not included in the analysis.

To further highlight the potentialities of the proposed approach, it is useful to compare the observed and the estimated response values. In particular, if the best model for a given group is used to predict the response variable for the units belonging to another group, results worsen as much as the groups differentiate with respect to the best quantiles as shown in Fig. 5. The figure is structured in

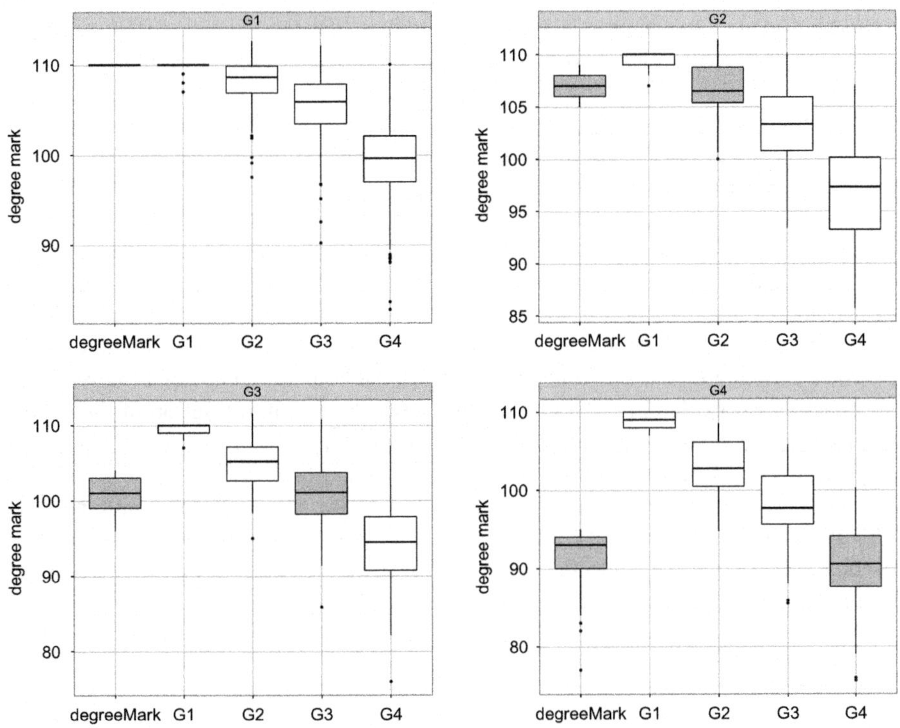

Fig. 5 Observed response distribution compared with the estimated distributions using the reference quantile of each group. Each panel depicts the degree mark of the students belonging to the group represented in the *grey label*. The two matching boxplots (observed and estimated) are colored in *grey*

four panels, one for each group. For example, the left-top panel refers to students belonging to group $G1$. From the third step of the proposed strategy (Sect. 3.3), the reference quantile for such a group is equal to 0.800. The left-most boxplot depicts the observed degree mark distribution in $G1$; the others show the estimated degree mark distributions obtained exploiting the reference quantiles associated to each group. It is evident how the estimated degree mark matches the observed degree mark using the model based on the reference quantile equal to 0.800. The above mentioned peculiarity of $G1$ (highest performance students) explains the flattened shape of the two matching boxplots. From the analysis of the other panels, it is evident how, in each panel, the observed matches with the estimated distribution obtained using the reference quantile associated to the specific group. To facilitate the reading, the two matching boxplots (observed and estimated) are colored in grey.

4 Concluding Remarks

The proposed approach provides a clustering of units according to the conditioned distribution of the dependent variable estimated through QR. It can represent a valuable tool to cluster units taking into account the dependence structure in the data. The underlying idea leading our approach relies on the expected observation that the dependence structure is affected by the features of the involved units. Obtaining a partition starting from the QR conditional distributions allows to group units where the effect played by the covariates on the dependent variable is similar.

The main strengths of the proposed approach are represented by the use of the whole sample to estimate the group dependence structures and the association of each group with a specific conditional quantile. The former point enables to easily test the statistical significance of the differences among the group. The latter provides a characterization of each group through the identification of a reference quantile describing the impact on the specific location of the dependent variable played by the features of the units belonging to the considered group. Finally, as the approach is embedded in the regression framework, the interpretation of the results can exploit the well-known rules of any linear model.

Avenues for further developments concern: *(i)* a proper comparison with the clusterwise linear regression, with whom we share the goal to simultaneously identify a partition of the data and the related model; *(ii)* a testing of the robustness of the method with respect to the number of groups, the distribution of the variables involved in the model and the model complexity through a simulation study.

References

1. Bruzzese, D., Vistocco D.: DESPOTA: DEndrogram slicing through a pemutation test approach. J. Classif. (in press)
2. Davino, C., Furno, M., Vistocco, D.: Quantile Regression: Theory and Applications. Hoboken, NJ, Wiley (2013)
3. Davino, C., Vistocco, D.: The evaluation of University educational processes: a quantile regression approach. Statistica **3**, 267–278 (2007)
4. Davino, C., Vistocco, D.: Quantile regression for the evaluation of student satisfaction. Italian J. Appl. Stat. **20**, 179–196 (2008)
5. DeSarbo, W.S., Cron, W.L.: A maximum likelihood methodology for clusterwise linear regression. J. Classif. **5**, 249–282 (1988)
6. Gordon, A.D.: Classification: Methods for the Exploratory Analysis of Multivariate Data. Chapman & Hall, London (1981)
7. Gould W.W.: sg70: Interquantile and simultaneous-quantile regression. Stata Tech. Bull. **38**, 14–22. Reprinted in Stata Tech. Bull. Reprints **7**, 167–176 College Station, TX: Stata Press (1997)
8. Grun, B., Leisch, F.: Fitting finite mixtures of linear regression models with varying & fixed effects in R. In: Rizzi, A., Vichi, M. (eds.) In: COMPSTAT 2006 - Proceedings in Computational Statistics, vol. 853–860. Physica Verlag, Heidelberg, Germany (2006)
9. Gurrutxaga, I., Albisua, I., Arbelaitz, O., Martn, J. I., Muguerza, J., Prez, J.M., Perona, I.: SEP/COP: An efficient method to find the best partition in hierarchical clustering based on a new cluster validity index. Pattern Recogn. **43**(10), 3364–3373 (2010)
10. Hartigan, J. A.: Clustering Algorithms. Wiley, New York (1975)
11. Hennig, C.: Identifiability of models for clusterwise linear regression. J. Classif. **17**, 273–296 (2000)
12. Kim, M., Ramakrishna, R.S.: New indices for cluster validity assessment. Pattern Recogn. Lett. **26**(15), 2353–2363 (2005)
13. Koenker, R.: Quantile Regression. Econometric Society Monographs. Cambridge University Press, Cambridge (2005)
14. Koenker, R., Basset, G.W.: Regression quantiles. Econometrica **46**, 33–50 (1978)
15. Koenker, R., Dorey, W.V.: Algorithm AS 229: computing regression quantiles. J. R. Stat. Soc. Ser. C (Appl. Stat.) **36**(3), 383–393 (1987)
16. Milligan, G.W.: A monte carlo study of thirty internal criterion measures for cluster analysis. Psychometrika **46**(2), 187–199 (1981)
17. Parzen, M.I., Wei, L., Ying, Z.: A resampling method based on pivotal estimating functions. Biometrika **81**, 341–350 (1994)
18. Scott, D.W.: On optimal and data-based histograms. Biometrika **66**, 605–610 (1979)
19. Spath, H.: Algorithm 39: clusterwise linear-regression. Computing **22**, 367–373 (1979)
20. Spath, H.: Correction to algorithm 39: clusterwise linear-regression. Computing **26**(3), 275 (1981)
21. Sturges, H.A.: The choice of a class interval. J. Am. Stat. Assoc. **21**, 6566 (1926)
22. Tibshirani, R., Walther, G., Hastie, T.: Estimating the number of clusters in a data set via the gap statistic. J. R. Stat. Soc. B **83**(2), 411–423 (2001)
23. Wu, K.-L., Yang, M.-S., Hsieh, J.-N.: Robust cluster validity indexes. Pattern Recogn. **42**(11), 2541–2550 (2009)
24. Zhen, Z., Yan, L., Nan, K.: Clusterwise linear regression with the least sum of absolute deviations - An MIP approach. Int. J. Oper. Res. **9**(3), 162–172 (2012)

Nonmetric MDS Consensus Community Detection

Carlo Drago and Antonio Balzanella

Abstract Community detection methods for the analysis of complex networks are increasingly important in modern literature. At the same time it is still an open problem. The approach proposed in this work is to adopt an ensemble procedure for obtaining a consensus matrix from which to perform a nonmetric MDS approach and then a clustering algorithm which allows to get a consensus partition of the nodes. The simulation study offers some interesting insights on the procedure because it shows that it is possible to understand the key nodes and the stable communities by considering different algorithms. The proposed approach is still applied to real data related to a network of patents.

Keywords Community detection • Complex networks • Nonmetric multidimensional scaling

1 Community Detection in Complex Networks

A network is a set of items defined vertices (or nodes) which can be connected with edges [7]. In this context a complex network can be defined as a network which is characterized by complex topological features. For example, modularity or the possibility to be divided into parts characterized by high density connections between vertices. Random graphs for example are not characterized by these features which generally happen in real networks. Usually, the connections in complex networks are not random but there are patterns which can be discovered and analyzed. In order to discover the properties of networks we need to consider their statistical features [2, 7]. Typical methods to analyze complex networks are:

C. Drago (✉)
Università degli Studi "Niccolò Cusano" Telematica Roma, via Don Carlo Gnocchi 3, 00166 Rome, Italy
e-mail: carlo.drago@unicusano.it

A. Balzanella
Department of Political Science, Second University of Naples, Viale Ellittico 31, 81100 Caserta, Italy

I. Morlini et al. (eds.), *Advances in Statistical Models for Data Analysis*, Studies in Classification, Data Analysis, and Knowledge Organization,
DOI 10.1007/978-3-319-17377-1_11

the Freeman degree centrality; the betweenness centrality, which represents the centrality for each vertex (node) in the network; the Closeness; the Density, which measures the ratio between the sum of vertices and possible vertices in a network; the Network centralization.

A complex network has a community structure if it can be divided into groups of nodes which are particularly dense in terms of within connections and sparse in terms of connections between the groups [4]. Detecting communities in a network is particularly relevant for applicative reasons; in fact, networks can usually be partitioned in community groups based on some attributes like location or occupation. At the same time there are important cases in which the communities can behave as independent departments in the network and to exhibit important functions [4].

Several algorithms and methods have been proposed in the literature. Among them, we recall the statistical methods based on the hierarchical clustering, the divisive methods such as the GirvanNewman algorithm, the modularity-based methods, the spectral algorithms, and finally the methods based on statistical inference like the Blockmodeling [4]. The effectiveness of these methods depends on the topological structure of the network so that each method seems to perform better in some situation than in others [6]; moreover, different methods performed on the same network can provide different partitions.

In explorative frameworks where no a priori information is available on the communities in the network, the choice of the right method can be unfeasible. In order to deal with this challenge, we propose an ensemble of community detection algorithms to find a consensus partition which allows to capture the most of information coming from the single community detection methods in the ensemble.

2 Community Detection Ensembles

We consider, as input, a network represented by an undirected graph $G = (V, E)$ where $V = (v_1, \ldots, v_i, \ldots, v_n)$ is the set of nodes of the network and E carries nonnegative values representing the presence of a connection between a pair of nodes. We consider the following set of methods in order to obtain a partitioning of the network into homogeneous communities: edge.betweenness, walktrap, fastgreedy, spinglass, leading.eigenvector, multilevel, infomap, label.propagation, optimal modularity [3]. Finally we consider also blockmodeling ex post as additional method.

We get, as output of each method, a partition $P^m = (C_1^m, \ldots, C_k^m, \ldots, C_K^m)$ where $m = 1, \ldots, M$ is the index of the community detection method and C_k^m is the set of nodes included in the k-th community for the method m-th.

Similarly to [1, 8], our ensemble method consists in building a consensus matrix $\mathbf{A} = [a_{i,j}]$ (with $i, j = 1, \ldots, n$) in which each cell $a_{i,j}$ (with $i \neq j$) records the number of times in which each couple of nodes is allocated to the same community of a local partition while the diagonal entries $a_{i,j} = M$ (with $i = j$) record the number of community detection methods. In this sense, a value of $a_{i,j}$ equal to the number M of methods in the ensemble indicates a full consensus in allocating the nodes (v_i, v_j) to the same community; on the contrary, the value 0 of $a_{i,j}$ indicates that no method allocates (v_i, v_j) to the same community. Finally, intermediate values of $a_{i,j}$ reveal that there is not a strong consensus in allocating the corresponding nodes to the same community as consequence of the differences among the methods in the ensemble.

3 Nonmetric Multidimensional Scaling for Community Detection

In this section, we consider the obtained consensus matrix $\mathbf{A}$ as a similarity matrix. This is motivated by the following assumptions: (1) we get a high value of $a_{i,j}$ only if a lot of community detection methods in the ensemble consider the two corresponding nodes (v_i, v_j) so similar to be very often allocated to the same community; (2) if $a_{i,j}$ records a low value, the two corresponding nodes are considered very dissimilar by a lot of members of the ensemble; (3) intermediate values account for nodes which are considered similar by some algorithm and dissimilar by others. This involves an intermediate value of similarity.

The similarity matrix $\mathbf{A}$ can be transformed into a dissimilarity one $\mathbf{D}$ by $\mathbf{D} = M\mathbf{1}-\mathbf{A}$, where M is the number of community detection algorithms used for building the consensus matrix and $\mathbf{1}$ is a (n, n) matrix of ones.

We propose to use the matrix $\mathbf{D}$ as input for a nonmetric multidimensional scaling algorithm (MDS) in order to reach two aims: the first one is to show, graphically, the proximity among nodes; the second is to get a consensus partition of nodes.

As it is well known, the main objective of nonmetric MDS consists in finding an arrangement of the input data objects into a low-dimensional space so that the new distances reflect as closely as possible the rank order of the data. It is worth noting that a two-dimensional space is often enough for getting a good approximation of the original distances and an easy-to-read graphical representation.

The motivation behind the choice of nonmetric MDS rather than the classic metric MDS is that each community detection algorithm acts as a voter in selecting which pairs of nodes should be allocated together in the same community and which pairs are so different to be allocated to different communities. So, the values in $\mathbf{D}$ reflect the ordering relations in nodes proximity.

Let n be the number of nodes in the network and let the dissimilarity between the nodes v_i and v_j be given by $d_{i,j}$. By means of nonmetric MDS, the nodes of the network are gathered in an nxp matrix $\mathbf{X}$, where $p \leq n$ is the dimensionality of the nodes in the new space. In nonmetric MDS, only the rank order of entries in the proximity matrix $\mathbf{D}$ is considered rather than the magnitude of the proximity.

The research of a matrix $\mathbf{X}$ such that the distances between its rows match as closely the order relations in $\mathbf{D}$ can be performed in several ways. A common approach consists in finding both a nonparametric monotonic relationship between the dissimilarities in **D** and the Euclidean distances between the items in **X** and the low-dimensional coordinates of each item in **X**. In order to obtain this relationship, we optimize the Kruskals stress function[5]:

$$Stress = \sqrt{\frac{\sum_{i,j}\left(d_{i,j}-\hat{d}_{i,j}\right)^2}{\sum_{i,j} d_{i,j}^2}} \tag{1}$$

where $\hat{d}_{i,j}$ are the estimated Euclidean distances among the items in **X**.

By means of MDS we get a representation of the nodes which can be plotted easier and which reflects the proximity relations resulting from the community detection algorithms in the ensemble.

In order to reach our second aim, we still use the output of MDS to get a consensus partition of the nodes. We use the well-known K-means on the low-dimensional coordinates in **X** to provide a partition P in K clusters of the nodes taking into account the Euclidean proximities estimated by the MDS.

To interpret these results we consider a simulation study based on different synthetic networks.

4 Analysis Based on Simulated Networks

In these examples we consider different types of simulated networks and different structures. In particular we consider different classes of networks: networks characterized by the preferential attachment, networks characterized by being based on the Preferential Attachment structure, Small World Networks, and finally the Barabasi Game BA structure [3]. In order to compare the results among the different methods and the consensus method we are proposing, we consider a network size of 40 nodes. For all the networks we obtain the results for each community detection algorithm. At this point we are able to compare the different results. So we start to consider the nonmetric MDS, then the K-Means. The number of the communities extracted is obtained by the majority of the number of the communities obtained by the community detection methods. In that sense we obtain the final consensus community structure (Figs. 1, 2, 3, and 4). It is important to note that we compare the consensus community structure with the original data by using the adjusted Rand Index. The reason of using the adjusted Rand Index is the need to assess the structure of the communities with the original communities extracted by the different community detection algorithms. The final results are interesting: in fact we obtain that it is very important to decide the right number of clusters in the K-means procedure in order to find the appropriate number of communities. At the same time it is important to consider that the method extracts not only communities

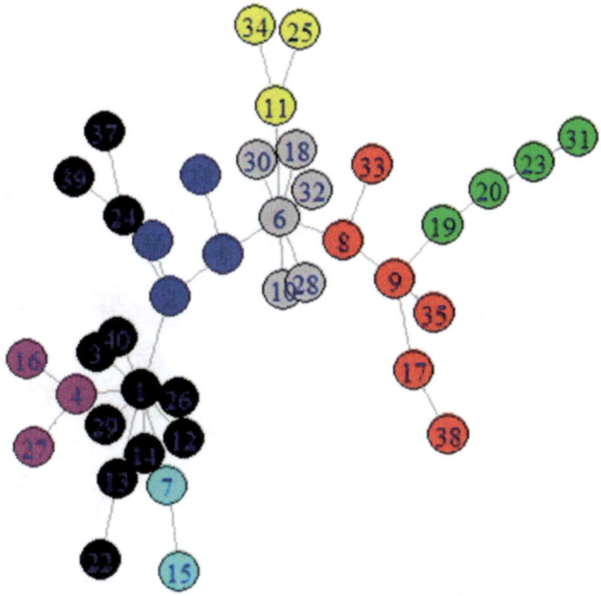

Fig. 1 Consensus partition for experiment 1

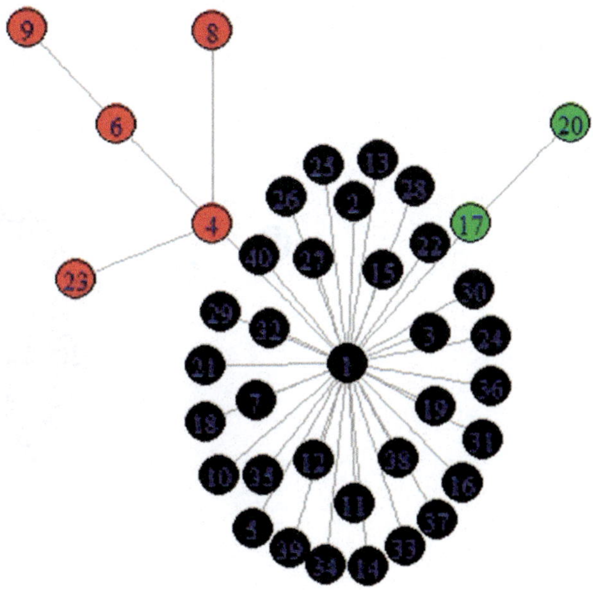

Fig. 2 Consensus partition for experiment 2

but it is able to identify some similarities between the structure of the communities as well. In Table 1 we show the adjusted rand index, computed on each evaluated network, between the consensus partition and the partitions obtained by the methods in the ensemble. We can see that the Adjusted Rand Index in the first case shows a good level of agreement between the results of the ensemble community detection

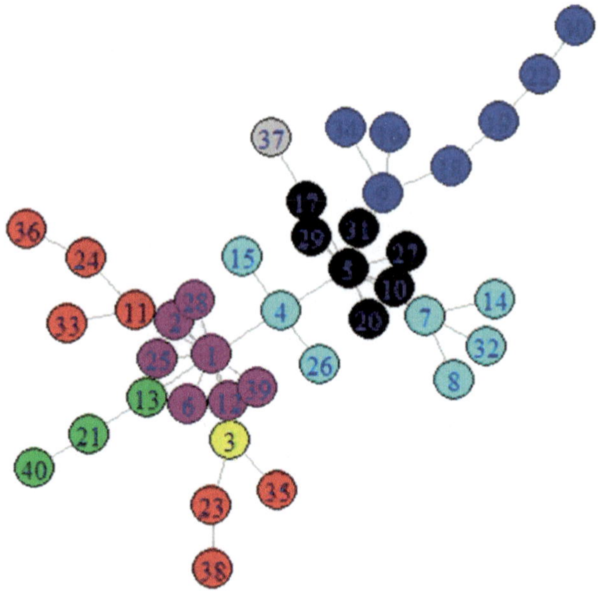

Fig. 3 Consensus partition for experiment 3

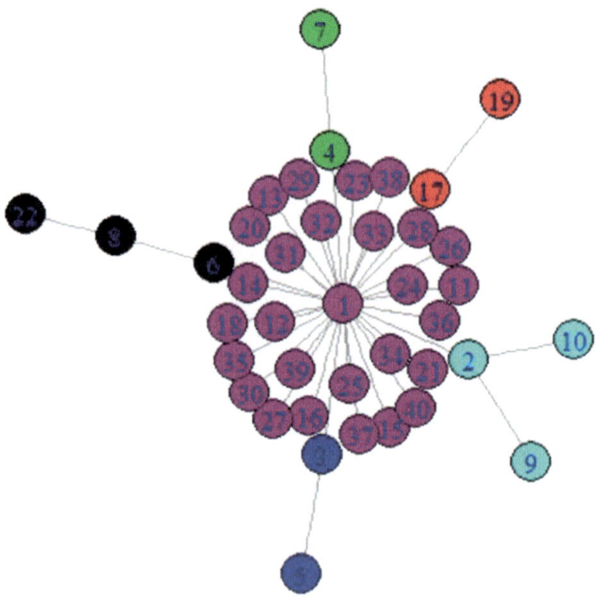

Fig. 4 Consensus partition for experiment 4

Table 1 Corrected rand index computed on the ensemble partition and the partition obtained by each community detection methods

	Experiment 1	Experiment 2	Experiment 3	Experiment 4
Edge betweenness	0.614	1	0.289	0.894
Walktrap	0.678	1	0.253	0.894
Fastgreedy	0.726	1	0.289	0.894
Leading eigenvector	0.614	1	0.407	0.506
Multilevel community	0.726	1	0.289	0.894
Infomap community	0.678	1	0.253	0.894
Label propagation	0.685	1	0.331	0.260

and the methods in the ensemble. In fact a level around 0.70 shows that there are a number of common nodes well captured by all the different algorithms. However, there can be some results which are due to single characteristics of the algorithms which cannot find any consensus in the other algorithms. In every case the method is able to find a solution which tends to minimize the risk of adopting the wrong community detection method, by finding a solution which "mediates" between the different solutions.

In the second case there is an Adjusted Rand Index of 1 for all the methods in the ensemble. It is important to note that we have detected the corrected number of communities (which are considered in the K-Means algorithm). The third experiment shows the case of a low Adjusted Rand Index (around 0.30). In this case we find a disagreement among the different methods which does not allow to find a unique satisfying solution. At the same time we can observe, visually, that the solution allows to find some similarities in the nonconnected groups of nodes. In this case there are some stable structures in these nonconnected groups. Finally in the experiment case we find a strong consensus between the different algorithms and the obtained solution, with the exception of algorithms 4 and 7. The latest, provides communities which are very different from those discovered by the other algorithms, while the consensus partition holds the information according to a majority scheme.

The communities can be considered as the group of nodes allocated to the same cluster by the K-Mean algorithm. In this sense the nodes which are part of the same cluster are stable groups detected by the procedure. The nodes which are no part of these stable groups can be detected by considering the graph and observing the nodes which are amidst other group of nodes (stable). In these cases we can detect situations in which different methods tend to have different outcomes for these nodes. This is a relevant information in the analysis of the network.

Now we consider a real case based on real data.

5 Application

The application is related to a joint patent application network obtained by a new dataset of innovative firms operating in Italy. The source of data is the OECD REGPAT database in which data are a subset of the original network. So at the end of the procedure we obtain 216 nodes. Each node represents a single different company and each vertex represents a common patenting project of two nodes. So, from the original matrix, we consider seven community detection methods: edge.betweenness, walktrap, fastgreedy, leading.eigenvector, multilevel, infomap, and label.propagation. These methods detect the different communities which are collected in the consensus matrix. Then we start the nonmetric MDS in order to find the distribution of the nodes in the axis X-Y identified by the procedure. Finally we use the clustering algorithm of K-Mean in order to find the stable communities. Due to a priori information, the number of classes is 40. The final interpretation of the results is that we can detect some communities which have the structure of a node very central (representing firms very innovative) and the other nodes representing companies which participate the common projects (Fig. 5).

In this case it is possible to see from the results we are able to understand the community structure of relevant group of nodes. At the same time here we can confirm that this tool is also useful to understand whether there are some differences in the algorithms. In fact when we are not able to detect a unique solution whether it is more likely to have similarities between not connected nodes. Also the result can show some similarities on the network structure which is effectively captured by a clustering algorithm and is not simply observed by a community detection one. In fact in some cases the similar groups cannot be connected but they show

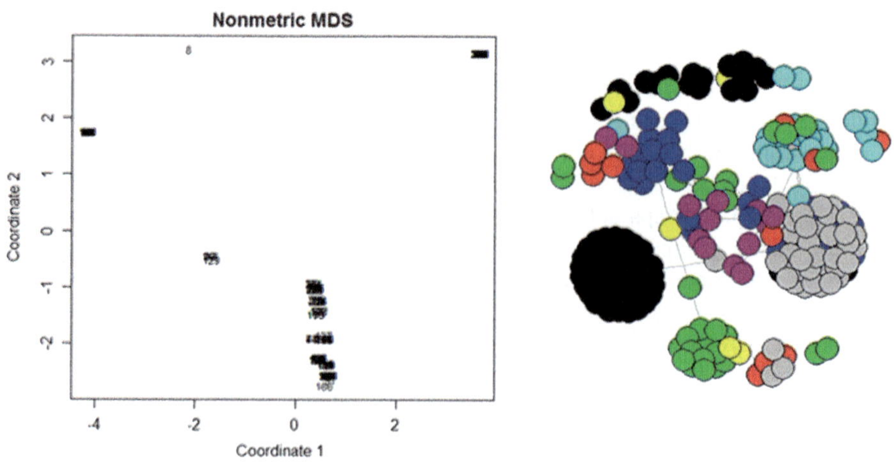

Fig. 5 MDS results plotted on the first two axes and k-means results on the two-dimensional data points from MDS

relevant similarities. In this case we are able to find this pattern. At the same time it is important to confirm the fact it is necessary to find the right number of K (obtained by majority voting of the algorithms) in order to detect the more relevant partitions using the a priori information given by the other methods.

6 Conclusions

In this work we have considered a new approach in order to perform community detection. In particular this method is useful in order to detect the different taxonomies of nodes in a network. So we can have nodes which are particularly unstable (so they participate in different communities) and nodes which are particularly stable. This information is explored more in depth by analyzing the nonmetric MDS procedure which maps the different nodes in the space and provide a more simple way to interpret the original network and allows to identify the relevant patterns (Fig. 5).

Acknowledgements The authors wish to thank Ivan Cucco for proving the data related to the joint patent application network.

References

1. Balzanella, A., Verde, R.: Summarizing and detecting structural drifts from multiple data streams. In: Giusti, A., Ritter, G., Vichi, M. (eds.) Classification and Data Mining, vol. XVIII, 26, pp. 105–112. Springer, Berlin (2013)
2. Barthelemy, M.: Betweenness centrality in large complex networks. Eur. Phys. J. B Condensed Matter Complex Syst. **38**(2), 163–168 (2004)
3. Csardi G., Nepusz T.: The igraph software package for complex network research. Int. J. Complex Syst. **1695**, 1–9 (2006)
4. Fortunato, S.: Community detection in graphs. Phys. Rep. **486**(3), 75–174 (2010)
5. Kruskal, J.B.: Nonmetric multidimensional scaling: a numerical method. Psychometrika **29**, 115–129 (1964)
6. Leskovec, J., Lang, K.J., Mahoney, M.: Empirical comparison of algorithms for network community detection. In: Proceedings of the 19th International Conference on World Wide Web, pp. 631–640. ACM, New York (2010)
7. Newman, M.E.: The structure and function of complex networks. SIAM Rev. **45**(2), 167–256 (2003)
8. Strehl, A., Ghosh J.: Cluster ensembles a knowledge reuse framework for combining multiple partitions. J. Mach. Learn. Res. **3**, 583–617 (2002)

The Performance of the Gradient-Like Influence Measure in Generalized Linear Mixed Models

Marco Enea and Antonella Plaia

Abstract A gradient-like statistic, recently introduced as an influence measure, has been proven to work well in large sample, thanks to its asymptotic properties. In this work, through small-scale simulation schemes, the performance of such a diagnostic measure is further investigated in terms of concordance with the main influence measures used for outlier identification. The simulation studies are performed by using generalized linear mixed models (GLMMs).

Keywords Diagnostics • GLMM • Gradient statistic • Outliers

1 Introduction

Generalized linear mixed models (GLMMs) [10] are useful extensions of both linear mixed models and generalized linear models in order to assess additional components of variability due to latent random effects. For this reason these models have received growing attention during the past decades. Unfortunately, the model estimates may heavily depend on a small part of the dataset or even on a particular observation or cluster. Therefore, the identification of potentially influential outliers is an important step beyond estimation in GLMMs. In the literature, two major approaches for detecting influential observations can be found. The first one is **the local influence approach**, which develops diagnostic measures by using the curvature of the influence graph of an appropriate function. The second one, **the deletion approach**, develops a diagnostic measure by assessing a chosen quantity change that is induced by the exclusion of individual data points from an analysis. However, since the observed-data likelihood function in a GLMM involves intractable integrals, the development, as well as the evaluation, of deletion diagnostic measures involving the information matrix is rather difficult. On the grounds of the measure suggested by Cook [4], Enea and Plaia [6] derive a diagnostic measure

M. Enea (✉) • A. Plaia
Dipartimento di Scienze Economiche Aziendali e Statistiche, University of Palermo, Palermo, Italy
e-mail: marco.enea@unipa.it; antonella.plaia@unipa.it

I. Morlini et al. (eds.), *Advances in Statistical Models for Data Analysis*, Studies in Classification, Data Analysis, and Knowledge Organization,
DOI 10.1007/978-3-319-17377-1_12

which does not require the information matrix, while maintaining the same large sample behaviour. Their proposal can be considered the analogue, in the study of influence, of the gradient statistic, recently introduced by Terrell [13] and further studied by Lemonte [8].

In this work, through well-tested GLMM-based simulation studies using small size samples, we assess the performance of this gradient-like measure in terms of concordance with the most used influence measures: the *likelihood displacement*, the *(generalized) Cook's distance* and the *(total) local influence*. The paper is structured as follows: first we define the GLMM for notation purposes in Sect. 2 and then we recall the three diagnostic measures above-mentioned and the gradient-like measure in Sect. 3, by providing some further computational details (w.r.t. [6]) for its calculation. The simulation studies are reported and discussed in Sect. 4 whereas Sect. 5 reports the conclusions. The appendix reports a simplified version of the R code we used to calculate the above-mentioned measures.

2 The GLMM

Let y_{ij} be the response of the jth observation, $j = 1, \ldots, n_i$, in the ith cluster, $i = 1, \ldots, N$. The GLMM is defined by the following equation:

$$g(\mu_{ij}) = g(E[y_{ij}|\boldsymbol{b}_i, \boldsymbol{x}_{ij}, z_{ij}]) = \eta_{ij} = \boldsymbol{x}'_{ij}\boldsymbol{\beta} + z'_{ij}\boldsymbol{b}_i, \tag{1}$$

where g is a link function, $\boldsymbol{x}_{ij}$ and z_{ij} are covariate arrays, $\boldsymbol{\beta}$ is the vector of fixed-effect parameters, $\boldsymbol{b}_i$ is assumed to be $N(0, G)$, with G unstructured. The *marginal likelihood* is

$$L(\boldsymbol{\beta}, G, \phi) = \prod_{i=1}^{N} f_i(\boldsymbol{y}_i|\boldsymbol{\beta}, G, \phi) = \prod_{i=1}^{N} \int \prod_{j=1}^{n_i} f_{ij}(y_{ij}|\boldsymbol{b}_i, \boldsymbol{\beta}, \phi) f(\boldsymbol{b}_i|G) d\boldsymbol{b}_i, \tag{2}$$

where

$$f(\boldsymbol{b}_i|G) = \frac{1}{\sqrt{(2\pi)^p|G|}} \exp\left\{-\boldsymbol{b}'_i G^{-1}\boldsymbol{b}_i/2\right\}, \tag{3}$$

$$f_{ij}(y_{ij}|\boldsymbol{b}_i, \boldsymbol{\beta}, \phi) = \exp\left\{\frac{y_{ij}\theta_{ij} - \psi(\theta_{ij})}{a_{ij}(\phi)} + c_{ij}(y_{ij}, \phi)\right\}, \tag{4}$$

with $\psi()$, $a_{ij}()$ and $c_{ij}()$ known functions and ϕ dispersion parameter. Parameter estimation of model (1) is usually performed via marginal log-likelihood maximization and by integrating out the random effects $\boldsymbol{b}_i$.

3 Influence Diagnostics

Let $\boldsymbol{\zeta} = (\boldsymbol{\beta}', \boldsymbol{\delta}')'$, with $\boldsymbol{\delta}$ representing the variance/covariance components. For the GLMM, the three most used influence measures, computable for a single observation, cluster or more generally for its subset M_i, are:

the *log-likelihood displacement* [5]

$$LD_{M_i} = 2\{l(\mathbf{y}|\hat{\boldsymbol{\zeta}}) - l(\mathbf{y}|\hat{\boldsymbol{\zeta}}_{(M_i)})\}, \tag{5}$$

the *Cook's distance* [3]

$$CD_{M_i} = (\hat{\boldsymbol{\zeta}} - \hat{\boldsymbol{\zeta}}_{(M_i)})'\{-H\}(\hat{\boldsymbol{\zeta}} - \hat{\boldsymbol{\zeta}}_{(M_i)}), \tag{6}$$

and the *Cook's total influence measure* [4]

$$C_{M_i} = 2\left|\boldsymbol{\Delta}'_{M_i} H^{-1} \boldsymbol{\Delta}_{M_i}\right|, \tag{7}$$

where H is the Hessian matrix of the log-likelihood relative to parameter $\boldsymbol{\zeta}$. Here $\boldsymbol{\Delta}'_{M_i} = \boldsymbol{s}_i - \boldsymbol{s}_{i(M_i)}$, where $\boldsymbol{s}_i = (\boldsymbol{s}'_{i\boldsymbol{\beta}}, \boldsymbol{s}'_{i\boldsymbol{\delta}})'$ is the subvector of the difference between the contribution to the score function of cluster i and the score function for such cluster without set M_i. Of course, if interest is only in the influence of the ith cluster, it will be sufficient to consider $\boldsymbol{\Delta}_{M_i} = \boldsymbol{s}_i$. Both H and $\boldsymbol{\Delta}_{M_i}$ are calculated at $\boldsymbol{\zeta} = \hat{\boldsymbol{\zeta}}$. Notice that we use the "total", as opposed to the "local", influence measure in the sense that (7) may be considered the deletion diagnostic subcase of [4], initially proposed to construct influence curves.

Now, let $\hat{\boldsymbol{\zeta}}_{(M_i)}$ be the estimate of $\boldsymbol{\zeta}$ when subset M_i is deleted. Since $\hat{\boldsymbol{\zeta}}_{(M_i)} \approx \hat{\boldsymbol{\zeta}} - H^{-1}_{(M_i)} \boldsymbol{\Delta}_{(M_i)}$, and by considering that $H^{-1}_{(M_i)}$ can be approximated by H^{-1}, as done by Zhu et al. [16], we have

$$\hat{\boldsymbol{\zeta}} - \hat{\boldsymbol{\zeta}}_{(M_i)} \approx H^{-1} \boldsymbol{\Delta}_{(M_i)}. \tag{8}$$

By pre-multiplying both members of (8) by $\boldsymbol{\Delta}'_{(M_i)}$, it becomes

$$\boldsymbol{\Delta}'_{(M_i)}(\hat{\boldsymbol{\zeta}} - \hat{\boldsymbol{\zeta}}_{(M_i)}) \approx \boldsymbol{\Delta}'_{(M_i)} H^{-1} \boldsymbol{\Delta}_{(M_i)}. \tag{9}$$

Note the similarity between the first member of (9) and the gradient statistic $\boldsymbol{\Delta}_0'(\hat{\xi} - \xi_0)$ [13]. Such a statistic is asymptotically χ^2 distributed, although it is not a quadratic form and might assume negative values for small sample sizes. By considering that $\sum_{i=1}^{N} \boldsymbol{\Delta}_{M_i} = 0$ and given that $\boldsymbol{\Delta}_{(M_i)} = \sum_{j \neq i} \boldsymbol{\Delta}_{M_j} = -\boldsymbol{\Delta}_{M_i}$, (9) becomes $\boldsymbol{\Delta}_{M_i}'(\hat{\boldsymbol{\xi}}_{(M_i)} - \hat{\boldsymbol{\xi}}) \approx \boldsymbol{\Delta}_{M_i}' H^{-1} \boldsymbol{\Delta}_{M_i}$. Finally we have

$$C_{M_i} \approx C_{M_i}^a = 2|\boldsymbol{\Delta}_{M_i}'(\hat{\boldsymbol{\xi}}_{(M_i)} - \hat{\boldsymbol{\xi}})|, \tag{10}$$

which is a measure of influence, because of the distance $\hat{\boldsymbol{\xi}}_{(M_i)} - \hat{\boldsymbol{\xi}}$, for which the use of the information matrix is no more necessary. Notice that if the M_ith subset is influential, $\hat{\boldsymbol{\xi}}_{(M_i)} - \hat{\boldsymbol{\xi}}$ will be large and imply a low accuracy of (8). However, if the aim of the approximation is just to detect influential data structure, as it is also discussed in Sect. 4, such an accuracy could be no more necessary since the higher $\hat{\boldsymbol{\xi}}_{(M_i)} - \hat{\boldsymbol{\xi}}$ the higher $C_{M_i}^a$ [5, p. 182].

A general expression of $\boldsymbol{s}_i$ for family (4) can be found in [11]. In particular, for GLMMs having $\phi = 1$, such as binomial or Poisson, it results that:

$$\boldsymbol{s}_{i\beta} - \boldsymbol{s}_{i\beta(M_i)} = X_i'(\boldsymbol{y}_i - \boldsymbol{\eta}_i) - X_{i(M_i)}'(\boldsymbol{y}_{i(M_i)} - \boldsymbol{\eta}_{i(M_i)}), \tag{11}$$

whereas the vector $\boldsymbol{s}_{i\delta} - \boldsymbol{s}_{i\delta(M_i)}$ of derivatives with respect to the elements of G is calculated from the following matrix D:

$$D = \frac{1}{2}\hat{G}^{-1}\{E[\boldsymbol{b}_i\boldsymbol{b}_i'|\boldsymbol{y}_i, \hat{\xi}] - E[\boldsymbol{b}_i\boldsymbol{b}_i'|\boldsymbol{y}_{i(M_i)}, \hat{\xi}]\}\hat{G}^{-1}. \tag{12}$$

In particular, the derivatives with respect to G_{jk}, $j \neq k$, will be the sum of the corresponding off-diagonal elements in D, whereas the derivatives with respect to G_{jj} will correspond to the jjth elements in D. The quantity in the curly brackets is the difference between the vectors of empirical Bayes (EB) estimates of the second moment of $\boldsymbol{b}_i$, based on the complete sample and on the sample without the observations in set M_i, respectively. To calculate these two quantities consider that $E[\boldsymbol{b}_i\boldsymbol{b}_i'|\boldsymbol{y}_i, \hat{\xi}] = VAR[\boldsymbol{b}_i|\boldsymbol{y}_i, \hat{\xi}] + E[\boldsymbol{b}_i|\boldsymbol{y}_i, \hat{\xi}]^2$. However calculating the EB means and variances can be time consuming. It is simpler and preferable calculating the EB modes and their variances [7, p. 234]. Actually, EB modes and their variances belong to the standard output of software to fit GLMMs like, for example, `glmer` in `lme4` R package [1].

4 Simulation Studies

4.1 Simulation Scheme 1

By following well-tested simulation schemes [9, 14, 15], a small-scale simulation study is performed from the following model: $y_{ij}|b_i \sim \text{Poisson}(\mu_{ij})$, $b_i \sim N(0, \sigma^2)$, $\log(\mu_{ij}) = x_{ij}\beta + b_i$, where $j = 1, \ldots, n$, $i = 1, \ldots, 10$, with equal sample size n in each cluster i. The single variable x_{ij} is chosen as j/n, while $\beta = 1, \sigma^2 = 0.1, 0.2, 1.0$ and $n = 30, 100, 200$. Both 100 and 1000 replications are considered for each combination of σ^2 and n. All the models are estimated by adding an intercept. The aim is to investigate the performance of $C^a_{M_i}$ by assessing its concordance with C_{M_i} in terms of proportion of correct identification of: (a) the cluster with the largest C_i; (b) the two clusters with the largest and the second largest C_i. Table 1 shows the results of the simulation. Observe that the proportions of correct identification are at least 83 % (a) and 64 % (b) using 100 replications and 86.1 % (a) and 62.4 % (b) using 1000 replications, which can be considered good results.

4.2 Simulation Scheme 2

By using the same parameters of the previous simulation scheme, in the second simulation study we generated 101 datasets, picked the one with median log-likelihood value and repeated the procedure 100 times. This scheme is aimed at assessing the pairwise concordance among LD_i, CD_i, $C_i/2$ and $C^a_i/2$, on 100 "typical" datasets [15]. Table 2 shows the pairwise concordance percentages of the cluster with the largest influence and the two clusters with the largest and the second largest influence.

Table 1 Proportion of correct identification of (a) the cluster with the largest C_i and (b) the two clusters with the largest and the second largest C_i, using C^a_i, for the simulation scheme 1

		$n = 30, \sigma^2$			$n = 100, \sigma^2$			$n = 200, \sigma^2$		
		0.1	0.2	1.0	0.1	0.2	1.0	0.1	0.2	1.0
Replications		(%)	(%)	(%)	(%)	(%)	(%)	(%)	(%)	(%)
100	(a)	89.0	83.0	90.0	87.0	95.0	90.0	88.0	91.0	94.0
	(b)	64.0	67.0	73.0	66.0	74.0	71.0	73.0	75.0	80.0
1000	(a)	87.7	88.0	86.1	89.7	92.4	87.6	89.2	92.2	89.7
	(b)	62.4	68.9	68.9	67.8	73.9	72.3	64.9	75.8	75.9

Table 2 Pairwise concordance percentages of (a) the cluster with the largest influence and (b) the two clusters with the largest and the second largest influence, using LD_i, CD_i, $C_i/2$ and $C_i^a/2$ for simulation scheme 2

n	σ^2		$C_i^a/2$, CD_i (%)	$C_i^a/2$, $C_i/2$ (%)	$C_i^a/2$, LD_i (%)	CD_i, $C_i/2$ (%)	CD_i, LD_i (%)	$C_i/2$, LD_i (%)	All four (%)
30	0.1	(a)	55	82	96	41	52	86	41
		(b)	21	53	87	10	22	53	10
100	0.1	(a)	68	94	98	64	67	94	64
		(b)	31	80	92	24	31	80	23
200	0.1	(a)	73	92	96	65	69	96	65
		(b)	59	65	92	37	56	69	37
30	0.2	(a)	78	86	96	65	74	90	65
		(b)	48	69	94	32	47	72	32
100	0.2	(a)	80	94	95	74	75	95	72
		(b)	58	71	89	37	55	68	34
200	0.2	(a)	83	94	98	77	83	94	77
		(b)	65	76	93	50	65	75	49
30	1	(a)	90	87	96	77	94	83	77
		(b)	67	66	93	39	69	61	39
100	1	(a)	92	87	98	79	92	87	79
		(b)	75	72	95	55	77	71	54
200	1	(a)	87	91	98	80	89	89	79
		(b)	75	78	98	59	77	76	58

Notice that $C_i^a/2$ shows high concordance rates with LD_i and $C_i/2$. Overall, for small σ^2 values there are small concordance rates among the four influence measures. This means that for this parameter setting the use of only one influence measure is not sufficient for outlier detection. From this view, the proposal of an additional diagnostic tool such as the gradient-like measure is advantageous.

Figure 1 shows some results from the simulated "typical" datasets, obtained by varying σ^2 and n, when all four measures are concordant. Observe that, even though the measures appear to be concordant in detecting the most influential clusters, for such clusters they can also provide values with different magnitude. Further, for small-size samples, we suggest to use C_i^a rather than $C_i^a/2$, in order to better highlight the most influential cluster.

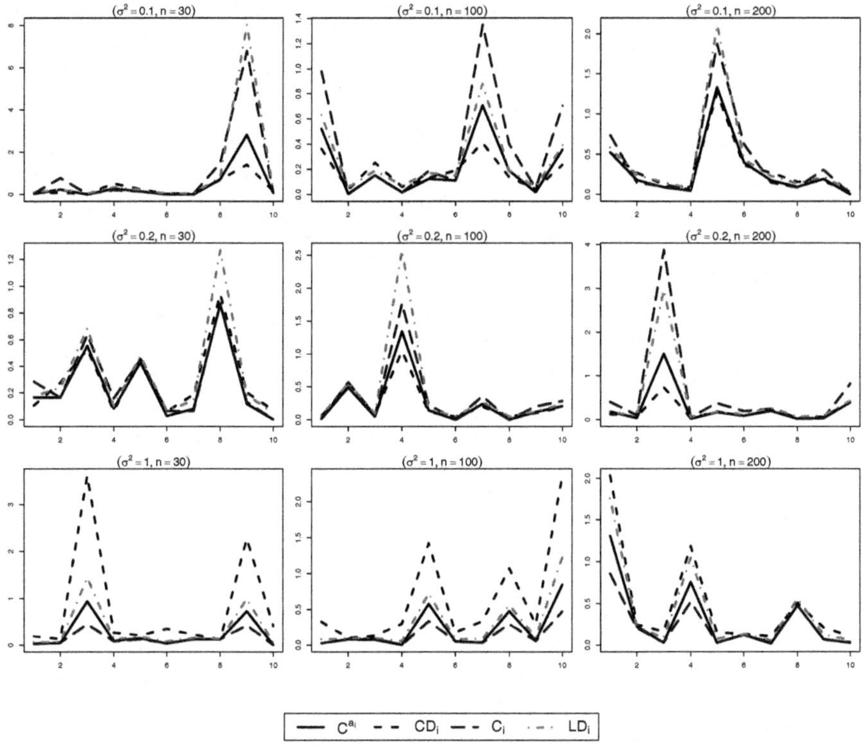

Fig. 1 Some cluster-oriented diagnostics from the second simulation scheme, by varying n and σ^2, for the case of concordance

5 Conclusions

In this work we have analysed the small-sample behaviour of the gradient-like influence measure, proposed by Enea and Plaia [6], by using Poisson-normal random intercept model. Such diagnostic measure resulted to be concordant, in cluster-level outlier detection, not only with C_{M_i}, from which it represents a direct approximation but also with CD_{M_i} and above all with LD_{M_i}. Although we have used the gradient-like measure in the context of the GLMMs, it may be used, as well as LD_{M_i}, CD_{M_i} and C_{M_i}, to carry out an influence diagnostics on any model.

Appendix: The following R [12] code allows

The following code allows to perform cluster-level influence diagnostics from an object returned by `glmer` for binomial or Poisson random intercept models. Currently, the code works under lme4 version 0.999999-2. At time of writing, package `lme4` was updated to version 1.0-5, but some bugs, concerning the conditional variances of the random effects, are not fixed yet. Further, as it has been explained by Enea and Plaia [6], the information matrix, which is necessary to perform the diagnostics using C_i and CD_i, can be obtained from package `glmmML` [2], which uses the same estimation method and provides the same estimates of `lme4`. A more complete code allowing diagnostics at the observation level, for random intercept/slopes models and for specified parameter subsets, here not reported due to space limits, can be requested to the authors.

```
influence.mer <- function(obj,H=NULL){
  options(warn=-1)
  parf <- obj@fixef
  nparf <- length(parf)
  oneresp <- is.null(ncol(obj@frame[[1]]))
  Y <- (if (oneresp) obj@frame[[1]] else obj@frame[[1]][,1])
  m <- if(oneresp) rep(1,length(Y)) else rowSums(obj@frame[[1]])
  nobs <- as.vector(table(obj@flist[,ncol(obj@flist)]))
  iclus <- obj@flist[,ncol(obj@flist)]
  clus <- levels(iclus)
  nclus <- length(clus)
  logLik1 <- logLik(obj)[1]
  delta <- VarCorr(obj)[[1]]
  names(delta) <- "delta"
  psi <- c(parf,delta)
  bi <- ranef(obj,postVar=TRUE)[[1]]
  Di <- c()
  for (i in 1:nclus) Di[i]<-(attributes(bi)$postVar[,,i]+bi[i,]^2)/(2*delta^2)
  E <- Y-fitted(obj)*m
  logLik2 <-c()
  offset <- if (length(obj@offset)>0) exp(obj@offset) else rep(1,length(Y))
  sDelta <- matrix(,nclus,nparf)
  Dpsi <- matrix(,nclus,length(psi))
  for (j in 1:nclus){
     yes <- (iclus==clus[j])
     sDelta[j,] <- crossprod(obj@X[yes,],E[yes])
     newobj <- update(obj,data=obj@frame[!yes,])
     deltai <- VarCorr(newobj)[[1]]
     Dpsi[j,] <- psi-c(fixef(newobj),deltai)
     logLik2[j] <- logLik(update(obj,data=obj@frame,start=list(ST=newobj@ST,
                    fixef=fixef(newobj)),control=list(maxFN=0,maxIter=0)))[1]
  }
  Delta <- cbind(sDelta,Di)
  DD <- Delta*Dpsi
  sGD <- 2*abs(DD)
  GD <- 2*abs(rowSums(DD))
  colnames(sGD) <- colnames(Delta) <- colnames(Dpsi) <- names(psi)
  Ci <- if (!is.null(H)) 2*diag(abs(Delta%*%solve(H)%*%t(Delta))) else NULL
  CDi <- if (!is.null(H)) diag(Dpsi%*%H%*%t(Dpsi)) else NULL
  return(list("GDi"=GD,"LDi"=2*abs(logLik1-logLik2),"Ci"=Ci,"CDi"=CDi))
}
library(lme4)
library(glmmML)
```

```
library(mvtnorm)
simul.pois  <- function(j,n,param){ #create an artificial data set
  pa <- as.vector(rmvnorm(j,c(0,0),matrix(a,2,2)))
  clus <- kronecker(1:j,rep(1,n))
  x <- rep((1:n)/n ,j)
  resp <- rpois(n*j,lambda=exp(param[1]+param[2]*x+cbind(kronecker(diag(j),
            rep(1,n)),kronecker(diag(j),(1:n/n)))%*%pa ))
  data.frame(clus,x,resp)
}
a <- c(1,0.5,0.5,1) #for variance/covariance components
dad <- simul.pois(j=10,n=30,param=c(1,-1,a))
m0 <- glmer(resp ~ x + (1|clus),data=dad, family=poisson, x=TRUE)
m0b <- glmmML(resp ~ x, cluster=clus,data=dad, family=poisson)
r0 <- influence.mer(obj=m0,H=solve(m0b$variance))
r01 <- r0
r01$Ci <- r01$Ci/2
r01$GDi <- r01$GDi/2 #GDi is the Gradient-like influence measure
r01 <- do.call("cbind",r01)
matplot(r01,lty=1:4,type="l",col=1:4,ylab="influence",xlab="cluster index")
legend("topright",c("GDi/2","LRi","Ci/2","CDi"),lty=1:4,col=1:4)
```

References

1. Bates, D., Maechler, M., Bolker, B.: lme4: Linear mixed-effects models using S4 classes. R package version 0.999999-2. http://CRAN.R-project.org/package=lme4 (2013)
2. Böstrom, G., Holmberg, H.: glmmML: Generalized linear models with clustering. R package version 0.82-1. http://CRAN.R-project.org/package=glmmML (2011)
3. Cook, R.D.: Detection of influential observations in linear regression. Technometrics **19**, 15–18 (1977)
4. Cook, R.D.: Assessment of Local Influence. J. R. Stat. Soc. B Met. **4**(2), 133–169 (1986)
5. Cook, R.D., Weisberg, S.: Residuals and Influence in Regression. Chapman and Hall, London (1982)
6. Enea, M., Plaia, A.: Influence diagnostics for meta-analysis of individual patient data using generalized linear mixed models. In: Vicari, D., Okada, A., Ragozini, G., Weihs, C. (eds.) Analysis and Modeling of Complex Data in Behavioral and Social Sciences. Studies in Classification, Data Analysis, and Knowledge Organization. Springer, New York (2014)
7. Fahrmeier, L., Tutz, G.: Multivariate Statistical Modelling Based on Generalized Linear Models. Springer, New York (1994)
8. Lemonte, A.J.: On the gradient statistic under model misspecification. Stat. Prob. Lett. **83**, 390–398 (2013)
9. McCulloch, C.E.: Maximum likelihood algorithm for generalized linear mixed models: applications to clustered data. J. Am. Stat. Assoc. **92**, 162–170 (1997)
10. McCulloch, C.E., Searle, S.R.: Generalized, Linear, and Mixed Models. Wiley, New York (2001)
11. Ouwens, M.J.N.M., Tan, F.E.S., Berger, M.P.F.: Local influence to detect influential data structures for generalized linear mixed models. Biometrics **57**(42), 1166–1172 (2001)
12. R Core Team: R: A language and environment for statistical computing. R Foundation for Statistical Computing, Vienna (2012) [ISBN 3-900051-07-0]
13. Terrell, G.R.: The gradient statistic. Comput. Sci. Stat. **34**, 206–215 (2002)

14. Xiang, L., Tse, S.-K., Lee A. H.: Influence diagnostics for generalized linear mixed models: applications to clustered data. Comput. Stat. Data Anal. **40**, 759–774 (2002)
15. Xu, L., Lee, S., Poon, W.: Deletion measures for generalized linear mixed models. Comput. Stat. Data Anal. **51**, 1131–1146 (2006)
16. Zhu, H., Lee, S., Wei, B., Zhou, J.: Case-deletion measures for models with incomplete data. Biometrika. **88**(3), 727–737 (2001)

New Flexible Probability Distributions for Ranking Data

Salvatore Fasola and Mariangela Sciandra

Abstract Recently, several models have been proposed for analysing the ranks assigned by people to some object. These models summarize the liking feeling towards the object, possibly with respect to a set of explanatory variables. Some recent works have suggested the use of the *Shifted Binomial* and of the *Inverse Hypergeometric* distribution for modelling the approval rate, while *mixture models* have been considered for taking into account the uncertainty in the ranking process. We propose two new probability distributions, the *Discrete Beta* and the *Shifted-Beta Binomial*, which ensure much flexibility and allow the joint modelling of the scale (approval rate) and the shape (uncertainty) parameters of the rank distribution.

Keywords Discrete Beta • Ranking data • Shifted-Beta Binomial

1 Introduction

Ranking data arise when n individuals are asked to order a set of K objects, or *item*, from the most to the least preferred. The response vector will be one of the possible permutations of the first K integers, assuming that ties cannot occur. Ranking data are generally arranged in an $n \times K$ matrix $R = \{r_i^k\}$, where the generic entry r_i^k represents the rank assigned by the i-th individual to the k-th item.

Ranking data modelling has received a lot of attention in the literature, and many models have been proposed over the years, such as order statistics models [7], distance-based models [9], paired-comparison models (e.g. Bradley-Terry model) and multistage models [2].

When the interest is to summarize the liking feeling towards a given item k, the response variable becomes univariate and can be denoted with R^k. Given the discrete nature of R^k, the class of generalized linear models has found large applicability in the study of ranking data, in particular the application of proportional odds

S. Fasola (✉) • M. Sciandra
Dipartimento di Scienze Economiche, Aziendali e Statistiche, Università di Palermo, Palermo, Italy
e-mail: salvatore.fasola@unipa.it; mariangela.sciandra@unipa.it

I. Morlini et al. (eds.), *Advances in Statistical Models for Data Analysis*, Studies in Classification, Data Analysis, and Knowledge Organization,
DOI 10.1007/978-3-319-17377-1_13

models [1]. The most recent proposals for modelling the approval rate of an item rely on the *Shifted Binomial* [3] and the *Inverse Hypergeometric* [4] r.v., but often these distributions are not sufficiently flexible to fit the empirical rank distributions. At this aim, mixtures of Discrete Uniform and Shifted Binomial r.v. (MUB models) have been proposed to deal with both the selection mechanism and uncertainty in the ranking process [5]. However, the discrete distributions existing in literature are not able to assume also "J" and "U" (say convex) shapes, as discussed by Punzo and Zini [10].

The aim of this paper is to introduce two new probability distributions which are more flexible in shape and preserve simplicity and interpretability of the relevant parameters.

The paper is organized as follows: Sect. 2 presents the *Discrete Beta* and the *Shifted-Beta Binomial* distributions, while Sect. 3 describes some inferential results about model estimation. Section 4 reports two applications on two real data sets, and finally Sect. 5 is devoted to discussion and future work.

2 Two New Flexible Distributions

2.1 The Discrete Beta Distribution

The first model we propose aims at exploiting the flexibility in shape of the Beta distribution to improve fitting performances for ranking data. Since the support of the Beta distribution is continuous, a suitable transformation is required to meet the discrete nature of ranks.

Let X be a Beta r.v. with parameters α and β and $f_X(x;\alpha,\beta)$ its *p.d.f.* Our proposal consists in splitting the support of X into K intervals of the same width and considering their respective (integrated) probabilities. At this end, a vector of $K-1$ equally spaced thresholds $x_j = j/K, j = 1, 2, \ldots, K-1$ can be defined; if we consider the discrete set of probabilities

$$P_j = F_X(x_j;\alpha,\beta) - F_X(x_{j-1};\alpha,\beta) \qquad j = 1, 2, \ldots, K,$$

where $x_0 = 0$, $x_K = 1$ and $F_X(x;\alpha,\beta)$ is the distribution function of X, it can be associated to ranks assuming

$$\text{Prob}(R_i^k = r_i^k) = P_{r_i^k}\,. \tag{1}$$

A latent variable interpretation can be given to the proposed model by assuming the ranks to be induced by a continuous latent r.v. X. In fact, if $X \sim B(a,b)$ and $R_i^k = j$ when $x_{j-1} \leq X_i < x_j$, the proposed model will follow. Usual approaches fix parameters of the distribution of X (e.g. a standardized version) and estimate thresholds, as in *proportional odds models* [1]. On the contrary, we fix thresholds and estimate the parameters of $f_X(x)$. This allows to have only two model parameters regardless of K, as in the MUB model [5]. Moreover, the choice

Fig. 1 Discrete Beta distribution with $\alpha = 2$, $\beta = 3$ and $K = 5$

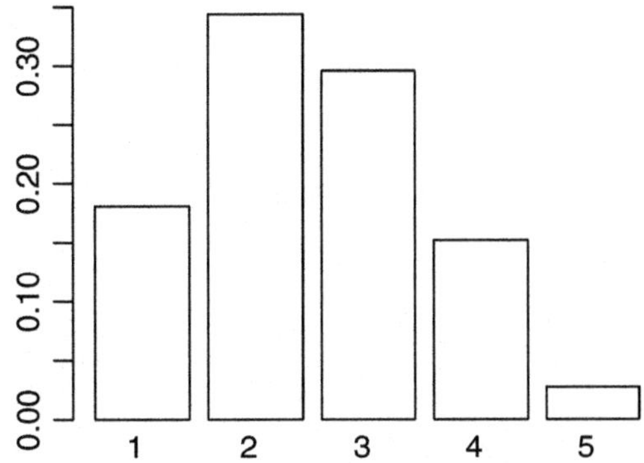

of equally spaced thresholds makes the discrete distribution to reflect the flexible shapes of the underlying Beta, especially when K is large, preserving its ease of interpretation in terms of the value assumed by a and b. For example, given a Beta r.v. with $\alpha = 2$ and $\beta = 3$ and $K = 5$, the associated thresholds are $x_1 = 0.2$, $x_2 = 0.4$, $x_3 = 0.6$ and $x_4 = 0.8$, and the probability of observing, for example, rank 4, corresponds to the probability that the Beta r.v. falls in the fourth interval, e.g. between 0.6 and 0.8. Figure 1 illustrates such distribution.

The expected value of R_i^k is

$$E(R_i^k) = K - \sum_{j=1}^{K-1} F_X(x_j; \alpha, \beta) .$$

This quantity tends to 1 (maximum liking feeling) as α tends to 0, because $f_X(x; \alpha, \beta)$ distributes the total probability mass in the close proximity of 0. Similarly, the expected value tends to K (minimum liking feeling) as β tends to 0. When $\alpha = \beta$ the distribution of X is symmetric, and, of course, $E(R_i^k) = (K+1)/2$. The variance also reflects the variance of the underlying Beta, and its expression is

$$V(R_i^k) = K^2 - \sum_{j=1}^{K-1}(1 + 2j)F_X(x_j, \alpha, \beta) - E(R_i^k)^2 .$$

In the two extreme scenarios ($\alpha \to 0$ or $\beta \to 0$), the variance tends to 0.

2.2 The Shifted-Beta Binomial Distribution

Let now R_i^k follow a *Shifted Binomial* [3] distribution

$$\text{Prob}(R_i^k = r_i^k) = \binom{K-1}{r_i^k - 1} \psi_k^{r_i^k - 1}(1 - \psi_k)^{K - r_i^k} , \tag{2}$$

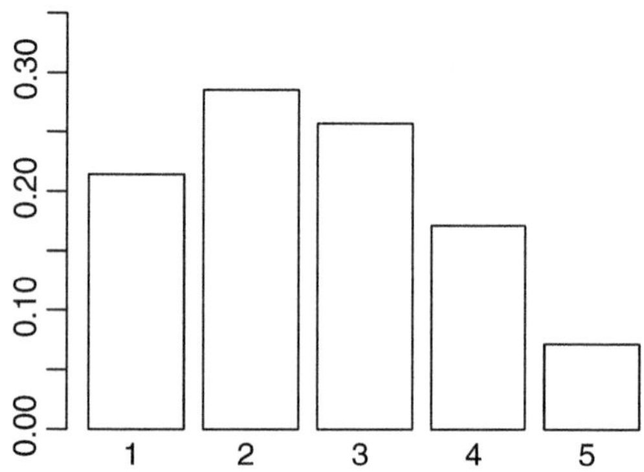

Fig. 2 Shifted-Beta Binomial distribution with $\alpha = 2$, $\beta = 3$ and $K = 5$

where ψ_k is the *disliking indicator*. Our proposal consists in using a rather natural generalization of (2), where ψ_k is assumed to be the realization of a Beta r.v. $\Psi_k \sim B(\alpha, \beta)$. This assumption leads to a *Shifted-Beta Binomial* distribution:

$$\text{Prob}(R_i^k = r_i^k) = \binom{K-1}{r_i^k - 1} \frac{B(\alpha + r_i^k - 1, \beta + K - r_i^k)}{B(\alpha, \beta)} . \tag{3}$$

When $\alpha = 1$, it reduces to the *Inverse Hypergeometric* model discussed in D'Elia [4]. As before, the model is ruled by the parameters of the underlying continuous Beta distribution. Despite the two proposed distributions are substantially different in their *p.d.f.*, they are similar in shape when using the same parameter values. For example, using the parameters of the previous example ($\alpha = 2$, $\beta = 3$), the Shifted-Beta Binomial distribution assumes the shape in Fig. 2; note similarities with the distribution in Fig. 1.

The expected value and the variance of R_i^k are, respectively,

$$E(R_i^k) = (K-1)\frac{\alpha}{\alpha+\beta} + 1 ,$$

$$V(R_i^k) = \frac{\left[E(R_i^k) - 1\right]\left[K - E(R_i^k)\right]}{K-1} \frac{\alpha + \beta + K - 1}{\alpha + \beta + 1} .$$

It is easy to note how the expected value tends to 1 as α tends to 0, while it tends to K when β tends to 0; it reduces to $(K+1)/2$ when $\alpha = \beta$. Once again the variance tends to 0 in the extreme scenarios.

3 Some Inferential Results

Estimates of α and β for models (1) and (3) can be derived via numerical maximization of the likelihood function:

$$L(\alpha, \beta; \boldsymbol{r}^k) = \prod_{i=1}^{n} \prod_{j=1}^{K} \text{Prob}(R_i^k = r_i^k)^{I(r_i^k = j)} . \tag{4}$$

The expressions of the mean and the variance for the two proposed distributions are quite complex and difficult to treat mathematically. Nevertheless, given the shape similarities between the Beta and the induced distributions, we propose to model the summary measures of the continuous version and use them as rough summary measures for the discrete versions. In particular we define

$$E(X) = E(\Psi) = \frac{\alpha}{\alpha + \beta}$$

as a *disliking indicator*, related to the scale of the distribution, and

$$\frac{E(X)[1 - E(X)]}{V(X)} - 1 = \frac{E(\Psi)[1 - E(\Psi)]}{V(\Psi)} - 1 = \alpha + \beta$$

as an *accuracy indicator*, related to the shape. The accuracy substantially reflects the degree of agreement between the judges in ranking the item.

An attractive reparameterization assumes

$$\eta = \text{logit}\left(\frac{\alpha}{\alpha + \beta}\right), \quad \gamma = \log(\alpha + \beta). \tag{5}$$

The use of this reparameterization makes the estimation process more stable, unconstrained (both α and β are strictly positive) and allows the joint modelling of the scale and the shape of the rank distribution. Another important feature concerns the possibility to introduce covariates in the model, for example assuming that two (possibly equal) vectors $\boldsymbol{x}_i$ and z_i have a linear effect on η and γ:

$$\eta_i = \boldsymbol{x}_i^T \boldsymbol{\theta}, \quad \gamma_i = z_i^T \boldsymbol{\lambda} . \tag{6}$$

Parameter interpretation reminds usual logit or log-linear models; if $\boldsymbol{x} = z = x$ we have

$$\theta = \log\left[\frac{\alpha(x_i + 1)/\beta(x_i + 1)}{\alpha(x_i)/\beta(x_i)}\right],$$

which resembles a log-odds ratio and

$$\lambda = \log\left[\frac{\alpha(x_i + 1) + \beta(x_i + 1)}{\alpha(x_i) + \beta(x_i)}\right],$$

which resembles a log-rate ratio. Of course, alternative reparameterizations could be allowed.

4 Two Real Applications

4.1 APA President Election

The two proposed models are here applied to the ranks assigned by $n = 5738$ members of the American Psychological Association (APA) to the five candidates during the election of the president in 1980. The complete data set is reported in Diaconis [6]; we focus on preferences expressed towards the third candidate, due to the particular shape assumed by the relevant observed rank distribution (Fig. 3).

The Discrete Beta model yields $\hat{\alpha} = 0.64$ and $\hat{\beta} = 0.70$ (AIC=18146.95), the Shifted-Beta Binomial model gives $\hat{\alpha} = 0.55$ and $\hat{\beta} = 0.61$ (AIC=18143.55). The disliking indicator estimate is about 0.48 for both the proposed models; this indicates that the intermediate rank is approximately the expected value of the distribution. The accuracy indicator is 1.34 for the Discrete Beta and 1.16 for the Shifted-Beta Binomial model. If we consider 2 as reference value (when α and β are both lower than one, the distribution assumes a convex shape and, then, is highly heterogeneous) the degree of agreement between the APA members appears to be low.

Despite the goodness of fit is not fully satisfactory (*Chi-squared* test statistic is highly significant), as Fig. 3 shows, the proposed models perform much better than the MUB model (AIC=18261.22), since they are able to fit convex distributions.

4.2 Computer Game Platforms

In this example, a model with explanatory variables is considered. The Discrete Beta model is applied to the ranks assigned by $n = 91$ students to six different platforms for computer games (see Fok, Paap and van Dijk, 2012)[8]. The explanatory variables are the *age* of the respondent, the number of *hours* spent on gaming per week and a dummy *own* indicating whether the platform is currently owned. We focus on preferences expressed towards PC, since the model fits quite well. Table 1

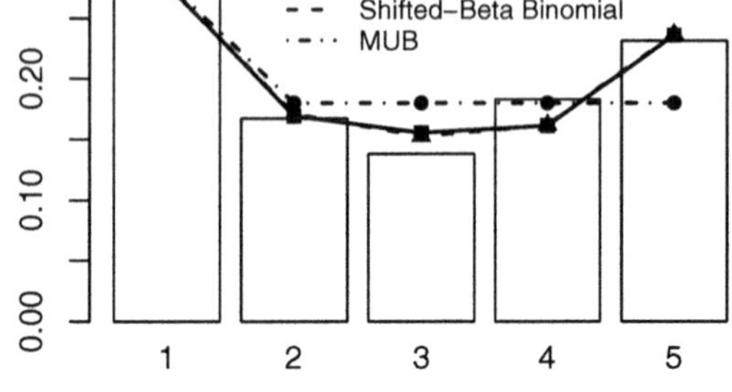

Fig. 3 Observed and fitted distributions of the ranks assigned to the third candidate for the election of the president of APA in 1980

Table 1 Disliking indicator (logit scale)

	Estimate	s.e	χ^2	p-value
Intercept	3.03	1.44	4.40	0.04
Age	-0.11	0.07	2.62	0.11
Hours	-0.09	0.03	8.89	0.00
Own	-1.39	0.39	12.56	0.00

Table 2 Accuracy indicator (log scale)

	Estimate	s.e	χ^2	p-value
Intercept	0.82	1.84	0.20	0.65
Age	-0.01	0.08	0.02	0.89
Hours	0.21	0.06	13.98	0.00
Own	-0.94	0.62	2.31	0.13

summarizes the output for η (the logit of the disliking indicator), Table 2 summarizes the output for γ (the log of the accuracy indicator); standard error estimates are derived from a numerical (observed) information matrix.

For a typical student with mean age (20.23), playing for a mean number of hours (3.88) without really owning a PC, the disliking indicator is about 0.63, greater than the intermediate scenarios. For the same profile the accuracy rate is 4.24, therefore there is quite a good agreement among that class of students in ranking PC (if, as before, 2 is taken as reference value).

As far as the explanatory variables are concerned, age does not appear to have a significant influence on the two indicators. As expected, students who actually own a PC have a lower disliking feeling, as well as students spending much time on gaming. In addition, the more the time spent playing PC, the higher would be the agreement between students in ranking PC, since the variance of the distribution of the ranks tends to reduce significantly for students who play more.

5 Conclusions

Two new flexible probability distributions for modelling the ranks assigned to an item have been proposed. These models are easy to interpret in terms of *disliking* feeling towards an item and *accuracy* rate for the corresponding distribution of ranks. These distributions result to be particularly useful when the rank distribution assumes a convex shape and represent also a good alternative for concave, monotonic and uniform distributions. The fitting process estimates the two summary measures simultaneously, through the use of a link function which maps parameters on an unbounded range of variation. Future work should focus on a more detailed description of the behaviour of the two proposed distributions with respect to parameter values; besides, a simulation study could be useful to evaluate, more rigorously, the fitting performances with respect to the existing competitors.

Acknowledgements We thank Marcello Chiodi for his suggestions and comments.

References

1. Agresti, A.: Categorical Data Analysis. Wiley, New York (2002)
2. Critchlow, D., Fligner, M., Verducci, S.: Probability models on rankings. J. Math. Psychol. **35**, 294–318 (1991)
3. D'Elia, A.: A shifted binomial model for rankings. In: Nunez-Antón, V., Ferreira, E. (eds.) Proceedings of the 15th International Workshop on Statistical Modelling, pp. 412–416. Servicio Editorial de la Universidad del Pais Vasco, Bilbao (2000)
4. D'Elia, A.: Modelling ranks using the inverse hypergeometric distribution. Stat. Model. **3**, 65–78 (2003)
5. D'Elia, A., Piccolo, D.: A mixture model for preference data analysis. Comput. Stat. Data Anal. **49**, 917–934 (2005)
6. Diaconis, P.: A generalization of spectral analysis with application to ranked data. Ann. Stat. **17**, 949–979 (1989)
7. Dwass, M.: On the distribution of ranks and of certain rank order statistics. Ann. Math. Stat. **28**, 424–431 (1957)
8. Fok, D., Paap, R., Van Dijk, B.: A rank-ordered logit model with unobserved heterogeneity in ranking capabilities. J. Appl. Econ. **27**, 831–846 (2012)
9. Lee, P.H., Yu, P.L.H.: Distance-based tree models for ranking data. Comput. Stat. Data Anal. **54**, 1672–1682 (2010)
10. Punzo, A., Zini, A.: Discrete approximations of continuous and mixed measures on a compact interval. Stat. Pap. **53**, 563–575 (2012)

Robust Estimation of Regime Switching Models

Luigi Grossi and Fany Nan

Abstract It is well known that generalized-M (GM) estimators for linear models are consistent and lead to a small loss of efficiency with respect to least squares (LS) estimator. When they are extended to threshold models the consistency of GM estimators is guaranteed only under certain objective functions. In this paper we explore, in a simulation experiment, the loss of consistency of GM-SETAR estimator under different objective functions, time-series length, parameter combinations and type of contaminations. Finally the best robust estimator is applied to study the dynamic of electricity prices where regime switching and high spikes are widely observed features.

Keywords GM estimator • Nonlinear models • Outliers

1 Introduction

Threshold auto regressive (TAR) models are quite popular in the nonlinear time-series literature. This popularity is due to the fact that they are relatively simple to specify, estimate, and interpret. The sampling properties of the estimators and test statistics associated with TAR models have been studied by Hansen [7]. It is very well known that time series can be contaminated by outliers which can dramatically influence parameter estimates (see, for example, [4]). In the class of nonlinear models, studies addressed to robustifying this kind of models are very few, although the problem is very challenging, particularly when it is not clear whether aberrant observations must be considered as outliers or as generated by a real nonlinear process. van Dijk [15] derived an outlier robust estimation method for the parameters in smooth threshold auto regressive (STAR) model, based on the principle of generalized maximum likelihood type estimation. Battaglia and Orfei [2] focused on outlier detection and estimation through a model-based approach when the time series is generated by a general nonlinear process. A general model

L. Grossi (✉) • F. Nan
University of Verona, Verona, Italy
e-mail: luigi.grossi@univr.it; fany.nan@univr.it

I. Morlini et al. (eds.), *Advances in Statistical Models for Data Analysis*,
Studies in Classification, Data Analysis, and Knowledge Organization,
DOI 10.1007/978-3-319-17377-1_14

able to capture nonlinearity, structural changes, and outliers has been introduced by Giordani et al. [6]. The authors suggest to employ the state-space framework which allows to estimate the coefficients of several nonlinear time-series models and simultaneously take into account the presence of outliers and structural breaks. The method seems quite effective in modeling macro-economic time series.

Apart from the previous methods which deal with the presence of outliers in very specific contexts, the issue of outliers in nonlinear time series models is far from being clearly solved. Chan and Cheung [3] extended the generalized M estimator method[1] to self-exciting threshold auto regressive (SETAR) models. Their simulation results show that the GM estimation is preferable to the LS estimation in the presence of additive outliers. As GM estimators have proved to be consistent with a very small loss of efficiency, at least under normal assumptions, the extension to threshold models, which are piecewise linear, looks quite straightforward. Despite this observation, a cautionary note [5] has been written to point out some drawbacks of GM estimator proposed by Chan and Cheung [3]. In particular, it is argued and shown, by means of a simulation study, that the GM estimator can deliver inconsistent estimates of the threshold even under regularity conditions. According to this contribution, the inconsistency of the estimates could be particularly severe when strongly descending weight functions are used.

Zhang et al. [16] demonstrate the consistency of GM estimators of autoregressive parameters in each regime of SETAR models when the threshold is unknown. The consistency of parameters is guaranteed when the objective function is a convex nonnegative function. A possible function holding these properties is the Huber ρ-function which is suggested to replace the polynomial function used in Giordani's [5] paper. However, the authors conclude that the problem of finding a threshold robust estimator with desirable finite-sample properties is still an open issue. Although a theoretical proof has been provided by the authors, there is not a well-structured Monte Carlo study to assess the extent of the distortion of the GM-SETAR estimator. From the analysis of the existing literature, there are at least three open issues which must be addressed: 1) What is the bias of SETAR robust estimators with respect to the LS estimator? 2) What is the best weight function to define the optimal robust estimator? 3) What are the forecasting performances of the different weight functions? Moreover, robust estimators of regime switching processes are not implemented within the most popular software platforms among statisticians, such as Matlab and R. In this paper we want to fill these gaps by presenting an extensive Monte Carlo study comparing LS and GM estimator under particular conditions. Both the simulation experiment and the analysis of real data rely on a library written in R which is available to the authors upon request. Finally, we propose an application of robust nonlinear estimators to the series of electricity prices following the results of the simulation experiment. It is indeed well known that, among the stylized facts which empirically characterize electricity prices, the

[1]For an overview about GM estimators see [1, Chap. 4] and [12, Chap. 8].

presence of sudden spikes is one of the most regularly observed and less explored features [9].

2 Robust SETAR Models

Given a time series y_t, a two-regime SETAR(p,d) model, is specified as

$$y_t = \begin{cases} \mathbf{x}_t\boldsymbol{\beta}_1 + \varepsilon_t, & \text{if} \quad y_{t-d} \leq \gamma \\ \mathbf{x}_t\boldsymbol{\beta}_2 + \varepsilon_t, & \text{if} \quad y_{t-d} > \gamma \end{cases} \tag{1}$$

for $t = 1, \ldots, N$, where y_{t-d} is the threshold variable with $d \geq 1$ and γ is the threshold value. The relation between y_{t-d} and γ states if y_t is observed in regime 1 or 2, $\boldsymbol{\beta}_j$ is the parameter vector for regime $j = 1, 2$ and $\mathbf{x}_t$ is the t-th row of the $(N \times p)$ matrix $\mathbf{X}$ comprising p lagged variables of y_t (and eventually a constant). Errors ε_t are assumed to follow an iid$(0, \sigma_\varepsilon)$ distribution.

In general the value of the threshold γ is unknown, so that the parameters to estimate become $\boldsymbol{\theta} = (\boldsymbol{\beta}_1', \boldsymbol{\beta}_2', \gamma, \sigma_\varepsilon)$. Parameters can be estimated by sequential conditional least squares: for a fixed threshold γ the model is linear, $\boldsymbol{\beta}_1$ and $\boldsymbol{\beta}_2$ can be estimated by ordinary least squares (OLS) and $\hat{\sigma}_\varepsilon = \sum_{t=1}^N r_t^2/N$, with $r_t = y_t - \mathbf{x}_t\hat{\boldsymbol{\beta}}$. The least square estimate of γ is obtained by minimizing the residual sum of squares $\gamma = \arg\min_{\gamma \in \Gamma} \sum_{t=1}^N r_t^2$ over a set Γ of allowable threshold values so that each regime contains at least a given fraction (ranging from 0.05 to 0.3) of all observations.

In the case of robust two-regime SETAR model, for a fixed threshold γ the GM estimate of the autoregressive parameters can be obtained by applying the iterative weighted least squares: $\hat{\boldsymbol{\beta}}_j^{(n+1)} = \left(\mathbf{X}_j'\mathbf{W}^{(n)}\mathbf{X}_j\right)^{-1} \mathbf{X}_j'\mathbf{W}^{(n)}\mathbf{y}_j$, where $\hat{\boldsymbol{\beta}}_j^{(n+1)}$ is the GM estimate for the parameter vector in regime $j = 1, 2$ after the n-th iteration from an initial estimate $\hat{\boldsymbol{\beta}}_j^{(0)}$, and $\mathbf{W}^{(n)}$ is a weight diagonal matrix, whose elements depend on a weight function $w(\hat{\boldsymbol{\beta}}_j^{(n)}, \hat{\sigma}_{\varepsilon,j}^{(n)})$ bounded between 0 and 1. The threshold γ can be estimated by minimizing the objective function $\rho(r_t)$ over the set Γ of allowable threshold values. Different weight functions have been proposed in the literature. The first method is described in [3]. Weights are calculated as

$$w(\hat{\boldsymbol{\beta}}_j, \hat{\sigma}_{\varepsilon,j}) = \psi\left(\frac{y_t - m_{y,j}}{C_y\hat{\sigma}_{y,j}}\right)\psi\left(\frac{y_t - \mathbf{x}_t\hat{\boldsymbol{\beta}}_j}{C_\varepsilon\hat{\sigma}_{\varepsilon,j}}\right),$$

where $m_{y,j}$ is a robust estimate of the location parameter (sample median) in the j-th regime. $\hat{\sigma}_{y,j}$ and $\hat{\sigma}_{\varepsilon,j}$ are robust estimates of the scale parameters σ_y and σ_ε, respectively, obtained by the median absolute deviation multiplied by 1.483. C_y and

C_ε are tuning constants fixed at 6.0 and 3.9, respectively. Although the choice of the values of the tuning constants could be calibrated, in this simulation experiment they are the same used in the study of Chan and Cheung [3]. This is to allow a fair comparison with their results. In this case, ψ is the Tukey bisquare weight function, defined as

$$\psi(u) = \begin{cases} (1-u^2)^2 & \text{if} \quad |u| \leq 1, \\ 0 & \text{if} \quad |u| > 1. \end{cases}$$

The objective function to minimize for the search of the threshold depends on Tukey bisquare weights. We use the same function described in [3].

For the second method, we follow [4]. The GM weights are presented in Schweppe's form $w(\hat{\boldsymbol{\beta}}_j, \hat{\sigma}_{\varepsilon,j}) = \psi(r_t)/r_t$ with standardized residuals $r_t = (y_t - \mathbf{x}_t\hat{\boldsymbol{\beta}}_j)/(\hat{\sigma}_{\varepsilon,j} w_x(\mathbf{x}_t))$ and $w_x(\mathbf{x}_t) = \psi(\mathrm{d}(\mathbf{x}_t)^\alpha)/\mathrm{d}(\mathbf{x}_t)^\alpha$. $\mathrm{d}(\mathbf{x}_t) = |\mathbf{x}_t - m_{y,j}|/\hat{\sigma}_{y,j}$ is the Mahalanobis distance and α is a constant usually set equal to 2 to obtain robustness of standard errors. The chosen weight function is the Polynomial ψ function as proposed in [11], given by

$$\psi(u) = \begin{cases} u & \text{if} \quad |u| \leq c_1, \\ \mathrm{sgn}(u) g(|u|) & \text{if} \quad c_1 \leq |u| \leq c_2, \\ 0 & \text{if} \quad |u| > c_2, \end{cases}$$

where $\mathrm{sgn}(u)$ is the signum function, $g(|u|)$ is a fifth-order polynomial such that $\psi(u)$ is twice continuously differentiable, and c_1 and c_2 are tuning constants, taken to be the square roots of the 0.99 and 0.999 quantiles of the $\chi^2(1)$ distribution ($c_1 = 2.576$ and $c_2 = 3.291$). The threshold γ is estimated by minimizing the objective function $\sum_{t=1}^{N} w(\hat{\boldsymbol{\beta}}, \hat{\sigma}_\varepsilon)(y_t - \mathbf{x}_t\hat{\boldsymbol{\beta}})^2$ over the set Γ of allowable threshold values.

The third method is based on the same methodologies of the second but with ψ as the Huber weight function, given by

$$\psi(u) = \begin{cases} -c & \text{if} \quad u \leq -c, \\ u & \text{if} \quad -c < u \leq c, \\ c & \text{if} \quad u > c, \end{cases}$$

where c is a tuning constant taken equal to 1.345 to produce an estimator that has an efficiency of 95 per cent compared to the OLS estimator if ε_t is normally distributed.

3 Simulation Experiment

To compare the performance of the three methods, we reproduce the simulation study of [3]. We generated time series from SETAR(1,d) models for fixed sample sizes of $N = 100, 500$, with 1000 replications, respectively, and $\sigma_\varepsilon^2 = 1$. Moreover, 18 parameter combinations for $\boldsymbol{\theta} = (\beta_1, \beta_2, \gamma, d)$ are considered. The series are contaminated following three schemes. For the single-outlier case, an additive outlier is located at $t = N/2$ with magnitude ω given by 0, 3, 4, 5 times the standard deviation of the process. For the three-outlier case, we fixed three outliers at $t = N/4, N/2$, and $N * 3/4$ with magnitude $-\omega, \omega$ and $-\omega$, respectively. The multiple-outlier case is applied only for series with $N = 500$: three outliers are fixed every 100 observations with the same scheme of the three-outlier case. For the first robust estimation method, following [3], the starting values β_1^0, β_2^0 of the parameters are calculated by four iterations with Huber weights with OLS estimates as initial points. For the second and third method the starting values are calculated by least median squares.[2]

In Table 1 we have summarized some results of the big Monte Carlo experiment which has been carried out to compare the performances of the robust GM estimator to the LS estimator by applying the three methods described previously and called "Tukey," "Polyn," and "Huber," respectively. Each row corresponds to a combination of parameters used to generate the trajectories of a SETAR process. The values reported in the table represent the ratio between robust and LS RMSEs (root mean square error): robust estimators are better than LS when the ratios are less than 1. For lack of space, we reported only 6 combinations out of the original 18. Moreover, we were not able to show the same results for 100 observations time series and for different contamination pattern (one single outlier and three-outlier case; contamination magnitude $\omega = 4$). According to what it has been proved by Zhang et al. [16] the robust estimator of the threshold parameter is very less efficient than the LS estimator in small samples. As a consequence, we found (results available upon request) that all three robust methods performed generally worse than the LS, at least for weak contamination patterns, that is in the single outlier case with small magnitude. The results reported in Table 1 refer to the most complex case, that is high sample size and multiple outliers. It is immediately clear that, while the method suggested by Chan and Cheung [3] based on the Tukey function does not show any significant improvement with respect to LS, the other two methods look to be competitive to LS in the estimation of the threshold. Moreover, the Polynomial and Huber functions are far better than the LS estimator in estimating autoregressive coefficients with a slight prevalence of the Polynomial method. These results confirm the theoretical results provided by Zhang et al. [16].

[2]Different starting values have been chosen deliberately to keep the first method as it was originally suggested by Chan and Cheung [3].

To give an overall idea of the results partially reported in Table 1, we have computed the average values of the RMSEs ratios of the robust estimators with respect to the LS estimator (Table 2) using all 18 simulated processes. The same ratios have been obtained in order to compare performances of each weight function to the others (Table 3). For instance, the first value in Table 2 (1.371) means that the RMSE obtained on the 18 simulated processes using the Polynomial weight function is on average 37.1 % higher than the RMSE of the LS estimator when the threshold is estimated on non-contaminated trajectories in accordance to the higher efficiency of LS. Thus, values greater than 1 mean that the analyzed estimator is worse than the compared estimator. From Table 2 we can conclude that all robust estimators are overperformed by the LS estimator when the parameters are estimated on non-contaminated series ($\omega = 0$). However, the Polynomial function is the only

Table 1 Ratios of the RMSE of the robust estimates to the LS estimates

		$\hat{\gamma}$			$\hat{\beta}_1$			$\hat{\beta}_2$		
	d	$\omega = 0$	3	5	$\omega = 0$	3	5	$\omega = 0$	3	5
Tukey (0,−0.5,−1)	1	1.17	0.98	0.86	1.84	0.46	0.41	3.02	0.81	0.41
Tukey (0,−1,−0.5)	1	1.18	0.77	0.72	3.06	0.42	0.26	1.83	0.72	0.47
Tukey (0,0.5,0.8)	1	1.31	1.36	1.44	1.66	0.67	0.76	3.35	0.87	0.48
Tukey (0,−0.5,0.8)	1	2.92	2.28	1.86	3.87	1.79	1.15	3.40	1.09	0.62
Tukey (0,0.3,0.8)	2	4.68	1.58	1.23	2.59	1.33	0.86	2.73	0.77	0.44
Tukey (−0.1,0.3,−0.8)	2	18.26	10.29	5.75	3.77	2.16	1.48	2.99	0.88	0.73
Polyn (0,−0.5,−1)	1	0.99	0.76	0.69	1.12	0.38	0.17	1.12	0.36	0.16
Polyn (0,−1,−0.5)	1	0.96	0.62	0.57	1.14	0.21	0.10	1.11	0.64	0.28
Polyn (0,0.5,0.8)	1	0.92	1.02	1.04	1.14	0.41	0.22	1.18	0.42	0.19
Polyn (0,−0.5,0.8)	1	1.28	1.06	0.93	1.38	0.78	0.49	1.07	0.45	0.21
Polyn (0,0.3,0.8)	2	1.74	0.65	0.39	1.18	1.04	0.60	1.10	0.43	0.19
Polyn (−0.1,0.3,−0.8)	2	2.57	1.21	0.59	1.08	0.75	0.38	1.08	0.41	0.15
Huber (0,−0.5,−1)	1	0.98	0.72	0.66	1.07	0.49	0.21	1.01	0.37	0.17
Huber (0,−1,−0.5)	1	0.92	0.63	0.58	1.09	0.29	0.14	1.03	0.62	0.28
Huber (0,0.5,0.8)	1	1.10	1.02	1.00	1.15	0.46	0.23	1.13	0.50	0.20
Huber (0,−0.5,0.8)	1	1.31	1.15	1.11	1.28	0.78	0.61	1.06	0.48	0.24
Huber (0,0.3,0.8)	2	2.04	0.72	0.51	1.31	1.04	0.71	1.14	0.63	0.26
Huber (−0.1,0.3,−0.8)	2	1.58	1.03	0.59	1.09	0.80	0.52	1.11	0.49	0.22

1000 MC simulations of time series with sample size 500, multiple-outlier case. First column reports the name of the weight function. The values of true parameters are in parentheses in the following order: γ, β_1, β_2

Table 2 Means of the 18 RMSEs ratios of the GM estimate to the LS estimate

	$\hat{\gamma}$			$\hat{\beta}_1$			$\hat{\beta}_2$		
	$\omega = 0$	3	5	$\omega = 0$	3	5	$\omega = 0$	3	5
Polyn	1.371	1.088	0.885	1.158	0.612	0.330	1.253	0.670	0.392
Huber	1.278	1.073	1.007	1.115	0.611	0.366	1.173	0.667	0.444
Tukey	4.152	2.939	2.079	3.085	1.109	0.765	3.121	1.485	1.057

1000 MC simulations of time series with sample size 500, multiple-outlier case. First column reports the name of the weight function

Table 3 Means of the 18 RMSEs ratios of the GM estimation

	$\hat{\gamma}$			$\hat{\beta}_1$			$\hat{\beta}_2$		
	$\omega = 0$	3	5	$\omega = 0$	3	5	$\omega = 0$	3	5
Polyn to Huber	1.067	1.013	0.887	1.038	0.966	0.902	1.060	0.962	0.874
Polyn to Tukey	0.519	0.504	0.491	0.439	0.586	0.419	0.437	0.520	0.399
Huber to Tukey	0.507	0.501	0.551	0.424	0.607	0.466	0.414	0.554	0.455

1000 MC simulations of time series with sample size 500, multiple-outlier case. First column reports the name of the weight function

one to overperform the LS estimator in the estimation of the threshold parameter when the magnitude of the contamination is high ($\omega = 5$). On the other hand, Polynomial and Huber functions are always far better than LS in the estimation of β_i on contaminated series. Tukey is better than LS only once. The comparison of the robust weight functions is shown in Table 3. The clear preference of Polynomial and Huber functions to the Tukey weights is strongly confirmed. Moreover, Polynomial reveals to be always better than Huber function with the exception of few cases related to the estimation of the threshold parameter where the two weight functions look to perform equally well.

4 Application: Italian Electricity Price

Prices fixed on deregulated electricity markets usually show changes in regime [8]. Another very well known stylized fact of electricity prices is the presence of isolated jumps as a consequence of sudden grid congestions which reflects immediately on prices because of lack of flexibility of the supply and demand curves [10]. This feature must be considered very carefully and robust techniques must be applied to avoid that few jumps could dramatically affect parameter estimates. Although many papers have applied quite sophisticated time-series models to prices and demand time series of electricity and gas very few have considered the strong influence of jumps on estimates and the need to move to robust estimators [13].

In this section, we apply LS and the three robust methods to estimate parameters of SETAR models on Italian electricity price data (*PUN, prezzo unico nazionale*).

Moreover, a comparison of prediction accuracy among the methods has been conducted.

The time series of prices used in the present work covers the period from January 1st, 2009 to December 31th, 2012 (35,064 data points, for $N = 1{,}461$ days): year 2012 has been left for out-of-sample forecasting. The data have an hourly frequency, therefore each day consists of 24 load periods with 00:00–01:00 am defined as period 1. Spot price is denoted as P_{tj}, where t specifies the day and j the load period ($t = 1, 2, \ldots, N; j = 1, 2, \ldots, 24$).

In this study, following a widespread practice in literature, each hourly time series is modeled separately.

Differences in load periods can cause significant variations in price time series. A first inspection, based on graphs, spectra, and ACFs (figures are not reported for lack of space but they are available upon request) for different hours, shows that the series have long-run behavior and annual dynamics, which change according to the load period. A common characteristic of price time series is the weekly periodic component (of period 7), suggested by the spectra that show three peaks at the frequencies $1/7$, $2/7$, and $3/7$, and a very persistent autocorrelation function.

We assume that the dynamics of log prices can be represented by a nonstationary level component L_{tj}, accounting for level changes and/or long-term behavior, and a residual stationary component p_{tj}, formally $\log P_{tj} = L_{tj} + p_{tj}$. To estimate L_{tj} we used the wavelet approach [14]. We considered the Daubechies least asymmetric wavelet family, LA(8), and the coefficients were estimated *via* the maximal overlap discrete wavelet transform (MODWT) method (for details, see [14]). Figure 1 shows $\log P_{tj}$ for hours 5 and 18, respectively, with the estimated nonstationary level component superimposed. The 2 h have been selected to give examples of peak (hour 18) and off-peak hours (hour 5).

After removing the long-term component, we estimated on the stationary time series p_{tj} the SETAR(p,d) model, as reported in Eq. (1). According to ACFs, a SETAR(7,1) model has been estimated over all the price series to highlight differences in the estimation given by different dynamics characterizing each load period.

We have to emphasize that the analyzed time series are very similar to the trajectories simulated in the previous section: large sample size and high contamination level. In this case Polynomial and Huber methods should perform better than both LS and Tukey. As a confirmation of the simulation experiment, Polynomial and Huber coefficients are very similar (tables not reported). Next step of the analysis will be to compare the forecasting performances of the robust methods with the forecasting performance of the LS estimator.

For comparing our robust/nonrobust SETAR models, we reproduced 366 one day-ahead forecasts $\hat{p}_{t+1}$ for each model estimated on a rolling window of 3 years. Comparisons are based on the predictions of the original spot prices, where the prediction of the long-term component L_{t+1} is obtained relying only on the information available in t. In particular, we set $\hat{L}_{t+1} = \hat{L}_t$, that is, we used the estimated value in t as a forecast for $t + 1$. Besides its simplicity, the motivation to

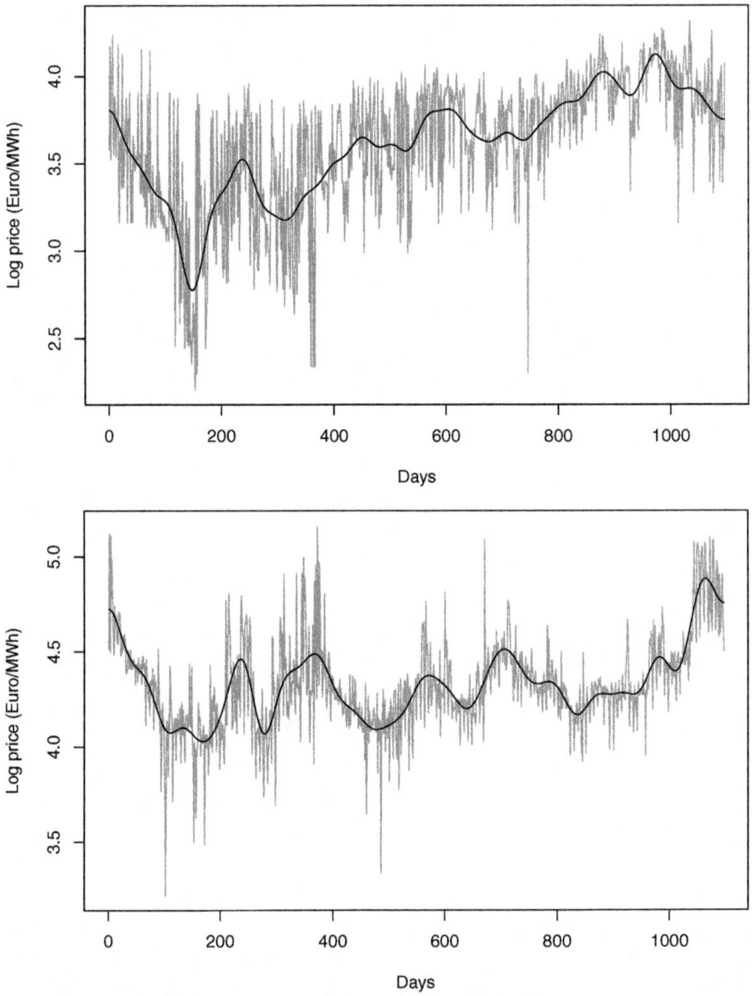

Fig. 1 $\log P_{tj}$ for hours 5, 18, respectively (from *top to bottom*), with the estimated nonstationary level component superimposed

use this method comes from the fact that the long-term component, by definition, should be basically the same for two contiguous days.

Forecasts have been compared in terms of MSE (mean square error) and MAE (mean absolute error) and the Diebold and Mariano test. These measures are based on the forecasting errors $e_{tj} = P_{tj} - \hat{P}_{tj}$ $(t = 1, 2, \ldots, N; j = 1, 2, \ldots, 24)$ for each method. We used the one-tailed Diebold and Mariano test (DM), whose null hypothesis is that the prediction accuracy of procedure (say) A is equal to or lower than that of procedure B. The test has been performed both with MSE and MAE.

Table 4 MSE and MAE of forecasts obtained from the four models on the four estimation periods and on the whole year. LS = least squares, POL = polynomial, HUB = Huber, TUK = Tukey

Hour 5, SETAR(7,1)								
	MSE				MAE			
Period	LS	POL	HUB	TUK	LS	POL	HUB	TUK
January–March	75.70	**69.50** (LS)	72.24	71.05	6.52	**6.13** (LS, HUB)	6.40	6.29
April-June	129.16	137.06	**128.26**	132.48	9.01	8.77	8.75	**8.56**
July–September	**54.29**	55.29 (HUB, TUK)	57.66	61.00	5.36	**5.32** (TUK)	5.44	5.51
October–December	47.44	**46.31**	47.34	49.98	5.12	**5.00**	5.08	5.18
Year	76.51	76.89	**76.25**	78.50	6.50	**6.30** (LS, HUB)	6.41	6.38
Hour 18, SETAR(7,1)								
	MSE				MAE			
Period	LS	POL	HUB	TUK	LS	POL	HUB	TUK
January–March	200.70	214.58	206.83	**195.80**	10.06	9.93	9.94	**9.80**
April–June	159.12	**148.20**	148.71	153.63	9.61	9.49	**9.41**	9.55
July–September	688.24	696.87	**686.51**	675.47	14.57	**13.49** (LS, HUB)	13.81 (LS)	13.66 (LS)
October–December	67.29	65.18	**64.35**	65.26	6.04	**5.71** (LS)	5.72 (LS)	5.72
Year	279.38	281.75	277.14	**273.07**	10.07	**9.65** (LS)	9.72 (LS)	9.68

The models whose forecasts are statistically worse than predictions of the model in the column are in parenthesis (1-tailed Diebold and Mariano test at 5 % significance level, MSE and MAE loss functions)

Table 4 shows MSE and MAE values on the whole year 2012 and on the four quarters. In parenthesis we reported the models whose forecasts are statistically worse than predictions of the model in the column considering the 1-tailed Diebold and Mariano test at 5 % significance level and MSE and MAE loss functions.

As can be seen, good results obtained from the simulation experiment are confirmed by the good forecasting performance of robust methods, with a slight preference of Polynomial on Huber's weights.

5 Conclusions

In this paper the statistical properties of different robust estimators for nonlinear time-series models have been examined. We have carried out an extensive Monte Carlo experiment to compare LS and GM estimators, with different weight functions, for SETAR models. The main result is that the bias in the threshold parameter estimator which has been observed in previous works seems to decrease when Huber and Polynomial weight functions are applied and when the sample size increases. From the estimation of parameters on the series of electricity prices we have observed that the application of robust estimators improves the prediction accuracy.

References

1. Andersen, R.: Modern Methods for Robust Regression. SAGE Publications, Chicago (2008)
2. Battaglia, F., Orfei, L.: Outlier detection and estimation in nonlinear time series. J. Time Ser. Anal. **26**(1), 107–121 (2005)
3. Chan, W.S., Cheung, S.H.: On robust estimation of threshold autoregressions. J. Forecast. **13**, 37–49 (1994)
4. Franses, P.H., van Dijk, D.: Non-linear Time Series Models in Empirical Finance. Cambridge University Press, Cambridge (2000)
5. Giordani, P.: A cautionary note on outlier robust estimation of threshold models. J. Forecast. **25**(1), 37–47 (2006)
6. Giordani, P., Kohn, R., van Dijk, D.: A unified approach to nonlinearity, structural change, and outliers. J. Econ. **137**(1), 112–133 (2007)
7. Hansen, B.E.: Inference in TAR models. Stud. Nonlinear Dyn. Econ. **2**(1), 1–14 (1997)
8. Huisman, R., Mahieu, R.: Regime jumps in electricity prices. Energy Econ. **25**(5), 425–434 (2003)
9. Janczura, J., Weron, R.: An empirical comparison of alternate regime-switching models for electricity spot prices. Energy Econ. **32**, 1059–1073 (2010)
10. Janczura, J., Trueck, S., Weron, R., Wolff, R.C.: Identifying spikes and seasonal components in electricity spot price data: a guide to robust modeling. Energy Econ. **38**, 96–110 (2013)
11. Lucas, A., van Dijk, R., Kloek, T.: Outlier robust GMM estimation of leverage determinants in linear dynamic panel data models. Discussion Paper 94–132, Tinbergen Institute (1996)
12. Maronna, R.A., Martin, R.D., Yohai, V.J.: Robust Statistics: Theory and Methods. Wiley, London (2006)
13. Nowotarski, J., Tomczyk, J., Weron, R.: Robust estimation and forecasting of the long-term seasonal component of electricity spot prices. Energy Econ. **39**, 13–27 (2013)
14. Percival, D., Walden, A.: Wavelet Methods for Time Series Analysis. Cambridge University Press, Cambridge (2000)
15. van Dijk, D.: Smooth transition models: extensions and outlier robust inference. Ph.D. thesis, Erasmus University Rotterdam, Rotterdam (1999)
16. Zhang, L.X., Chan, W.S., Cheung, S.H., Hung, K.C.: A note on the consistency of a robust estimator for threshold autoregressive processes. Stat. Probab. Lett. **79**, 807–813 (2009)

Incremental Visualization of Categorical Data

Alfonso Iodice D'Enza and Angelos Markos

Abstract Multiple correspondence analysis (MCA) is a well-established dimension reduction method to explore the associations within a set of categorical variables and it consists of a singular value decomposition (SVD) of a suitably transformed matrix. The high computational and memory requirements of ordinary SVD make its application impractical on massive or sequential data sets that characterize several modern applications. The aim of the present contribution is to allow for incremental updates of existing MCA solutions, which lead to an approximate yet highly accurate solution; this makes it possible to track, via MCA, the association structures in data flows. To this end, an incremental SVD approach with desirable properties is embedded in the context of MCA.

Keywords Correspondence analysis • Incremental methods • Singular value decomposition

1 Introduction

Multiple correspondence analysis (MCA) is a suitable dimension reduction method for the visual exploration of the association structure characterizing a set of categorical attributes [9]. Classic applications of MCA range from marketing to psychology, to social and environmental sciences. In the last decade, new frameworks of application emerged, which usually involve large/massive amounts of categorical data. For instance, Multiple and Simple CA have been effectively used for data preprocessing and feature space reduction [16, 21, 24], as well as for visualizing meaningful associations in high-dimensional data [10, 20]. Other

A.I. D'Enza (✉)
Department of Economics and Law, Università di Cassino e del Lazio Meridionale, Cassino, Italy
e-mail: iodicede@unicas.it

A. Markos
Department of Primary Education, Democritus University of Thrace, Thrace, Greece
e-mail: amarkos@eled.duth.gr

I. Morlini et al. (eds.), *Advances in Statistical Models for Data Analysis*, Studies in Classification, Data Analysis, and Knowledge Organization, DOI 10.1007/978-3-319-17377-1_15

examples include the continuous monitoring of typical product purchase combinations in market basket data, visualization of web-page visiting patterns via web-log analysis, tracking patient symptoms and behaviors over time, and monitoring of word associations that are present in data pulled on-the-fly from social networking sites. In all these examples, there is a high rate of data accumulation coupled with constant changes in data characteristics; the applicability of ordinary MCA is limited and requires a different approach.

MCA can be accomplished via an eigenvalue decomposition (EVD) or a singular value decomposition (SVD) of a suitably transformed data matrix. The application of ordinary EVD or SVD to large and high-dimensional data is infeasible because of the high computational and memory requirements: this aspect also limits the applicability of MCA on large data sets. In addition EVD/SVD, and hence MCA, are unsuitable for sequential data or data flows, i.e., when new data arrive, one needs to rerun the method with the original data augmented by the new data and the whole data structures being decomposed have to be kept in memory.

In the literature, there are several proposals aiming to overcome the EVD or SVD-related limitations when the full data set is not available from the start, as in data flows. Such approaches are based on incremental updates of existing EVD/SVD solutions according to new data (see [1] for an overview). The solution obtained from the starting data block has to be incrementally updated each time new data comes in.

The aim of the present contribution is to extend the use of MCA as a visual tracking tool of evolving association structures. To this end, we propose a block-based MCA algorithm to deal with incremental updates of existing solutions, which we refer to as "Live" MCA and leads to approximate, albeit accurate, solutions.

The paper is organized as follows: In Sect. 2 we briefly recall MCA as a dimension reduction method for categorical data. Section 3 reviews the literature on incremental eigen-decomposition methods. An incremental modification of MCA for tracking association structures is proposed in Sect. 4. In Sect. 5, we provide experimental results on synthetic data to investigate the convergence and accuracy of Live MCA, as compared to ordinary MCA. In Sect. 6 we illustrate a real-world application on data gathered from a social networking service. The paper concludes in Sect. 7.

2 MCA as a Matrix Decomposition Technique

This section provides a brief introduction to MCA from a matrix decomposition viewpoint. Let $\mathbf{Z}$ be a $n \times Q$ binary matrix, where n is the number of observations and Q the total number of categories that characterize q categorical variables. The general element is $z_{ij} = 1$ if the ith statistical unit is characterized by the jth category, $z_{ij} = 0$ otherwise; let $\mathbf{P} = \frac{1}{n \times q}\mathbf{Z}$ be the correspondence matrix, where $n \times q$ is the grand total of $\mathbf{Z}$. The core step of MCA is the matrix decomposition of the standardized residual matrix $\mathbf{S}$, defined as follows:

$$\mathbf{S} = \mathbf{D}_{\mathbf{r}}^{-1/2} \left(\mathbf{P} - \mathbf{r}\mathbf{c}^{\mathsf{T}}\right) \mathbf{D}_{\mathbf{c}}^{-1/2}, \tag{1}$$

where $\mathbf{r}$ and $\mathbf{c}$ are the row and column margins of $\mathbf{P}$, respectively; $\mathbf{D}_{\mathbf{r}}$ and $\mathbf{D}_{\mathbf{c}}$ are diagonal matrices with values in $\mathbf{r}$ and $\mathbf{c}$. The MCA solution can be obtained via the SVD of $\mathbf{S} = \mathbf{U}\boldsymbol{\Sigma}\mathbf{V}^{\mathsf{T}}$, where $\mathbf{U}$ is an $n \times Q$ orthonormal matrix with left singular vectors on columns, $\boldsymbol{\Sigma}$ is a diagonal matrix containing the Q singular values, and $\mathbf{V}$ is a $Q \times Q$ matrix of right singular vectors. The jth singular value corresponds to the standard deviation of data along the direction of jth singular vector, $j = 1, \ldots, Q$. The principal coordinates of the statistical units are $\mathbf{F} = \mathbf{D}_{\mathbf{r}}^{-1/2}\mathbf{U}\boldsymbol{\Sigma}$, whereas $\mathbf{G} = \mathbf{D}_{\mathbf{c}}^{-1/2}\mathbf{V}\boldsymbol{\Sigma}$ are the attribute coordinates.

3 Enhanced Eigen Decompositions

The SVD and related EVD lie at the heart of several multivariate methods, applicable to continuous, categorical, and compositional data. Both techniques have been applied to a wide spectrum of fields, ranging from signal processing and control theory to pattern recognition and time-series analysis. In cases where the data size is too large to fit in memory, or if the full data set is not completely available from the beginning, as in the case of data flows, the SVD/EVD application is infeasible. Therefore, it may be advantageous to perform the computations as the data become available. The so-called incremental methods aim to update (or downdate) an existing SVD or EVD solution when new data is processed. These methods can be applied to sequential data blocks without the need to store past data in memory.

Incremental EVD/SVD approaches that operate on streaming data are plentiful in the literature. A popular class of algorithms is based on sequential decomposition, that is, they seek to find the best subspace estimate each time a new data block arrives, but without performing the full EVD/SVD at each step [1]. Most of the early work introduced algorithms for sequential EVD or SVD in the fields of computer vision and signal processing [4–6, 11, 15, 17]. Although some of these methods are quite efficient, they only allow for a single column update, do not take into account the mean information or, equivalently, assume that the data is inherently zero mean, and have some potential numerical instability.

More recently, Brand [2, 3] proposed an efficient and stable incremental SVD algorithm with block update, which can also handle missing or uncertain values; however, this method assumes the sample mean is fixed when updating the eigenbasis. Hall et al. [12] presented an integrated approach for merging and splitting subspaces, using stable incremental computations of both EVD and SVD. This approach allows for multiple column (or block) update and downdate and takes into account the mean information. Similar methods to that of [12] were proposed by Fidler et al. [7] and Ross et al. [22]. In the former method, the mean information is preserved, but block update is not considered. The latter approach extended the sequential Karhunen–Loeve algorithm proposed by Levy and Lindenbaum [15] to an incremental SVD algorithm with mean and block update; it goes beyond other approaches in that it has constant space and time complexity. A generic approach, unifying some of the previous methods for approximating the dominant SVD, can be found in [1].

In this paper, we utilize the incremental SVD procedure proposed by Ross et al. [22], in order to provide an incremental MCA algorithm. The procedure allows to keep track of the data mean, which is a desirable property in the context of MCA, so as to simultaneously update the center of the low-dimensional space of the solution. Also, the method has a computational advantage over other approaches in that the decomposition can be computed in constant time regardless of data size. This property makes it appealing for an incremental MCA implementation in the case of data flows.

4 Block-Wise MCA

In order to describe an incremental or block-wise MCA scheme we first introduce some necessary definitions. An *eigenspace* is a collection of the quantities needed to define the result of a matrix eigen decomposition, as it involves eigenvalues (singular values), eigenvectors (singular vectors), data mean, and size.

In particular, with respect to the SVD, for an $n_1 \times Q$ matrix $\mathbf{X}_1$ and an $n_2 \times Q$ matrix $\mathbf{X}_2$, we can specify two eigenspaces as

$$\Omega_1 = (n_1, \mu_1, \mathbf{U}_1, \boldsymbol{\Sigma}_1, \mathbf{V}_1) \text{ and } \Omega_2 = (n_2, \mu_2, \mathbf{U}_2, \boldsymbol{\Sigma}_2, \mathbf{V}_2).$$

The aim of incremental decomposition is to obtain an eigenspace Ω_3 for the concatenated matrix: $\begin{bmatrix}\mathbf{X}_1\\ \mathbf{X}_2\end{bmatrix}$, using uniquely the information in Ω_1. The total number of statistical units and the global data mean can be easily updated: $n_3 = n_1 + n_2$ and $\mu_3 = \frac{n_1\mu_1 + n_2\mu_2}{n_3}$.

Setting the problem in an MCA framework, it is necessary to derive the matrix to be decomposed and the data mean. This is because, in the case of MCA, variables are transformed according to the margins of each data block. In particular, we first express the standardized residual matrix of Eq. (1) in covariance matrix form:

$$\mathbf{S} = \underbrace{\frac{\mathbf{Z}}{Q\sqrt{n}}\mathbf{D}_{\mathbf{c}}^{-1/2}}_{\mathbf{X}_1} - \mathbf{1}_n \underbrace{\frac{1}{\sqrt{n}}\mathbf{1}_{\mathbf{c}}^{\mathsf{T}}\mathbf{D}_{\mathbf{c}}^{1/2}}_{\mu_1^{\mathsf{T}}}, \tag{2}$$

where $\mathbf{X}_1 = \frac{1}{Q\sqrt{n}}\mathbf{Z}\mathbf{D}_\mathbf{c}^{-1/2}$ is the $n_1 \times Q$ row-wise centered matrix of the first data block and $\mu_1 = \frac{1}{\sqrt{n}}\mathbf{D}_\mathbf{c}^{1/2}\mathbf{1}$ is the data mean. For an incoming data block, we obtain the corresponding $n_2 \times Q$ matrix $\mathbf{X}_2$. In order to take into account the varying mean, the vector $\sqrt{\frac{nm}{n+m}}(\mu_2 - \mu_1)$ is added to $\mathbf{X}_2$.

In order to obtain the eigenspace Ω_3 of the super matrix $\begin{bmatrix}\mathbf{X}_1\\ \mathbf{X}_2\end{bmatrix}$, we adopt and briefly describe the incremental SVD approach proposed by Ross et al. [22].

Lemma 1 *Given the SVD of* $\mathbf{X}_1 = \mathbf{U}_1\boldsymbol{\Sigma}_1\mathbf{V}_1^\mathsf{T}$,

$$\begin{bmatrix}\mathbf{X}_1\\ \mathbf{X}_2\end{bmatrix} = \begin{bmatrix}\mathbf{U}_1 & \mathbf{0}\\ \mathbf{0} & \mathbf{I}\end{bmatrix}\begin{bmatrix}\boldsymbol{\Sigma}_1 & \mathbf{0}\\ \mathbf{L} & \mathbf{H}\mathbf{Q}^\mathsf{T}\end{bmatrix}\begin{bmatrix}\mathbf{V}_1\\ \mathbf{Q}\end{bmatrix},$$

where $\mathbf{L} = \mathbf{X}_2\mathbf{V}_1^\mathsf{T}$, $\mathbf{Q}$ *is a result from the* $\mathbf{QR}$*-decomposition of* $\mathbf{H} = \mathbf{X}_2 - \mathbf{L}\mathbf{V}_1$, *and* $\mathbf{I}$ *is the identity matrix.*

Proof Let $\mathbf{L} = \mathbf{X}_2\mathbf{V}_1$ be the projection of $\mathbf{X}_2$ onto the orthogonal basis $\mathbf{V}_1$ and $\mathbf{H} = \mathbf{X}_2 - \mathbf{L}\mathbf{V}_1$ the orthogonal component of $\mathbf{L}$. Apply a $\mathbf{QR}$-decomposition to $\mathbf{H}$ to obtain $\mathbf{Q}$. Thus, $\mathbf{H} = \mathbf{X}_2 - \mathbf{L}\mathbf{V}_1^\mathsf{T} \Leftrightarrow \mathbf{X}_2 = \mathbf{H} + \mathbf{L}\mathbf{V}_1 \Leftrightarrow \mathbf{X}_2 = \mathbf{H}\mathbf{Q}^\mathsf{T}\mathbf{Q} + \mathbf{L}\mathbf{V}_1$. Therefore,

$$\begin{bmatrix}\mathbf{U}_1 & \mathbf{0}\\ \mathbf{0} & \mathbf{I}\end{bmatrix}\begin{bmatrix}\boldsymbol{\Sigma}_1 & \mathbf{0}\\ \mathbf{L} & \mathbf{Q}\end{bmatrix}\begin{bmatrix}\mathbf{V}_1\\ \mathbf{Q}\end{bmatrix} = \begin{bmatrix}\mathbf{U}_1\boldsymbol{\Sigma}_1\mathbf{V}_1\\ \mathbf{L}\mathbf{V}_1 + \mathbf{H}\mathbf{Q}^\mathsf{T}\mathbf{Q}\end{bmatrix} = \begin{bmatrix}\mathbf{X}_1\\ \mathbf{X}_2\end{bmatrix}.$$

Apply the SVD to the matrix $\begin{bmatrix}\boldsymbol{\Sigma}_1 & \mathbf{0}\\ \mathbf{L} & \mathbf{H}\mathbf{Q}^\mathsf{T}\end{bmatrix}$ to obtain $\mathbf{U}_m\boldsymbol{\Sigma}_m\mathbf{V}_m^\mathsf{T}$.

Finally, $\mathbf{U}_3 = \begin{bmatrix}\mathbf{U}_1 & \mathbf{0}\\ \mathbf{0} & \mathbf{I}\end{bmatrix}\mathbf{U}_m$, $\boldsymbol{\Sigma}_3 = \boldsymbol{\Sigma}_m$, $\mathbf{V}_3 = \mathbf{V}_m\begin{bmatrix}\mathbf{V}_1\\ \mathbf{Q}\end{bmatrix}$.

In each update, the new row and column margins are given by $\mathbf{D}_\mathbf{r}^{(3)} = n_3^{-1}$ and $\mathbf{D}_\mathbf{c}^{(3)} = \left(n_1\mathbf{D}_\mathbf{c}^{(1)} + n_2\mathbf{D}_\mathbf{c}^{(2)}\right)n_3^{-1}$, respectively. Thus, $\mathbf{D}_\mathbf{c}^{(3)}$ is set to be the average of the "local" margins or the margins of the merged data blocks. Finally, row and column principal coordinates are given by $\mathbf{F}_3 = \left(\mathbf{D}_\mathbf{r}^{(3)}\right)^{-1/2}\mathbf{U}_3\boldsymbol{\Sigma}_3$ and $\mathbf{G}_3 = \left(\mathbf{D}_\mathbf{c}^{(3)}\right)^{-1/2}\mathbf{V}_3\boldsymbol{\Sigma}_3$, respectively.

Since the whole data matrix is unknown and the global margins are approximated by the local margins, the Live approach leads to an approximate MCA solution. An investigation of the convergence properties of the Live approach will be provided in Sect. 5.1.

5 Experimental Results

In this section, numerical experiments are presented to empirically study: *(i)* the convergence rate of the "local", $\mathbf{c}^*$, towards the "global", $\mathbf{c}$, margins and *(ii)* the accuracy of Live MCA compared to ordinary MCA. The convergence of the margins is assessed in terms of the mean absolute discrepancy of $\mathbf{c}^*$ from $\mathbf{c}$. The accuracy of Live MCA is measured in terms of the similarity between the ordinary MCA and Live MCA configurations in principal coordinates, computed on the same data set. In particular, the similarity measure is the R index, which equals $\sqrt{1-m^2}$, where m^2 is the symmetric orthogonal Procrustes statistic [14]. The index ranges from 0 to 1 and can be interpreted as a correlation coefficient; it was calculated using the function *protest* of the R package **vegan** [19]. MCA was applied using the **ca** package [18] and figures were produced using the package **ggplot2** [23].

5.1 Convergence

In order to study the convergence rate of Live MCA to ordinary MCA solutions, a 1 million rows data set was generated that acts as a population, and four categorical variables were considered. The probability of occurrence of the different categories changes in four equally sized blocks of rows. The number of categories per variable was randomly generated between 2 and 5. The global margins $\mathbf{c}$ were considered as the reference quantities. The starting and the incoming data blocks were randomly sampled with replacement from the "population" data set. Different sizes for the starting $sbs \in [750, 1000, 1250, 1500]$, and for the updating data blocks, $ubs \in [750, 1000, 1250, 1500]$, were considered. At each update, the mean absolute discrepancy was computed between $\mathbf{c}^*$ and $\mathbf{c}$. Figure 1 shows the convergence rate of the margins over 250 updates; the upper part of the figure shows the convergence rates for varying starting data block size, with fixed updating block size ($ups = 250$); the lower part of the figure shows the convergence rates for varying updating block size and fixed starting data block size ($sbs = 1000$). Both figures show that the discrepancy drops down considerably and converges towards zero (below 0.001), after only few updates. In particular, the size of the starting data does not have a relevant effect on the convergence rate, except for the very first updates, when for smaller size the discrepancy is higher. The size of the incoming blocks plays a more important role in the convergence rate: in fact, smaller sized data blocks require more updates for the discrepancy to drop down.

Fig. 1 Average absolute convergence of the block-based margins and the global margins over 250 updates. The *upper part* of the figure shows results for varying starting block size (sbs) and fixed updating block size (ubs); in the *lower part*, updating block size varies while starting block size is fixed

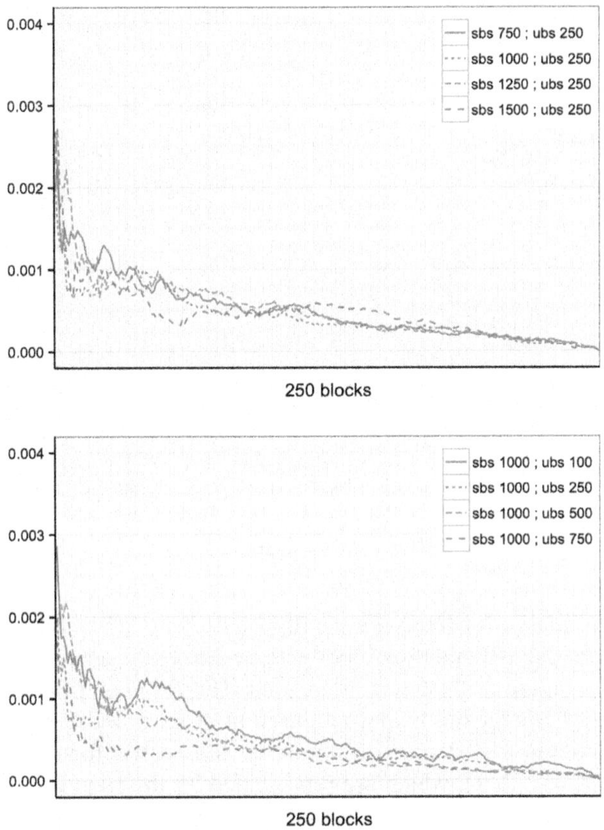

5.2 Accuracy

The accuracy of Live MCA was investigated with a similar experimental setup to that of the previous section. In particular, the same "population" data set is considered to sample the data blocks from, and each experiment is defined by the following parameters:

- n = 100,000, number of total rows (units) analyzed (that is, the row sum of the starting and updating blocks)
- $Q = 240$, number of total columns (attributes) for $q = 70$ variables
- $sbs \in \{100, 200, \cdots, 1000\}$, size of the starting data block
- $ubs \in \{60, 120, \cdots, 600\}$, size of the updating data block

The aim of this experiment was to examine whether different values of *sbs* and *ubs* affect the accuracy of Live MCA. In particular, different starting block sizes were considered for fixed updating block size. The size of *ubs* was fixed to the optimal size in terms of computational complexity [15]: $ubs = (Q - q) / \sqrt{2} = 120$.

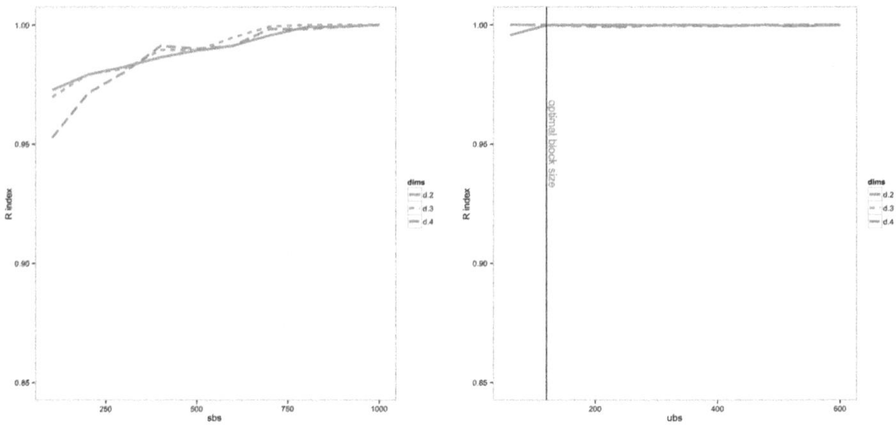

Fig. 2 Similarity between ordinary and Live MCA configurations for varying sizes of starting blocks (*left*) and updating blocks (*right*), the number of updates depends on *sbs* and *ubs*. (**a**) Protocol (*left*): cols = 240, sbs = 100 to 1000, ubs = 120. (**b**) Protocol (*right*): cols = 240, sbs = 1000, ubs = 60 to 600

Similarly, for fixed $sbs = 1000$, different values of *ubs* were considered to check whether the updating block size penalizes the accuracy of Live MCA.

Figure 2 illustrates the results. The R index, plotted on the vertical axis, shows how similar the final configuration of Live MCA is to the ordinary MCA configuration. The degree of similarity is assessed for different number of dimensions ($d = 2, 3, 4$), which are represented by different lines, and for different values of *sbs* and *ubs*, respectively, on the left-and right-hand side of Fig. 2.

With respect to the left-hand side of Fig. 2, the accuracy is always high ($R > 0.95$), and it increases with the starting data size; in fact, from $sbs = 750$ and on, $R \simeq 1$ irrespective of dimensionality. The right-hand side of Fig. 2 shows that, in terms of accuracy, the updating block size is almost irrelevant: even for the optimal block size, which is indicated by a vertical line on the plot, the accuracy is $R \simeq 1$. All in all, the experiments demonstrated that, although Live MCA is approximate, its accuracy can be very high, since in most cases the discrepancy from a full MCA solution is negligible.

6 A Real-World Application: Monitoring Consumer Attitudes in Twitter

The proposed approach is eventually applied to a real-world data set. The data refers to a small corpus of messages or tweets mentioning seven major hotel brands. It was gathered by continuously querying and archiving the Twitter Streaming API service, which provides a proportion of the most recent publicly available tweets,

Fig. 3 Exact and Live MCA map of the attributes. *Different symbols* are used for each method

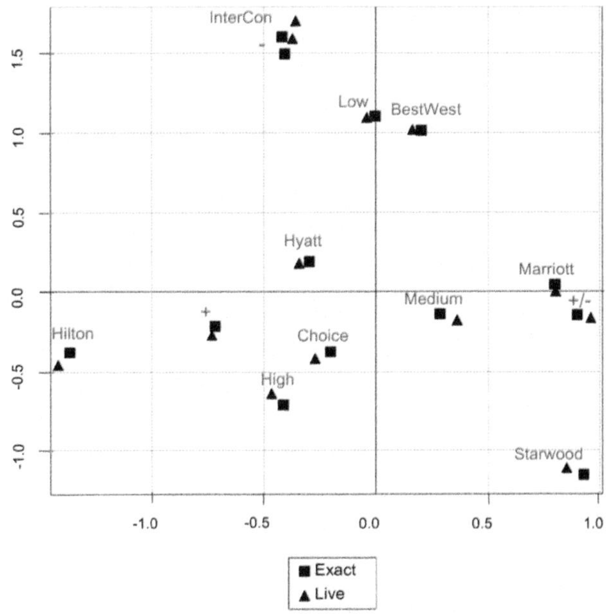

along with information about the user. The data was collected using the twitteR package in R [8]. A total of about 10,000 tweets were extracted within a time period of 6 days, from June 23rd to June 28th 2013. Only tweets in the English language were considered. Sentiment analysis was performed on each message to assess the corresponding user's sentiment towards a brand. Sentiment analysis is an active area of research involving complicated algorithms and subtleties. For the purposes of this toy example, we estimated a tweet's sentiment by counting the number of occurrences of "positive +," "neutral +/−," and "negative −" words. A third variable, user visibility or popularity, as measured by the number of followers each user had, was also included in the data set. The variable was categorized into three groups, "low," "medium," and "high."

The purpose of the present example is to show the evolving association structure of sentiments towards the brands as new data blocks are processed, using Live MCA. The first block for the incremental implementation consisted of 500 rows (tweets), and five equally sized blocks were consequently added to update the original solution.

In Figs. 3 and 4, we plot both solutions (Exact and Live MCA) on the same map for the attributes and tweets, respectively. These figures refer to the final configuration, however an animation is available at http://www.amarkos.gr/research/dynMCA that shows the evolution in time for the attribute configuration. In order to obtain a smooth animation, a morphing has been applied between one update and another; 25 frames were generated between each configuration and the next. The positions of the points tend to stabilize after the first two updates. With respect to the tweet map, points from different blocks are shown in different colors. Both

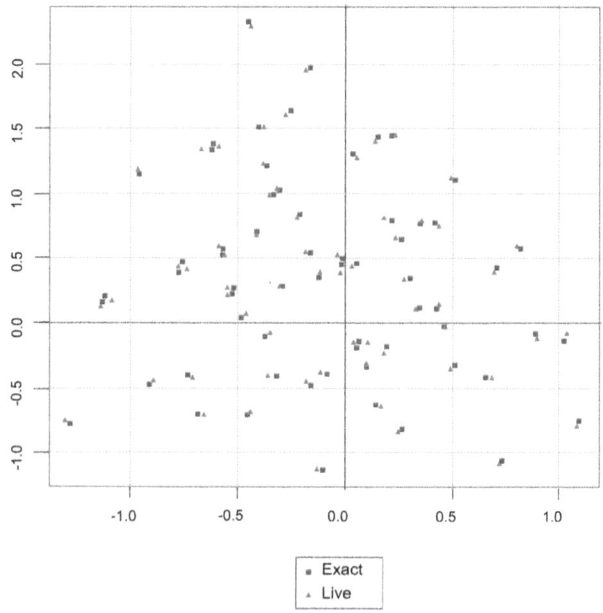

Fig. 4 Exact and Live MCA map of the tweets. *Different symbols* are used for each method

Live MCA configurations are characterized by a slightly larger scale of the points' coordinates over the axes, due to the use of local margins. The relative position of the points, however, stays the same, when compared to the ordinary solution. Therefore, the similarity between the ordinary and Live configurations is very high ($R = 0.997$ for both tweets and attributes).

7 Conclusions

An enhanced MCA implementation has been proposed that extends its applicability to modern big data problems and categorical data flows. Such implementations become then feasible, for instance, for continuous monitoring of word associations that are present in data pulled on-the-fly from social networking sites or for revealing and visualizing web-page visiting patterns via web-log analysis. Since such an implementation leads to an approximation of the ordinary MCA solution, we conducted a series of experiments to study the discrepancy between Live and ordinary MCA, as well as the convergence of Live MCA. In general, we conclude that (a) the larger the size of the starting data block, the more representative the margins of the block will be of the final data matrix and (b) the size of the updating data block does not significantly affect the accuracy of the solution. For further theoretical and empirical evaluation of the procedure, see [13].

An interesting perspective would be to study the relationships between the proposed incremental MCA scheme based on incremental SVD and others, e.g., based on stochastic approximation. We defer consideration of these possibilities to future work in order to keep the focus on our main contribution. Another idea is to extend the applicability of the proposed implementations to the case when new data is characterized by a partially overlapping set of attributes, that is, when the original data space dimensionality differs between updates.

References

1. Baker, C., Gallivan, K., Van Dooren, P.: Low-rank incremental methods for computing dominant singular subspaces. Linear Algebra Appl. **436**(8), 2866–2888 (2012)
2. Brand, M.: Fast online svd revision for lightweight recommender systems. In: Proceedings of SIAM International Conference on Data Mining, pp. 37–46 (2003)
3. Brand, M.: Fast low-rank modifications of the thin singular value decomposition. Linear Algebra Appl. **415**(1), 20–30 (2006)
4. Chahlaoui, Y., Gallivan K., Van Dooren, P.: An incremental method for computing dominant singular spaces. In: Berry, M.W. (ed.) Proceedings of the Computational Information Retrieval Conference, pp. 53–62. SIAM, Philadelphia (2001)
5. Chandrasekaran, S., Manjunth, B.S., Wang, Y.F., Winkeler, J., Zhang, H.: An eigenspace update algorithm for image analysis. Graph. Model. Image Process. **59**(5), 321–332 (1997)
6. DeGroat, R.D., Roberts, R.: Efficient, numerically stablized rank-one eigenstructure updating. IEEE Trans. Acoust. Speech Sig. Process. **38**(2), 301–316 (1990)
7. Fidler, S., Skocaj, D., Leonardis, A.: Combining reconstructive and discriminative subspace methods for robust classification and regression by subsampling. IEEE Trans. Pattern Anal. **28**(3), 337–350 (2006)
8. Gentry, J.: twitteR: R based Twitter client. http://cran.r-project.org/web/packages/twitteR/ (2011)
9. Greenacre, M.J.: Correspondence Analysis in Practice. Chapman and Hall/CRC, London (2007)
10. Greenacre, M., Hastie, T.: Dynamic visualization of statistical learning in the context of high-dimensional textual data. J. Web Semant. **8**, 163–168 (2010)
11. Gu, M., Eisenstat, S.C.: A stable and efficient algorithm for the rank-one modification of the symmetric eigenproblem. SIAM J. Matrix Anal. Appl. **15**, 1266–1276 (1994)
12. Hall, P., Marshall, D., Martin, R.: Adding and subtracting eigenspaces with eigenvalue decomposition and singular value decomposition. Image Vis. Comput. **20**, 1009–1016 (2002)
13. Iodice D'Enza, A., Markos, A.: Low-dimensional tracking of association structures in categorical data. Stat. Comput. (on-line, April, 2014)
14. Jackson, D.A.: PROTEST: A Procrustean randomization test of community environment concordance. Ecoscience **2**, 297–303 (1995)
15. Levy, A., Lindenbaum, M.: Sequential Karhunen-Loeve basis extraction. IEEE Trans. Image Process. **9**(8), 1371–1374 (2000)
16. Lin, L., Shyu, M.L.: Weighted association rule mining for video semantic detection. Int. J. Multimed. Data Eng. Manag. **1**(1), 37–54 (2010)
17. Murakami, H., Kumar, B.V.: Efficient calculation of primary images from a set of images. IEEE Trans. Pattern Anal. Mach. Intell. **4**(5), 511–515 (1982)
18. Nenadić, O., Greenacre, M.J.: Correspondence analysis in R, with two- and three-dimensional graphics: the ca package. J. Stat. Softw. **20**, 1–13 (2007)
19. Oksanen, J., Kindt, R., Legendre, P., O'Hara, B., Simpson, G.L., Solymos, P., et al.: Vegan: Community ecology package (2008)
20. Petrović, S., Bašic, B.D., Morin, A., Zupan, B.: Textual features for corpus visualization using correspondence analysis. Intell. Data Anal. **13**(5), 795–813 (2009)

21. Pham, N.K., Morin, A., Gros, P., Le, Q.T.: Intensive use of correspondence analysis for large scale content-based image retrieval. Stud. Comp. Intell. **292**, 57–76 (2010)
22. Ross, D., Lim, J., Lin, R.S., Yang, M.H.: Incremental learning for robust visual tracking. Int J. Comput. Vis. **77**, 125–141 (2008)
23. Wickam, H.: ggplot2: An implementation of the Grammar of Graphics. R package version 0.8.2 (2009)
24. Zhu, Q., Lin, L., Shyu, M.L., Chen, S.C.: Effective supervised discretization for classification based on correlation maximization. In: Proceedings of the IEEE International Conference on Information Reuse and Integration, pp. 390–395. IEEE, New York (2011)

A New Proposal for Tree Model Selection and Visualization

Carmela Iorio, Massimo Aria, and Antonio D'Ambrosio

Abstract The most common approach to build a decision tree is based on a two-step procedure: growing a full tree and then prune it back. The goal is to identify the tree with the lowest error rate. Alternative pruning criteria have been proposed in literature. Within the framework of recursive partitioning algorithms by tree-based methods, this paper provides a contribution on both the visual representation of the data partition in a geometrical space and the selection of the decision tree. In our visual approach the identification of the best tree and of the weakest links is immediately evaluable by the graphical analysis of the tree structure without considering the pruning sequence. The results in terms of error rate are really similar to the ones returned by the classification and regression trees (CART) procedure, showing how this new way to select the best tree is a valid alternative to the well-known cost-complexity pruning.

Keywords Classification and regression trees • Model selection • Pruning • Visual representations

1 Tree-Based Recursive Partitioning Methods and Tree Model Selection: An Overview

Recursive partitioning tree procedures have been the subject of extensive research in the past. Specially tree-based methods have been proposed for both prediction and exploratory purposes. Hierarchical segmentation obtained by decision trees can be seen as a stepwise procedure performed according to some optimization criteria, which provides a progressive sequence of partitions of an initial set of objects, described by some explanatory variables (either numerical or/and categorical) and a response variable [10], via a top down criterion.

C. Iorio (✉) • M. Aria • A. D'Ambrosio
Department of Economics and Statistics/Department of Industrial Engineering, University of Naples Federico II, 80138 Napoli, Italy
e-mail: carmela.iorio@unina.it; massimo.aria@unina.it; antdambr@unina.it

I. Morlini et al. (eds.), *Advances in Statistical Models for Data Analysis*,
Studies in Classification, Data Analysis, and Knowledge Organization,
DOI 10.1007/978-3-319-17377-1_16

Several methods have been proposed over the years. The oldest tree-based method was automatic interaction detector (AID) proposed by Morgan and Sonquist [16]. Goal of AID is to grow regression trees through binary splitting rules that provide recursive reduction in unexplained sum of squares.

Messenger and Mandell [13] and Morgan and Messenger [15] extended AID for categorical outcome according to the so-called *theta criterion* (THAID, THeta AID). A descendant of AID and THAID is CHAID (CHi-square AID), introduced by Kass [11]. CHAID uses Chi-square splitting criterion to classify a categorical response variable.

Quinlan [17, 18] developed an iterative algorithm, known as ID3. The input is a table of objects and each object induces a decision tree. Leaves of decision tree indicate the class to which the objects belong. ID3 uses the entropy criteria for splitting nodes. An extension of ID3 is C4.5. It utilizes a normalized entropy measure, known as Gain Ratio, which expresses the proportion of information induced by any split [3].

One of the most popular tree-based techniques is classification and regression trees (CART) developed by Breiman et al. [6]. Induction of decision trees is typically performed in two steps. In the first step, a training set (used to grow the tree) is recursively divided into subgroups according to splitting criteria expressed in terms of decrease in impurity. Often, the criterion used to split is the Gini diversity index. The tree-growing step continues until some stopping rule is reached, such as all samples for a node belong to the same class. In literature there are several proposals for tree-growing step [4, 14, 21].

In the second step, called pruning, the tree is reduced to prevent "overfitting." Pruning generates a decision tree by simplifying the tree structure by removing some of the branches of the fully expanded tree with the goal of improving the classification accuracy.

A generic internal node of a tree can be seen as a starting point for a sub-tree that will end with several leaves or terminal nodes. The data falling down in the leaves are evaluated via misclassification rate or expected value according to the nature of the response. As a consequence global badness of fit indices can be either the misclassification ratio or the mean squared error [6]. Alternative pruning criteria have been proposed in the literature [8].

The CART pruning procedure considers both the accuracy (evaluated by some error measure not necessarily coincident with the one used for the growing step) and the complexity (given by the number of terminal nodes) of the tree, introducing the so-called *cost-complexity measure* [6]. The algorithm works either by using a separate independent set of samples or by cross-validation. The goal is to produce the best sequence of pruned subtrees of the fully expanded tree. A complexity parameter needs to be defined. It represents both a penalty for any additional node and the cost associated to the removal of any terminal node belonging to a given branch. The optimal decision tree is based on the definition of a trade-off measure between the accuracy (*cost*) and the size (*complexity*) of the tree. Quinlan suggested two methods of pruning. The first one [18] is known as *reduced error pruning* and it prunes the nodes according to a bottom-up approach. It generates a sequence of subtrees and uses the test set to evaluate the performance of the tree. Since the misclassification rate on the training set is optimistically biased, Quinlan introduced a continuity correction for the binomial distribution, which might provide a more realistic error rate. This pruning method is known as *pessimistic error pruning*. The second pruning method is implemented in C4.5 [19]. It is known as *error based pruning* and produces a simplified tree structure. Cappelli et al. [7] proposed an alternative pruning method based on a so-called *impurity-complexity measure* which evaluates the accuracy of the classification directly through the impurity measure. Siciliano et al. [22] proposed a model-based tree growing that implicitly prunes the tree within the tree-growing step. This phase is based on the concept of retrospective split as well as on the recursive estimation of GLM.

2 Decision Tree Visualization

Fayyad et al. [9] stated that without proper visualization techniques, data mining models may not give the desired insight to help humans to understand the phenomena.

Different visual representations of decision tree have been proposed. Hierarchical and radial views are the two most popular graphs for decision tree [12]. Hierarchical view is the most natural way to display a decision tree. A decision tree is defined as a directed connected acyclic graph. A graph is a set of nodes. Scheduling information is placed in the nodes. In the internal node, the information represents the splitting rule; in the terminal node, it consists of prediction. Radial view is applied mainly in displaying object structures, such as organizations.

Node-link diagrams are the most familiar tree structures. This representation is poor in revealing the overall structure of a tree, such as its depth levels. In addition, it does not demonstrate node sizes.

A tree map [20] is another way to display all the partitions using area based plot, in which each terminal node is represented by a rectangle. The rectangular area of tree map corresponds to the full dataset. This area is partitioned recursively with an alternating horizontal and vertical partitioning directions until the terminal nodes are reached. The size of each sub-rectangle is proportional to the number of cases in the corresponding node. Tree map does not allow a relative comparison of groups within nodes.

A tree ring maps the hierarchies into circles and it displays both tree topology and node size. The most inner circle represents the root node. Tree maps and tree rings are space-filling visualization methods, since they make full use of the available space.

The icicle plot represents a tree node as a rectangle whose length is proportional to the number of records associated with it. This visualization is more space-efficient than node-link diagrams, since there are no links between nodes [5].

The basic idea of circle segment [2] visualization technique is to display the data dimension as a segment of a circle. If the data consists n dimension, the circle is partitioned into n segments, each segment represents one different attribute; each pixel inside a segment is a single value of the attribute.

Values of each attribute are sorted independently and assigned to a different color based upon its class. Another similar approach proposed by Ankerst et al. [1] uses a stacked bar representation instead of circle segments. In other words, bars representing an attribute are displayed horizontally and are displayed stacked upon each other. This technique is easily expandable to support many attributes. The circle segments method appears to start losing display granularity as the number of segments increases. Circle segments and similar techniques provide great visibility into multivariate classification techniques. They are a great aid in identifying obvious relationships between data values and classes and for identifying potentially weak relationships as well.

3 Visual Tree Model Selection

Our approach is based on the definition of a new way to represent the tree structure by a node-link diagram. Indeed the length of a path is proportional to the decrease of the error measure. The lower is the error in the descendant nodes, the longest is the length of the path.

Θ_{jz} indicates an *oriented path* starting from node j to node z. A path Θ_{jz} can be seen as a sequence of intermediate oriented paths θ_{mn} with node m directly connected with node n and $m < n$ such as $\Theta_{jz} = \langle \theta_{jk}, \ldots, \theta_{mn}, \ldots, \theta_{sz} \rangle$, with $j < \ldots < m < n < \ldots < z$.

We define the *depth of an oriented path* V_{jz} as

$$V_{jz} = \frac{e_j \cdot p_j - e_z \cdot p_z}{e_{root}} = \sum_{r=2}^{|H_{jz}|} \frac{e_{h^{(r)}} \cdot p_{h^{(r)}} - e_{h^{(r+1)}} \cdot p_{h^{(r+1)}}}{e_{root}}, \tag{1}$$

where:

- e_j, e_z, p_j and p_z are, respectively, the error measures and the proportion of cases in nodes j and z
- e_{root} is the error measure of root node of the tree
- H_{jz} is the vector containing the ordered list of nodes belonging to the path Θ_{jz}
- $h^{(r)}$ is the generic node at the position (r) of the ordered set of node H_{jz}, with $r = 1, \ldots, |H_{jz}|$

The generic path starting from the root node to the node z can be indicated as $\Theta_{\cdot z}$ and its depth measure $V_{\cdot z}$.

V_{jz} can be interpreted as a relative decrease of error measure from j to z which explains the predictive strength of the sequence of splits from the parent node j to the descendant node z.

V has some properties that allow to measure tree quality at each level:

- By definition $V_{\cdot root} = V_{\cdot 1}$ is equal to 0, since it is the depth of the path $\theta_{\cdot 1}$.
- $V_{\cdot j}$ is between 0 and 1, so the depth of a generic path is at most equal to 1.

As a consequence, it is possible to define a *relative error measure* of the node j as $\varphi_{\cdot j}$:

$$\varphi_{\cdot j} = 1 - V_{\cdot j} = \frac{e_j \cdot p_j}{e_{root}}. \tag{2}$$

The properties of V measure linked to this new way to represent a tree structure suggest us to define an alternative model selection procedure based on "tree graphical representation." Each internal node is an equally likely candidate to be the point which identifies a cut to the depth level of the structure. In this way, given a node t candidate a cutting point, for each path which span the ideal cutting line, the nodes departing from it will become terminal. At each potential cutting line is linked an error measure φ_T defined as

$$\varphi_T = \sum_{t \in T} \varphi_t, \tag{3}$$

where T is the set of terminal nodes of a generic sub-tree. The distribution of errors follows a typical descending trend over the training sample and a convex trend over the test or CV sample.

4 An Application of Visual Tree

In the framework of CART methodology we want to show how it is possible to use a visual model instead of the classic pruning procedure.

The main interpretative advantages of the visual tree are shown in an application on a real dataset. We have developed the analysis of Credit dataset (Decisia SPAD Repository). According to the numbering system developed by CISIA Software Informer, let k be the generic tth node, it generates descendant nodes numbered as $2k$ (on the left) and $2k + 1$ (on the right). By convention, node number 1 indicates the root node.

Figure 1 displays, in the upper row, the graphical representation of the CART approach. Both the maximum expanded tree and the decision tree are provided, respectively, on the left and the right sub-figures in the first row. Decision tree was selected via test-set procedure, the size of test sample is about 30 % of the entire dataset. As it can be noted, no information is contained in length of paths and in levels of tree (Fig. 1, first row, left side). In the classical visualization there is a lack of information about the goodness of split, the purity of nodes, and the goodness of the tree (Fig. 1, first row, right side). The second row of Fig. 1 shows

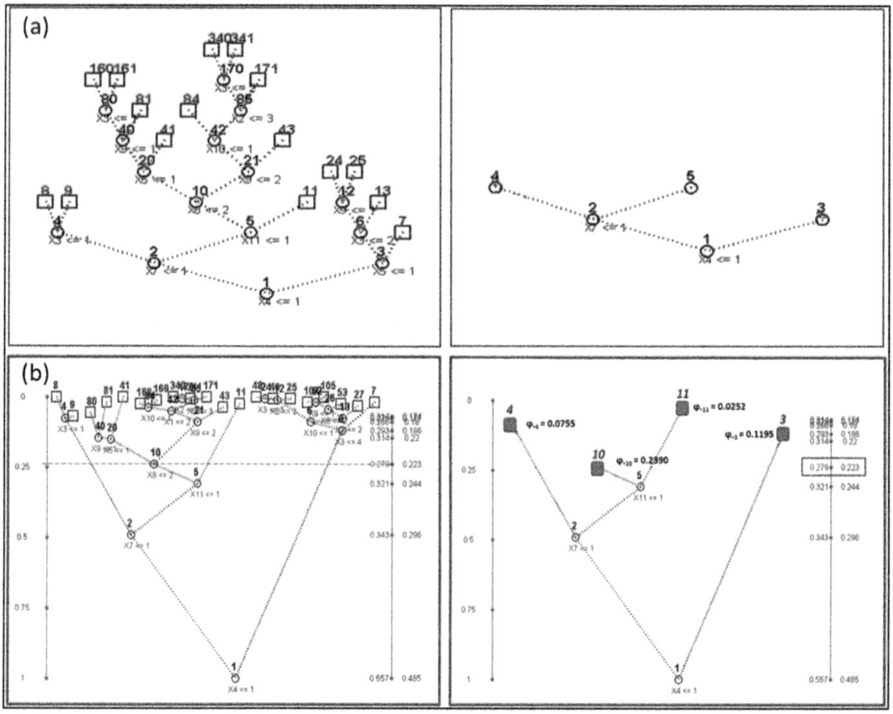

Fig. 1 Comparison of CART approach and visual tree model selection. (**a**) CART approach: exploratory classification tree (on the *left*) and decision tree (on the *right*). (**b**) Visual tree model selection: fully expanded tree and cutting sequence (on the *left*) and decision tree (on the *right*)

the visualization of the tree structure using our approach on the Credit dataset. The sub-figures of second row provide, respectively, the maximum expanded tree and the decision tree. The pruned tree was obtained by using the same test set used before. In the second row of Fig. 1, first plot, the left axis measures the relative error $\varphi_{.j}$ and the right axis measures the misclassification error linked to each cutting point calculated, respectively, on the training sample (right) and test sample (left). At first sight, this plot points out the relative importance of splits and the best cut level to obtain an optimal decision tree. By looking at the graph, as in a dendrogram of the hierarchical cluster analysis, we can decide an automatic cutting of the tree as a function of the error rate. As can be noted in this row, right plot, the tree structure highlights that the relative error measure of terminal nodes $3, 4, 11$ is close to 0, while it is higher for node 10. Note that overall relative error measure of the tree φ_T, as defined in Eq. (3), is proportional to the misclassification error on the training sample $R(T)$.

The visual decision tree shows the contribution of the nodes with a higher relative error measure to the tree badness of fit. The shorter is the length of a path, the higher is the contribution to the global error measure. By visual tree model selection we can

Table 1 Tree model comparison: visual pruning vs classical pruning (1000 bootstrap replications)

				Visual		Cart	
Dataset	Continue	Categorical	Response	Error	Size	Error	Size
Credit	2	11	Binary	0.2489	4.516	0.2467	4.676

see that at the first split there is a significant decrease of the error measure especially for node 3 that become immediately a terminal node (Fig. 1, second row, right side). The splits are the same for both trees obtained with the CART approach and the visual tree model selection, but in visual decision tree the depth of a path explains the predictive strength of a split to decrease the error measure.

This visual approach can be used to build trees both in supervised classification and in nonparametric regression.

We carried out a comparison by bootstrap and empirical evidence suggests how both procedures return similar outcomes. The results, reported in Table 1, show that the misclassification rate and the tree size are very similar. Both measures are referred to trees validated via test-set procedure. Moreover, in visual approach, the identification of the best tree and the weakest links is immediately evaluable by the graphical analysis of the tree structure without considering the pruning sequence.

5 Conclusion

The proposed visual tree model selection seems to be a valid alternative to the cost-complexity strategy to select decision trees. We propose a new tree structure visualization that allows to identify more discriminant splits, weakest links and help the user to catch the optimal substructure as decision tree.

References

1. Ankerst, M., Ester, M., Kriegel, H.P.: Towards an effective cooperation of the computer and the user for classificaton. In: Proceedings of the Sixth International Conference on Knowledge Discovery and Data Mining, Boston, pp. 178–188 (2000)
2. Ankerst, M., Keim, D.A., Kriegel, H.P.: Circle segments: a technique for visually exploring large multidimensional datasets. In: Proceedings of IEEE Visualization, Hot Topic Session, Sab Francisco (1996)
3. Apté, C., Weiss, S.: Data mining with decision trees and decision rules. Future Gener. Comput. Syst. **13**, 197–210 (1997)
4. Aria, M., Siciliano, R.: Learning from trees: two-stage enhancements. In: Proceedings of Classification and Data Analysis Group (CLADAG 2003), Cleub, pp. 22–24 (2003)
5. Barlow, S.T., Neville, P.A.: Comparison of 2-D visualization of hierarchies. In: Proceedings of the IEEE Symposium on Information Visualization, San Diego, pp. 131–138 (2001)

6. Breiman, L., Friedman, J.H., Olshen, R.A., Stone, C.J.: Classification and Regression Trees. Wadsworth International Group, Belmont (1984)
7. Cappelli, C., Mola, F., Siciliano, R.: An alternative pruning method based on the impurity-complexity measure. In: Rayne, R., Green, P. (eds.) Proceedings in Computational Statistics 13th Symposium, pp. 221–226. Springer, New York (1998)
8. Esposito, F., Malerba, D., Semeraro, G., Kay, J.: A comparative analysis of methods for pruning decision trees. IEEE Trans. Pattern Anal. Mach. Intell. **19**, 476–491 (1997)
9. Fayyad, U.M., Grinstein, G., Wierse, A.: Information Visualization in Data Mining and Knowledge Discovery. Morgan Kaufmann Publishers, San Francisco (2002)
10. Hastie, T., Tibshirani, R., Friedman, J.: The Elements of Statistical Learning. Springer, New York (2009)
11. Kass, G.V.: An exploratory technique for investigating large quantities of categorical data. J. Appl. Stat. **29**, 119–127 (1980)
12. Liu, Y., Salvendy, G.: Design and evaluation of visualization support to facilitate decision trees classifications. Int. J. Hum. Comput. Stud. **65**, 95–110 (2007)
13. Messenger, R., Mandell, L.: A modal search technique for predictive nominal scale multivariate analysis. J. Am. Stat. Assoc. **67**, 768–772 (1972)
14. Mola, F., Siciliano, R.: A fast splitting procedures for classification and regression trees. Stat. Comput. **7**, 208–216 (1997)
15. Morgan, J.N., Messenger, R.C.: THAID a Sequential Analysis Program for Analysis of Nominal Scale Dependent Variables. Survey Research Center, Institute for Social Research, University of Michigan, Ann Arbor (1973)
16. Morgan, J.N., Sonquist, J.A.: Problems in the analysis of survey data and a proposal. J. Am. Stat. Assoc. **58**, 415–434 (1963)
17. Quinlan, J.R.: Discovering rules by induction from large collections of examples. In: Michie, D. (ed.) Expert Systems in the Micro Electronic AgeSoftware Pioneers, pp. 168–201. Edinburgh University Press, Edinburgh (1979)
18. Quinlan, J.R.: Simplifying decision trees. Int. J. Man Mach. Stud. **27**, 221–234 (1987)
19. Quinlan, J.R.: C.4.5: Programs for Machine Learning. Morgan Kaufmann, San Mateo (1993)
20. Shneiderman, B.: Tree visualization with tree-maps: 2-d space. J. ACM Trans. Graphs (TOG) **11**, 92–99 (1992)
21. Siciliano, R., Aria, M.: TWO-CLASS trees for non parametric regression analysis. In: Fichet, B., Piccolo, D., Verde, R., Vichi, M. (eds.) Classification and Multivariate Analysis for Complex Data Structures. Series of Studies in Classification, Data Analysis and Knowledge Organizations, pp. 63–71. Springer, Heidelberg (2011)
22. Siciliano, R., Aria, M., D'Ambrosio, A.: Posterior prediction modelling of optimal trees. In: Brito, P. (ed.), Proceedings in Computational Statistics (COMPSTAT 2008), 18th Symposium, pp. 323–334. Springer, New York (2008)

Object-Oriented Bayesian Network to Deal with Measurement Error in Household Surveys

Daniela Marella and Paola Vicard

Abstract In this paper we propose to use the object-oriented Bayesian networks (OOBNs) architecture to model measurement errors in the Italian survey on household income and wealth (SHIW) 2008 when the variable of interest is categorical. The network is used to stochastically impute microdata for households. Imputation is performed both assuming a misreport probability constant over all the population and learning a Bayesian network for estimating such a probability. Finally, potentialities and possible extensions of this approach are discussed.

Keywords Categorical variable • Misreport probability • Mixed measurement model • Structural learning • Underreporting

1 Introduction

Measurement error is the difference between the value of a feature provided by the respondent and the corresponding true but unknown value. Together with nonresponse, measurement error is one of the main nonsampling error sources. The presence of measurement errors may severely affect the quality of survey results leading to erroneous conclusions.

Object-oriented Bayesian networks (OOBNs) have been recently proposed as a new tool to model and correct measurement errors. In particular, the measurement error in a categorical variable is described by a mixed measurement model implemented in an Bayesian network (BN); for details see [5]. The aim of the paper is to apply this model to 2008 survey on household income and wealth (SHIW), a bi-annual sample survey conducted by Banca d'Italia. Its main objective is to study the economic behavior of Italian households. Interviews are considered valid if there are

D. Marella (✉)
Dipartimento di Scienze della Formazione, Via del Castro Pretorio 20, 00185 Roma, Italy
e-mail: daniela.marella@uniroma3.it

P. Vicard
Dipartimento di Economia, Via Silvio D'Amico 77, 00145 Roma, Italy
e-mail: paola.vicard@uniroma3.it

I. Morlini et al. (eds.), *Advances in Statistical Models for Data Analysis*,
Studies in Classification, Data Analysis, and Knowledge Organization,
DOI 10.1007/978-3-319-17377-1_17

no missing items on the questions regarding income and wealth [6]. Therefore unit nonresponse and measurement errors are two major issues. In particular, financial assets in SHIW are affected by misreporting of financial amounts with a prevalence of underreporting. In our application we aim at correcting bond amount declared values. Bond amount, being a continuous variable, will be discretized in order to apply the mixed measurement model in [5].

The measurement error model parameters have been estimated using a validation sample. In particular, the probability of an error has been initially estimated by assuming that it is constant over all the population. This assumption in unrealistic, therefore a BN has then been used to model and predict the misreport probability on the basis of auxiliary information in a validation sample. Once estimated this probability, the overall measurement error model has been implemented and bonds microdata have been imputed in SHIW 2008.

In [6] measurement errors are modeled using propensities to misreport estimated on the validation sample. The propensities are then used to adjust for SHIW data. More specifically, first bond amounts are estimated by a logistic model using a vector of socioeconomic characteristics both at the household and at the head of household level as covariates together with the declared value. Then misreporting on the amount held is estimated through a separate model using a set of household characteristics and the declared amount. Our approach differs in the use of BNs allowing to: (a) exploit auxiliary variables directly and indirectly influencing the misreport probability; (b) model the measurement error generating process and predict microdata.

The paper is organized as follows. In Sect. 2 the mixed measurement error model is briefly described and implemented in a OOBN. In Sect. 3 the network is used to impute microdata in SHIW 2008 and the performance of the imputation is evaluated both assuming a misreport probability constant over all the population and learning a BN from the validation sample (Sect. 3.1). Finally in Sect. 4 potentialities and possible extensions of our approach for dealing with measurement error in sample surveys for continuous variables are discussed.

2 A OOBN Model for Measurement Error in SHIW 2008

In this paper we consider an ordered categorical variable X with K response categories whose frequencies p_k, $k = 1, \ldots, K$, are assumed known. When a measurement error occurs the observed category is different from the true category. Let $q_{i->j}$ be the intercategory transition probability from the true category i to the observed category j, where $\sum_{j=1}^{K} q_{i->j} = 1$. In order to estimate the $K(K-1)$ probabilities $q_{i->j}$, we could carry out an interview–reinterview study. Alternatively, the transition probabilities $q_{i->j}$ can be expressed by means of models characterized by a smaller number of parameters to be estimated. Here we use scalar models (see [9]) having the form $q_{i->j} = \lambda s_{i->j}$, where λ represents the error parameter and the nonnegative quantities $s_{i->j}$, $j \neq i$, specify the model.

A realistic and plausible representation of the measurement error generating process is the mixed measurement model

$$s^{mix}_{i->j} = (1-h)s^{prop}_{i->j} + hs^{MMT}_{i->j} \tag{1}$$

given by a mixture of the proportional model $s^{prop}_{i->j}$ and the one-T step model $s^{MMT}_{i->j}$. h is the mixture parameter taking values between 0 and 1 according to the relative importance of the one-T step model. The model $s^{prop}_{i->j}$ reflects the assumption that, whenever a measurement error occurs, the observed value j is generated at random from the population frequency distribution, i.e., regardless of the true value i. The model $s^{MMT}_{i->j}$ implies that, if an error takes place, the observed category j can only be a neighboring category or a category up to T steps away from the true category i. The one-T step model is flexible since it also allows for modeling an asymmetric error generating mechanism, which is common for financial variables.

By the mixed measurement model both accidental and deliberate errors can be described. Therefore the model (1) is completely general and can be applied to various contexts by suitably: (*i*) tuning the mixture parameter h; (*ii*) estimating the parameters (μ, α_t) where μ is the misreport probability and α_t is the probability that the difference between the observed and the true category is t, for $|t| = 1, \ldots, T$. Notice that μ is proportional to λ

$$\mu = \lambda\beta \quad \text{with} \quad \beta = (1-h)\left(1 - \sum_i p_i^2\right) + h.$$

A sensitivity analysis to model parameters has been performed in [5].

In order to automate and efficiently perform error detection and correction, the mixed measurement model (1) has been implemented in a OOBN. For an account on OOBNs, we refer to [2].

Figure 1 shows the main network (top-level) representing the overall measurement error process. In what follows, instance and regular nodes are indicated in teletype face, while bold face is used for network classes.

The observed category, i.e., that declared by the respondent, is represented by the standard node `oc`. The false (wrong) category is represented by an instance of network class **fc**. This means that the round-shaped rectangular node `fc` is a

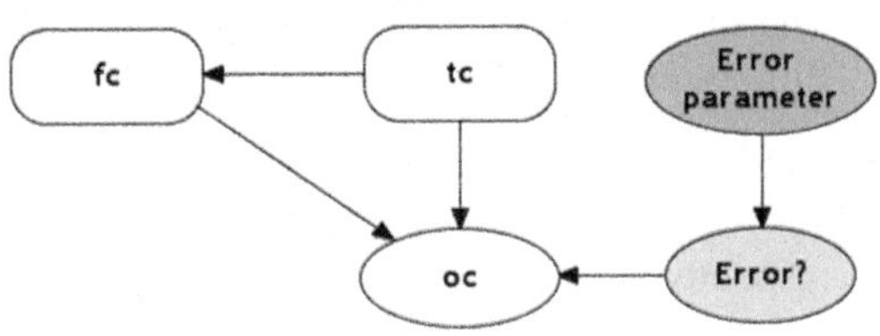

Fig. 1 Top-level network for the measurement error model for the respondent

BN itself encoding the mixed measurement model (1). For details and graphical model representation, see [5]. The true category, node `tc`, is represented by an instance of network class **tc** associated with the probability distribution of the variable of interest. The fact that the respondent may consciously or unconsciously report the wrong category is represented by the random node **Error?** associated with a Bernoulli distribution of parameter λ, i.e., the error parameter. Specifically, if the respondent declares the true category, coded as `Error?`=0, then the observed category coincides with the original one. If the respondent is wrong, coded as `Error?`=1, then the observed category is different from the original one and it is generated according to the mixed measurement model implemented in the false category instance node `fc`.

3 Using the Network to Impute Microdata

Given the OOBN in Fig. 1, SHIW 2008 microdata have been imputed by a two-step strategy. First the measurement model parameters (h, μ, and α_t, $|t| = 1, \ldots, T$) have been estimated using a validation sample; secondly the estimated measurement model in Fig. 1 has been used to impute microdata for units in SHIW 2008. Specifically, for each respondent in SHIW 2008 sample, the evidence, i.e., the corresponding observed value, is inserted and propagated throughout the network in Fig. 1 to estimate the probability distribution of the true value given the observed one. The individual true value is predicted by a random draw from such a distribution.

As far as the first step is concerned, the validation sample has been obtained by means of data collected by Banca d'Italia and a major Italian bank group on a sample of customers of the latter. In particular, the survey was carried out (independently from SHIW 2008) in 2003 on a sample of 1.681 households where at least one member was a customer of the bank group. Survey data had then been matched with the bank customers database containing the amount of the assets (stocks and bonds) actually held by the individuals selected in the sample; see [6] for details.

In this analysis we focus on (government and private) bonds. Since model (1) and its OOBN representation are developed for categorical variables, the true and the observed amount of bonds in the validation sample have been discretized. The discretization (in ten classes) has been performed using the *Chi*2 discretization algorithm, see [4]. In the rest of this paper the discretized distributions are treated as the original observed and true distributions.

The performance of the imputation procedure has been evaluated by means of the following two indicators:

1. Kullback–Leibler distance between the true and the observed distribution and between the true and the imputed distribution denoted by KL^{TO} and KL^{TI}, respectively

Table 1 Estimates of the measurement model parameters α_t, $t = -7, \ldots, 3$

α_{-7}	α_{-6}	α_{-5}	α_{-4}	α_{-3}	α_{-2}	α_{-1}	α_1	α_2	α_3
0.07	0.07	0.08	0.15	0.12	0.13	0.16	0.11	0.06	0.05

2. The percentage of correct imputations $\psi = \frac{1}{n^*} \sum_{i \in S^*} I_{x_i}(x_i^*) * 100$, where S^* is the subsample composed of units affected by measurement errors in the validation sample and I_{x_i} is the indicator function assuming the values 1 if the true value for unit i denoted by x_i is equal to the corresponding imputed value x_i^* and 0 otherwise

As seen in Sect. 2, the measurement model parameters are h, μ, and α_t, $|t| = 1, \ldots, T$. The value of the mixture parameter $h = 0.9$ has been determined through a sensitivity analysis. The probabilities α_t of a mismatch of length t between the observed and the true class have been estimated with the proportions of mismatches of length t in the validation sample. In our case, as expected, the comparison between observed and true amount class results in the prevalence of underreporting. Specifically, the maximal length of negative mismatches with a positive estimated probability is larger than that of positive mismatches, being equal to -7 and 3, respectively. The estimates of α_t, $t = -7, \ldots, 3$, are reported in Table 1. Regarding the misreport probability μ, it has been first assumed constant over all the population and estimated as the proportion of mismatches between the observed and the true value of bond amount class in the validation sample. It resulted $\hat{\mu} = 0.54$.

Having estimated the mixed measurement model parameters, the OOBN in Fig. 1 can be used to impute the bond amount class in SHIW 2008. For each unit in the SHIW 2008 sample, the observed bond amount class is inserted in the node `oc` in Fig. 1 and is propagated through the network. The updated distribution for the node `tc`, that is the probability distribution of the true value given the observed one, is obtained and a predicted value is drawn at random from it. Note that in this analysis we assumed that the underreporting behavior observed in the validation sample (collected in 2003) remained unchanged in 2008.

After having estimated the observed and the imputed distributions of bonds from SHIW 2008, such distributions have been compared with the true distribution estimated from the validation sample using the indicator 1. The distance between the true and the imputed distribution $KL^{TI} = 0.13$ is less than the distance between the true and the observed distribution $KL^{TO} = 0.47$. The bonds microdata have been also imputed in the validation sample in order to compute the performance indicator in 2. As a result, 11 % of data affected by measurement error in the validation sample are correctly reconstructed using the proposed imputation method. Therefore this procedure reveals a good performance when applied to categorical or previously discretized variables.

3.1 Learning the Error Parameter by a BN

The mixed measurement model in the above section is based on the hypothesis that the propensity to report incorrect bond amount class is the same over all the population. This is not a reasonable assumption since it is realistic to expect that high amount of bond owners will tend to underreport, while people having no or little amount of bond may overreport. Hence imputation results could be improved using information in the validation sample in order to model the misreport probability.

The idea is to integrate the error probability model and the measurement error model used for imputation. Note that these models refer to different samples: the first one can be estimated from the validation sample; the second one is applied to the sample to be corrected (the SHIW 2008 sample, in this case). Thanks to their hierarchical structure and modularity property [2], OOBNs provide a natural tool to deal with these models together, integrating them in a single system. Figure 2 shows the OOBN representation of the measurement error process in a single variable.

The round-shaped rectangles represent instances of network classes and are Bayesian networks themselves. In particular, the network class **Measurement Error Parameters** contains the BN model for the attitude to report incorrect information. From this the misreport probability, i.e. the error parameter μ of the measurement error model, is derived. In turn, as suggested by the arrow direction, this output constitutes an input for the network class **Measurement Error Model** whose expanded representation is the BN in Fig. 1.

As far as the estimation of the network **Measurement Error Parameters** is concerned, the variables in the validation sample described in Table 2 have been used. The network has been learnt using the NPC (*Necessary Path Condition*) algorithm [8], implemented in the software Hugin, allowing us to also take into account logical constraints such as the presence/absence of links or their directions. Here we impose that if node `Error?` is connected with any of the other variables, the direction has to be from these into `Error?`

The resulting network in Fig. 3 shows that `Error?` is directly influenced by the declared class of owned bond amount (`Observed Bond Class`), by the age of the respondent (`Age`), and by the residence geographic area (`Geo Area`). The evaluations given by the interviewer (`Comprens`, `Facil`, `Verored`) affect the probability to misreport only indirectly. Notice that the variable `Observed Bond Class` logically corresponds to the variable `oc` (observed category) in

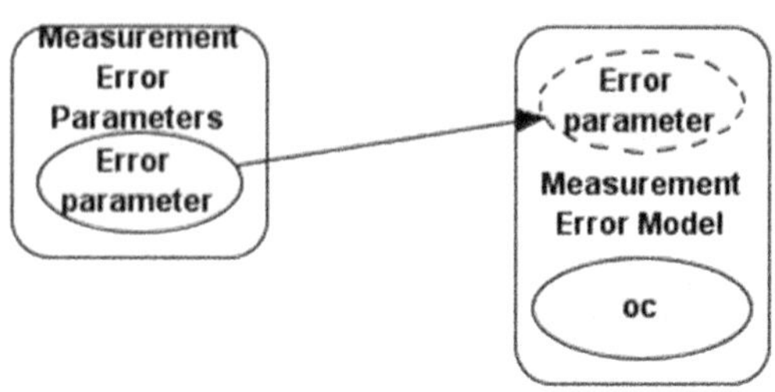

Fig. 2 OOBN representation of the measurement error process in a single variable

Table 2 Variables for the measurement error parameters network

Variable	States	Description
`Comprens`	{1–5, 6–8, 9–10}	Interviewee's level of understanding[a]
`Facil`	{1–5, 6–8, 9–10}	Easy for the interviewee to answer[a]
`Verored`	{1–5, 6–8, 9–10}	Reliability of the information on income[a]
`Age`	{2–38, 38–50, 50–65, 65–89}	Age of the respondent
`Geo Area`	{0, 1}	Living in the North-Center of Italy no/yes (0/1)
`Observed Bond Class`	10 classes	Class of owned bond amount
`Error?`	{0, 1}	Reporting a wrong value no/yes (0/1)

[a]Evaluation given by the interviewer

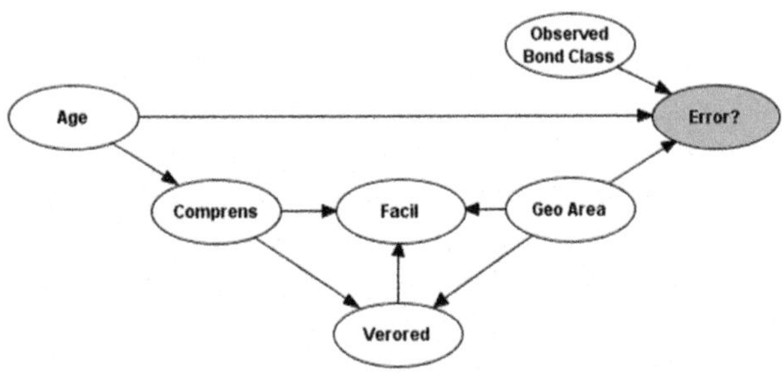

Fig. 3 BN structure for the error parameter (**Measurement Error Parameters** network class)

the measurement error model network in Fig. 1, although in this last network the observed category is measured on the sample to be edited.

Having a model for the error generating process derived from the validation sample, we can specify the different error probabilities for different configurations of the respondents, obtaining a more refined and flexible tool to correct our data. More specifically, for each respondent in SHIW 2008 the evidence given by (`Observed Bond Class`, `Age`, `Verored`) is inserted and propagated through the network in Fig. 3 and the corresponding misreport probability is derived. This probability is then fed into the measurement error model network in Fig. 1 together with the observed value of `oc` in order to produce the predicted value. The KL distance between the true and the imputed distribution decreases from 0.13 to 0.08 revealing that the imputation procedure accounting for auxiliary information formalized by a BN improves the results.

4 Conclusion and Discussion

In this paper we have seen that OOBNs can be fruitfully used to model measurement error problems. By the application to SHIW 2008 data we have shown that using auxiliary information to estimate the propensity to misreport helps improving the results. Our model deals with errors in a single variable. Further research is still needed to extend the analysis to the multivariate case and account for auxiliary information also in the true value prediction phase. Other problems at this research stage may limit the application of BNs to measurement error correction as well as to official statistics. Among them we mainly refer to: (*i*) the use of hybrid BNs where continuous and discrete variables are considered; (*ii*) the necessity to take into account the complexity of sampling design when BNs are applied to sample surveys.

Regarding the first point, it is well known that mixtures of Gaussian distributions can approximate any probability distribution. Then it should be possible to solve any hybrid BN by first approximating it by a mixture of Gaussian BNs and then using the Lauritzen algorithm [3] to solve this mixture (see [7]).

As far as the second point is concerned, design complexity should be taken into account in survey analysis by an appropriate use of sampling weights in order to obtain unbiased estimates. Some results in a likelihood approach are in [1]. Furthermore, when auxiliary variables are used to improve either parameter estimates or imputation procedure, BNs should be learnt suitably accounting for the sampling design features.

References

1. Ballin, M., Scanu, M., Vicard, P.: Estimation of contingency tables in complex survey sampling using probabilistic expert systems. J. Stat. Plan. Inference **140**, 1501–1512 (2010)
2. Koller D., Pfeffer A.: Object-Oriented Bayesian Networks. In: Proceedings of the Thirteenth Annual Conference on Uncertainty in Artificial Intelligence, pp. 302–313 (1997)
3. Lauritzen, S.L.: Propagation of probabilities, means and variances in mixed graphical association models. J. Am. Stat. Assoc. **87**, 1098–1108 (1992)
4. Liu, H., Setiono, R.: Chi2: Feature selection and discretization of numeric attributes. In: Proceedings of The Seventh International Conference with Artificial Intelligence (1995)
5. Marella, D., Vicard P.: Object-oriented Bayesian network for modeling the respondent measurement error. Commun. Stat. Theory Methods **42**(19), 3463–3477 (2013)
6. Neri, A., Ranalli, M.G.: To misreport or not to report? The case of the Italian survey on household income and wealth. Stat. Trans. **12**, 281–300 (2011)
7. Shenoy, P.P.: Inference in Hybrid Bayesian Networks using mixtures of Gaussians. In: Prooceedings of the 22nd Conference in Uncertainty in Artificial Intelligence (2006)
8. Steck, H.: Constraint-based structural learning in Bayesian networks using finite data sets. Ph.D. thesis, Department of Computer Science, University of Munich (2001)
9. Vicard, P., Dawid, A.P.: A statistical treatment of biases affecting the estimation of mutation rates. Mutat. Res. **547**, 19–33 (2004)

Comparing Fuzzy and Multidimensional Methods to Evaluate Well-Being in European Regions

Maria Adele Milioli, Lara Berzieri, and Sergio Zani

Abstract We suggest a new criterion based on fuzzy sets theory in order to evaluate well-being in European regions at NUTS 2 level. With reference to the various domains of this vague and multidimensional concept, a subset of 16 variables available in Eurostat database is selected. After a fuzzy transformation, the variables are aggregated into a fuzzy synthetic indicator, considering different weighting criteria. For each region the fuzzy indicator value, in the range [0, 1], may be interpreted as a membership degree to the subset of the areas with the highest well-being. The results are compared with the ones obtained by principal component analysis (PCA) and *k*-means cluster analysis applied to the same dataset. Furthermore, the relationships of the fuzzy indicator with GDP per capita and with human development index (HDI) are highlighted. The advantages and the drawbacks of the suggested approach are discussed.

Keywords Cluster analysis • Composite indicators • Fuzzy sets • Membership function • Principal components

1 Introduction

Is increasing GDP per capita a symptom of better life conditions? "Yet Gross Domestic Product measures everything, in short, except that which makes life worthwhile". (Speech excerpt by Robert F. Kennedy, 1968.) The growing interest in the "beyond GDP" ideas has resulted in the construction of several alternative measures of economic development and social progress (e.g. [5]). Well-being and quality of life are the most recurrent terms used to describe these concepts, but in the literature non equivalent definitions and specifications are considered.

M.A. Milioli (✉) • S. Zani
Department of Economics, University of Parma, 43125 Parma PR, Italy
e-mail: mariaadele.milioli@unipr.it; sergio.zani@unipr.it

L. Berzieri
Comune di Parma, 43121 Parma PR, Italy
e-mail: l.berzieri@comune.parma.it

I. Morlini et al. (eds.), *Advances in Statistical Models for Data Analysis*,
Studies in Classification, Data Analysis, and Knowledge Organization,
DOI 10.1007/978-3-319-17377-1_18

"Human development, as an approach, is concerned with what I take to be the basic development idea: namely, advancing the richness of human life, rather than the richness of the economy in which human beings live, which is only a part of it" (Amartya Sen).

Well-being may concern either a single person's life situation (subjective well-being, see, e.g. [9]) or the living conditions of people in a certain area. The two main features of well-being are multidimensionality and vagueness: this latent concept cannot be directly measured, but it can be captured by means of a set of observable variables encompassing different domains. Composite indicators should ideally measure multidimensional concepts which cannot be captured by a simple variable [13]. Furthermore, it is possible to point out the gradual transition from poor to rich living conditions, considering increasing levels of well-being. The measures of well-being should be obtained using multidimensional analysis and fuzzy sets approach, providing a mathematical framework in which this vague concept can be studied.

Most of the researches on well-being are carried out at country level. The recent "Better Life Index" allows to compare well-being across countries, based on 11 topics identifying the areas of material living conditions and quality of life [15]. By narrowing down the analysis at sub-national level, a wide variety of situations emerge across and within the countries.

In this paper we propose the construction of fuzzy composite indicators in order to evaluate well-being in the European regions of the 27 member States, as defined in NUTS 2 (Nomenclature of Territorial Units for Statistics of second level).

Related recent studies on the measurement of the living conditions across European regions are: [2, 4, 16, 19].

The theoretical socio-economic framework that we consider is described in: [1, 8, 14, 20]. Well-being at territorial level may be determined by two main domains: material living conditions (or "economic welfare") that include income and wealth, consumption, jobs and earnings, housing; quality of life, defined as the set of non-monetary attributes of individuals and their opportunities and life chances (health status, education and skills, environmental quality, personal security, etc.). The framework also considers the sustainability over time of the socio-economic conditions and of the natural systems.

Well-being composite indicators are highly sensitive to the variables that are selected, to the methods and weights used in the aggregation: different choices may entail quite different results [18].

Starting from the previous conceptual models and the above mentioned considerations, in Sect. 2 we select a subset of variables available in Eurostat database at NUTS 2 level.

In Sect. 3 we describe the steps for the construction of a fuzzy composite indicator, assumed as a synthetic measure of well-being level in the regions.

In Sect. 4 we present the values of this indicator in the map of European regions and sketch the best and the worst areas. In Sect. 5 the fuzzy sets approach is compared with GDP per capita values and with the results of classical multidimensional methods for dimension reduction and classification of the units: principal component analysis (PCA) and k-means cluster analysis, applied to the same dataset.

In Sect. 6 we compare the values of the fuzzy indicator with the ones of human development index (HDI) at NUTS 2 level and we show their non-linear relation.

Concluding remarks in Sect. 7 highlight the additional information of the suggested approach with respect to the traditional ones.

2 The Selection of the Variables

The NUTS 2 classification subdivides the 27 European States into 271 regions. Source of the data is Eurostat's database. This classification corresponds in Italy to the administrative regions, with the exception of Trentino-Alto Adige, divided into the provinces of Trento and Bolzano.

First of all we have erased from the data set 5 units not belonging to European Union: HR1, HR2, HR3—candidate regions in Croatia; IS 00 Iceland (Efta Country); FI 1B Helsinki (new region). We have also deleted the following six regions in other continents: Guyane, Réunion, Martinique, Guadalupe (FR); Melilla, Ceuta (ES).

The selection of the variables has been done starting from the list of all available indicators at NUTS 2 level for European regions (reference year 2010), which is a strong limitation in the definition of the complex concept of well-being. Above all, there is a lack of suitable variables for describing at regional level the aspects of the sustainability, social connection, personal security and subjective well-being.

In order to avoid redundancy for the available domains, a variable selection procedure has been carried out. In most cases, the inclusion of all the variables in a statistical analysis is, at best, unnecessary and, at worst, a serious impediment to the correct interpretation of the data. If two variables are highly correlated, then one of them can often be deleted without the final result being greatly influenced. One way of achieving a simple interpretation is to reduce the number of variables, i.e. to select a subset of the variables to preserve as far as possible the original information. On this topic see, e.g. [17].

Using the criteria of the correlation matrix and PCA, a subset of 16 standardized variables has been selected, with respect to six domains: health and road accidents, wealth and free time, labor market, education, demography, environment. In Table 1 the list of the variables and their relationship, positive or negative, with the global well-being is presented. We point out that the set of the selected variables includes the three aspects considered in the HDI: life expectancy, education and GDP per capita (analysed in Sect. 6).

Table 1 Subset of well-being indicators used in the analysis

Well-being indicators	Relationship with well-being
Health and road accidents	
Life expectancy at birth	+
Victims in road accidents (on 100,000 residents)	−
Wealth and free time	
GDP at current market prices (100 = mean value)	+
Family disposable income (100 = mean value)	+
% free time weekly hours	+
Labour market	
Employment rate	+
Unemployment rate	−
Long-term unemployment rate	−
Differences between young and adult unemployment rate	−
Education	
% persons with tertiary education	+
Life-long learning	+
Demography	
Elderly rate	−
% under 10 years old	+
Fertility rate	+
Natural Change rate (mean 2006–2010)	+
Environment	
% land use for residential, commercial and industrial purpose	−

3 The Suggested Fuzzy Indicator

Fuzzy sets theory (e.g. [24]) provides an approach to deal with vague concepts as well-being or quality of life [3, 11]; poverty [7, 12], customer satisfaction [22, 23]. Using the fuzzy approach, the well-being of an area may be interpreted as a question of degree, showing the gradual transition from poor to rich regions: the measure of well-being can be expressed as membership degree to the subset A of the best areas.

Consider a set of n regions r_i ($i = 1, 2, \ldots, n$) and p manifest variables X_s ($s = 1, 2, \ldots, p$) reflecting the different aspects of well-being. Without loss of generality, let us assume that each variable is positively related with well-being. If a quantitative variable X_s shows negative correlation, we substitute it with a simple decreasing transformation, e.g. $f(x_{si}) = \max(x_{si}) - x_{si}$.

In order to define the membership function for each variable X_s it is necessary:

1. To identify the extreme situation such that $\mu_A(x) = 0$ (non-membership) and $\mu_A(x) = 1$ (full membership)
2. To define a criterion for assigning membership function values to the intermediate categories of the variable

For each standardized variable X (for simplicity of notation we omit index s), we choose an inferior (lower) threshold l and a superior (upper) threshold u, with l and u finite, and we define the m.f. $\mu_A(x_i)$ as follows:

$$\mu_A(x_i) = \begin{cases} 0 & x_i \leq l \\ \frac{x_i - l}{u - l} & l < x_i < u \\ 1 & x_i \geq u \end{cases}$$

We have chosen: *lower* threshold l = median of the variable; *upper* threshold u = 90th percentile. With this choice the regions with a value of the variable under the median do not belong to the subset A of the best regions, with reference to the considered aspect, and the regions with the 10 % highest values totally belong to the subset of the areas with the highest quality of life.

Among the steps of the construction of a composite index, weighting and aggregation criteria are the most difficult ones as they directly affect the quality and reliability of the results (e.g. [10, 18]). Let us consider the criteria for aggregating p fuzzy variables into a fuzzy composite indicator. A general aggregation function is the weighted generalized mean:

$$\mu_A(i) = \{\sum_{s=1}^{p} [\mu_A(x_{si})]^{\alpha} w_s\}^{1/\alpha}$$

where $w_s > 0$ is the normalized weight that expresses the relative importance of the variables X_s; $(\sum_{s=1}^{p} w_s = 1)$. For fixed arguments and weights, the function is monotonic increasing with α; if $\alpha \to -\infty$, then it becomes the intersection; if $\alpha \to +\infty$, then it is equal to the union. For $\alpha \to 0$ it becomes the weighted geometric mean.

The weighting criteria may be:

- Equal weights, which imply a careful selection of the variables in order to assure a balance of the different aspects of the latent phenomenon
- Factor loadings, obtained by PCA
- Subjective weights obtained by expert judgments, with reference to the importance of the different aspects

Obviously other thresholds, other functions (as the exponential or the cubic ones) and other weights may be considered and in the next section we test the sensitivity of the results obtained using different selection criteria.

4 Fuzzy Well-Being Levels in European Regions

We have calculated the values of a first fuzzy composite indicator with equal weights for the 16 variables and of a second fuzzy indicator with weights proportional to the factor loadings of the first principal component (see successive Sect. 5). The correlation between unweighted and weighted indicators is very high ($r = 0.992$) and therefore the classification of the regions obtained by the two criteria is very similar; so we will describe only the unweighted indicator. For this index we have considered also different upper thresholds: 80th and 79th percentile. The correlation with the indicator using 90th percentile is very high: 0.990 and 0.975, respectively. Therefore we present only the results with reference to the 90th percentile upper threshold.

The values of this fuzzy indicator have an interesting interpretation: a value equal to 0 corresponds to a region under the median for all the variables, a value equal to 1 identifies a region over the 90-th percentile for all the variables and a value in the open range (0, 1) may be assumed as membership degree of the region to the subset A of the best areas, i.e. as a fuzzy measure of well-being.

The top ten regions for well-being level are (fuzzy indicator value in brackets):

- Berkshire, Buckinghamshire and Oxfordshire, UK (0.86)
- Stockholm, SE (0.81)
- Noord-Holland, NL (0.80)
- Bedfordshire and Hertfordshire, UK (0.78)
- Zuid-Holland, NL (0.74)
- Flevoland, NL (0.72)
- Gloucestershire, Wiltshire and Bristol/Bath area, UK (0.70)
- Overijssel, NL (0.69)
- Hovedstaden, DK (0.69)

We highlight that no region presents a value equal to one of the fuzzy indicator, i.e. no region shows values greater than the 90-th percentile for all the 16 variables.

The regions with the worst conditions, all with zero values of the fuzzy well-being indicator, are:

- Yuzhen Tsentralen, BG
- Dytiki Makedonia, EL
- Nyugat-Dunántúl, HU
- Dél-Dunántúl, HU
- Dél-Alföld, HU
- Sud-Muntenia, RO

The complete list of the values of the fuzzy indicator may be requested to the first author.

In Fig. 1 we present the map of the values of the fuzzy composite indicator in European regions, according to a partition with five equal classes, based on percentiles. The map is obtained using the program GvSig (http://www.gvsig.org),

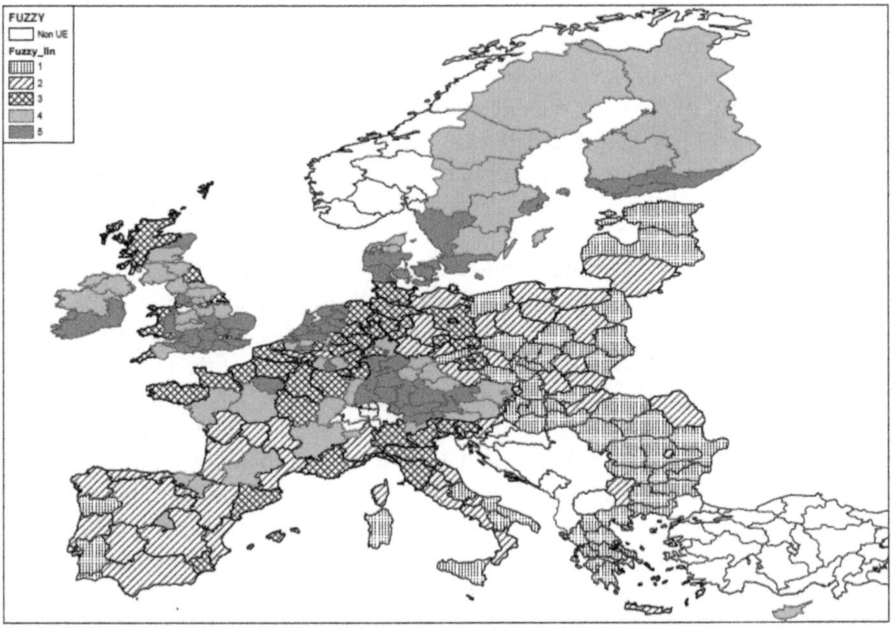

Fig. 1 Map of the European regions according to the values of the fuzzy composite indicator of well-being

a cartographic information system for visualizing the results. The map of the European regions shows the five classes of percentiles by different types of grids: the darkest areas correspond to the best regions.

The lowest levels of well-being are located in the East European countries and in a few regions in Portugal, Spain and southern Italy. The best areas are scattered in different countries of central and northern Europe.

5 Comparison with GDP Per Capita and With Other Multidimensional Indicators

It is interesting to compare the results of the fuzzy multidimensional approach with the traditional indicator of development, i.e. GDP per capita and with the results of other multidimensional methods, applied to the same set of 16 variables. The metropolitan areas of Brussels, Inner and Outer London show too high values for a few variables and may be considered as multidimensional outliers. They have been omitted in the following comparisons and therefore only 257 units are considered.

The correlation between the fuzzy indicator and GDP per capita is moderate ($r = 0.712$) and this restates that GDP is a poor and insufficient criterion for a global evaluation of living conditions of territorial units.

Another comparison method is the classification of the values of the fuzzy indicator values and GDP per capita into a contingency table, considering for each indicator the partition corresponding to five classes of percentiles (Table 2).

The regions in the same percentile class with the two criteria (main diagonal of the matrix) are 42.8 %, and Kendall's tau is 0.620. We highlight that the extreme regions (the worst and the best) are rated in a similar manner on the basis of the two criteria, whereas the regions in the middle of the range present more different classifications.

We have applied PCA to the same set of 16 variables of well-being. The first PC accounts for 37.5 % of the total variance and the second PC for 20.2 %. The percentage explained by the two PC is equal to 57.7 % and is superior to the threshold $0.95^{16} = 0.44$ (Cronbach's alpha = 0.844). The first PC is highly related to the variables measuring income and wealth, education, labour market and life expectancy; the second PC describes demographic domain. The linear correlation between the previous fuzzy indicator and the first PC is sufficiently high ($r = 0.932$) and also the rankings of the regions obtained by the two criteria are similar, but not equal (Spearman's rho = 0.950). The contingency table with reference to the fuzzy indicator values and the scores of the first PC, considering for each indicator the partition corresponding to five classes of percentiles (Table 3), shows that most of the regions (67.7 %) are in the same percentile class with the two criteria, i.e. the two indicators show similar but not equal results (Kendall's tau = 0.847).

Table 2 Contingency table of the values of the fuzzy indicator and GDP per capita

		Percentile classes of GDP per capita					
		1	2	3	4	5	Total
Percentile Classes of the Fuzzy indicator	1	35	14	1	1	0	51
	2	15	17	11	7	2	52
	3	1	13	14	15	8	51
	4	0	7	16	16	13	52
	5	0	1	9	13	28	51
Total		51	52	51	52	51	257

Table 3 Contingency table of the values of the fuzzy and the first PC indicators

		Percentile classes of the first PC					
		1	2	3	4	5	Total
Percentile classes of the fuzzy indicator	1	39	12	0	0	0	51
	2	11	32	9	0	0	52
	3	1	8	30	12	0	51
	4	0	0	12	31	9	52
	5	0	0	0	9	42	51
Total		51	52	51	52	51	257

Table 4 Five clusters of regions by *k*-means method

Cluster index	Number of regions	Fuzzy values average	First PC scores average	Second PC scores average
3	61	0.089	−1.240	0.178
1	29	0.143	−0.906	0.258
2	37	0.309	0.254	0.974
5	65	0.331	0.323	−1.286
4	65	0.491	1.100	0.437

Finally, we have applied *k*-means cluster analysis to the 16 standardized variables selecting five groups (for comparison reasons with the previous partitions), ranked according to the average of the values of the fuzzy indicator of the regions in each cluster. The average of the scores of the first and second PC is also presented (Table 4). The 65 regions in cluster n. 4 are the ones with the highest well-being measured by fuzzy and PC indicators.

6 Comparison with HDI

The comparison of the suggested well-being indicators with the results of other researches on this topic at sub-national level is not an easy task, as a consequence of the differences in the choice of variables, methods and territorial units.

We compare our results with the values of HDI computed by Bubbico and Dijkstra [6] for European regions at NUTS 2 level, with reference to 27 EU countries for the year 2007.

HDI is the average of three normalized indices, one in each dimension of human development:

- Life expectancy at birth
- Education
- GDP per capita (PPP US dollars)

The index presents values in the range [0, 100], where 0 is equal to the lowest level of human development and 100 to the highest. The HDI is usually calculated in order to compare the development of the nations all over the world (see, e.g. [21]).

Figure 2 shows the scatterplot with respect to HDI and the fuzzy composite indicator for the same 257 European regions examined in Sect. 5. The relationship between the two indices is moderate and non-linear: $r^2 = 0.597$ for the linear function and $r^2 = 0.720$ for the quadratic function (the cubic function shows a non-significant increase $r^2 = 0.722$). There are also a few bivariate outliers, very far from the curve. In the left side of the figure we can see: 201 = Acores (PT); 202 = Madeira (PT); 205 = Nord-Est of Romania; 208 = Bucuresti (RO) and 224 = Východné Slovensko (SK). Under the curve there is 15 = Brabant Wallon (BE)

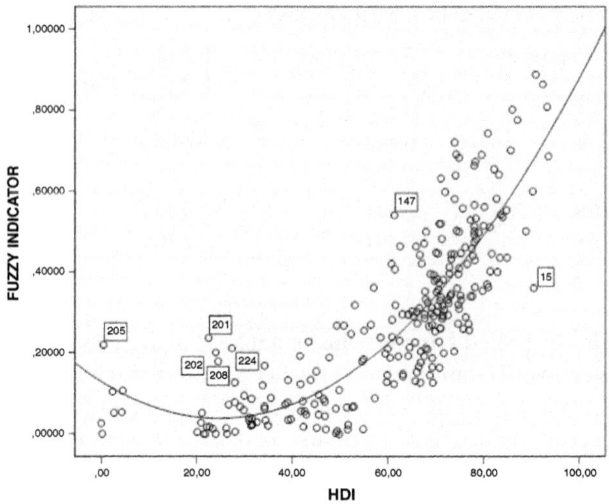

Fig. 2 Scatterplot of the 257 European region with reference to HDI and fuzzy indicator with superimposed quadratic function. The numbers correspond to regions that are bivariate outliers

and over the function 147 = Bolzano (IT). For the seven above mentioned regions the two criteria of well-being evaluation entail quite different results. Deleting these 7 outliers, we obtain a slight improvement in the goodness-of-fit for the quadratic function: $r^2 = 0.750$. The differences between the two indicators may be explained by the sets of variables (3 against 16), the transformation and aggregation criteria, the reference year (2007 and 2010).

7 Concluding Remarks

In this paper we have suggested a criterion based on fuzzy sets theory for the construction of well-being indices at sub-national level. Our fuzzy composite indicator is based on a set of variables describing the various domains of well-being and it presents values in the closed range [0, 1]. The great advantage of this index is its simple and interesting interpretation: a value equal to 0 corresponds to a region under the median for all the variables, a value equal to 1 identifies a region over the 90-th percentile for all the variables and a value in the open range (0, 1) may be assumed as membership degree of the region to the subset of the areas with highest well-being.

The application of the fuzzy indicator to the European regions at NUTS 2 level, considering a set of 16 variables, has pointed out new aspects and better explanations of well-being. The comparison with the results of PCA, applied to the same set of variables, has highlighted that the linear and rank correlation between the previous fuzzy indicator and the first PC are sufficiently high ($r = 0.932$; Spearman rho = 0.950), i.e. the rankings of the regions obtained by the two criteria are similar, but not equal. The correlation of these two composite indicators with GDP per capita is moderate ($r = 0.712$ and $r = 0.759$, respectively) and this confirms the inadequacy of such single variable for a complete description of well-being concept. The relation of the fuzzy indicator with HDI is non-linear ($r^2 = 0.720$ for the quadratic function) and there are a few regions that may be considered as bivariate outliers.

The shortcomings of the suggested approach are related to the following subjective choices in the various steps of the construction of a fuzzy composite indicator:

- Set (or subset) of variables
- Form of the membership function and lower and upper thresholds
- Weights of the variables
- Aggregation criterion.

The sensitivity and robustness of the results with respect to a few different choices in the previous steps have been examined in Sect. 4.

References

1. Allin, P., Hand D.J.: The Wellbeing of Nations: Meaning, Motive and Measurement, Environmetrics. Wiley, New York (2014)
2. Annoni, P., Weziak-Bialowolska, D.: Quality of Life at the Sub-national Level: An Operational Example for the EU. Publications Office of the European Union, Luxembourg (2012)
3. Baliamoune-Lutz, M.: On the measurement of human well-being: fuzzy set theory and Sen's capability approach. In: McGillivray, M., Clarke, M. (eds.) Understanding Human Well-being. United Nation University Press, New York (2006)
4. Berzieri, L., Milioli M.A., Zani S.: A Fuzzy Approach to Measure Well-Being in European Regions. CD, SIS Conference, Brescia (2013)
5. Bleys, B.: Beyond GDP: Classifying alternative measures for progress. Soc. Indic. Res. **109**, 355–376 (2012)
6. Bubbico, R.L., Dijkstra, L.: The European regional Human Development and Human Poverty Indices, Regional Focus, no. 2, European Union, Regional Policy (2011)
7. Cerioli, A., Zani, S.: A fuzzy approach to the measurement of poverty. In: Dagum, C., Zenga, M. (eds.) Income and Wealth Distribution, Inequality and Poverty, pp. 272–284. Springer, Berlin (1990)
8. CNEL-ISTAT, Rapporto BES 2013: il benessere equo e sostenibile in Italia, Roma (2013)
9. Diener, E.: Subjective well-being: the science of happiness and a proposal for a national index. Am. Psychol. **55**, 34–44 (2000)
10. Hsieh, C.-M.: Importance is not important: the role of importance weightings in QOL measures. Soc. Indic. Res. **109**, 267–278 (2012)
11. Lazim, M.A., Abu Osman, M.T.: A new Malaysian Quality of Life index based on fuzzy sets and hierarchical needs. Soc. Indic. Res. **94**, 499–50 (2009)
12. Lemmi, A., Betti. G. (eds): Fuzzy Set Approach to Multidimensional Poverty Measurement. Springer, New York (2006)
13. OECD: Handbook on Constructing Composite Indicators. OECD Publishing, Paris (2008)
14. OECD: Compendium of OECD Well-Being Indicators. OECD Publishing, Paris (2011)

15. OECD: Better Life Index. OECD Publishing, Paris (2013)
16. Okulicz-Kozaryn, A.: Income and well-being across European Provinces. Soc. Indic. Res. **106**, 371–392 (2012)
17. Pacheco J., Casado S., Porras S.: Exact methods for variable selection in principal component analysis: guide functions and pre-selection. Comput. Stat. Data Anal. **57**, 95–111 (2013)
18. Paruolo P., Saisana M. Saltelli A.: Ratings and rankings: voodoo or science? J. R. Stat. Soc. A **176**, **3**, 609–634 (2013)
19. Pittau, M.G., Zelli, R., Gelman, A.: Economic Disparities and Life Satisfaction in European Regions. Soc. Indic. Res. **96**, 339–361 (2010)
20. Stiglitz, J., Sen, A., Fitoussi, J.-P.: Report by the Commission on the Measurement of Economic Performance and Social Progress, Paris (2009)
21. UNDP: Human Development Report 2011. Palgrave Macmillan, New York (2011)
22. Zani, S., Milioli, M.A. Morlini, I.: Fuzzy Methods and Satisfaction Indices. In: Kennett, R.S., Salini, S. (eds.) Modern Analysis of Customer Surveys, pp. 439–456. Wiley, New York (2012)
23. Zani, S., Milioli, M.A. Morlini, I.: Fuzzy Composite Indicators: An Application for Measuring Customer Satisfaction. In: Torelli, N., Pesarin, F., Bar-Hen, A. (eds.) Advances in Theoretical and Applied Statistics, pp. 243–253. Springer, New York (2013)
24. Zimmermann, H.J.: Fuzzy Sets Theory and its Applications, 4th edn. Kluwer, Boston (2001)

Cluster Analysis of Three-Way Atmospheric Data

Isabella Morlini and Stefano Orlandini

Abstract Classification of meteorological time series is important for the analysis of the climate variability and climate change. The clustering of several years in groups that are homogeneous with reference to the amount of precipitation and to the atmospheric condition can aid in understanding the structure of precipitation and may be important in developing hydrological models. In this paper we propose a cluster analysis of multivariate time series based on a dissimilarity measure that considers the functional form of the data. The unit to be classified are 148 years, from 1861 to 2008, and the variables are the values of precipitation, the minimum temperature, and the maximum temperature in different occasions (days or months) in the province of Modena (Northern Italy).

Keywords Climate change • Clustering • Functional data analysis • Precipitation

1 Introduction

When studying climate change in a spatial area, we may search for typical patterns, common to some time periods, describing the underlying atmospheric process. The analysis and the comparison of these different patterns may give an insight into the long-period changes in meteorological variables, such as rain and temperature. These typical patterns may be thought of as centroids of homogeneous clusters, where the units to be classified are years over a long period of time and the variables are measurements of rain and temperature in different occasions (for example, days

I. Morlini (✉)
Department of Economics "Marco Biagi", University of Modena and Reggio Emilia, Via Berengario 51, 41100 Modena, Italy
e-mail: isabella.morlini@unimore.it

S. Orlandini
Department of Mechanical and Civil Engineering, University of Modena and Reggio Emilia, Strada Vignolese 905, 41125 Modena, Italy
e-mail: stefano.orlandini@unimore.it

I. Morlini et al. (eds.), *Advances in Statistical Models for Data Analysis*, Studies in Classification, Data Analysis, and Knowledge Organization,
DOI 10.1007/978-3-319-17377-1_19

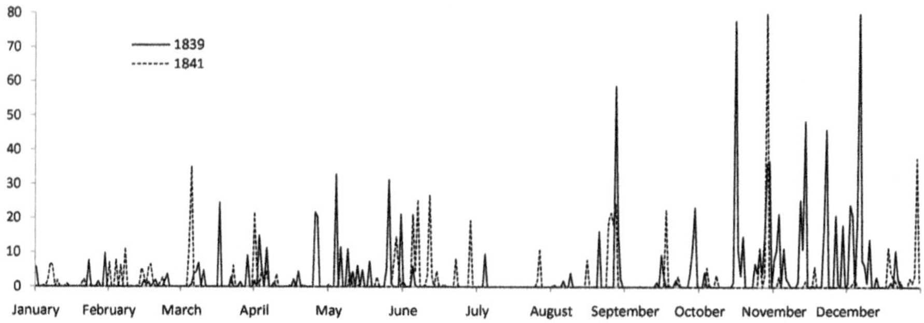

Fig. 1 Daily values of rainfall in Modena in the years 1839 and 1841

or months). Classification of these there-way data (unit × variables × occasions) should consider the functional form of the multivariate time series. Indeed, salient features of atmospheric measurements, such as extreme values, maxima or minima, may result shifted in the different series. The transformation of time, that is the warping function from one series to another, must be estimated, before computing the dissimilarity between pairs of series. This function permits a fruitful alignment of the two sequences of measurements. As an example, Fig. 1 reports the daily values (`explains`) of rainfall intensity in the years 1839 and 1841, in the province of Modena (Northern Italy). The two sequences show a great similarity, considering that both years have a peak around 30 mm in March, three days with more than 20 mm in the period May–June and, in particular, a very rare event such as a daily value near 80 mm in October. The timing of this very rare event is shifted of 13 days in the 2 years (it occurs the 16th of October in the year 1839 and the 29th of October in the year 1841). Cross-sectional similarities, which compare measurements gathered in the same day, produce pessimistic values for these two series. A more comprehensive similarity should align similar events that occur in nearby days. Even the simplest data analysis, such as computing a mean, can require that features be first aligned by a time transformation, a process that is called time-series registration.

Classical functional data analysis [2, 7, 8, 10] interpolates the sequences of values by a smooth curve and assumes that also the time-warping function is a smooth function, differentiable as the curves themselves. Suppose we have n observed values x_{ijt}, $i = 1, \ldots, n$, of variable j ($j = 1, \ldots, p$) at time t ($t = 1, \ldots, T$). In functional data analysis, the model usually assumed is

$$x_{ijt} = s_{ij}(w_{ij}(t)) + \varepsilon_{ijt}, \tag{1}$$

where s_{ij} is the smooth function underlying the time series i of variable j ($i = 1, \ldots, n$), ($j = 1, \ldots, p$), $w_{ij}(t)$ is the smooth time warping function, and ε_{ijt} is the error term. The errors are assumed independent and identically distributed. The function $w_{ij}(t)$ is subject to the following constraints:

1. $t_1 < t_2 \Longleftrightarrow w_{ij}(t_1) < w_{ij}(t_2)$
2. $w_{ij}(0) = 0$
3. $w_{ij}(T) = T$

To keep the notation simple, in (1) it is assumed that, for each variable j and for each time series i, both the number T and the timing of the sampled values x_{ij} are identical. However, many applications involve variation in locations and numbers of sampling points across replications and formula (1) may be adjusted for these cases. The smooth functions s_{ij} and w_{ij} depend on the time series i and on the variable j and each observation x_{ijt} is associated with the registered curve value $s_{ij}(w_{ij}(t))$. The simplest curve alignment procedure is a landmark registration. A landmark is a feature with a location that is clearly identifiable in all curves. The curves are aligned by transforming the physical time so that the location of the landmarks is the same for all curves. In case of a single landmark, if t_0 is the timing of this landmark in variable j and t_i is the timing of this landmark in curve i, then the time-warping function $w_{ij}(t)$ is specified by fitting a smooth function to the three points $(0, 0)$, (t_0, t_i), and (T, T). This function is as differentiable as the curves s_{ij} themselves. According to this definition, $w_{ij}(t_i) = t_0$, and all the registered functions defined as $y_{ij}(t) = s_{ij}(w_{ij}(t))$ will all automatically arrive at the landmark at the same time, namely t_0. Both the definition of multiple landmarks and their unequivocal identification in individual curves are problematic, especially in long time series of atmospheric data. For example, in Fig. 1, the timing of the rain peak in March may be either the 5th or the 17th. In October, t_0 may be either the 16th or the 29th. Moreover, these peaks may be not so visible in other years. It is evident that both the exact number and the daily locations of landmarks in rain and temperature time series are not objectively identifiable. As an alternative to landmark registration, in this paper we use the dynamic time warping algorithm (dtw). In its original formulation, the dtw estimates a "warping path" for aligning one series to another and minimizes a measure of "discrepancy" between the two series which is called dynamic time warping cost (dtwc). However, if we modify one of the constraints in the classical formulation, the algorithm estimates a path which is a discrete time warping function and minimizes, a cost which is a dissimilarity measure between the two registered series. The dtw algorithm doesn't require the estimate of the smooth curves interpolating the time series. However, we may estimate these curves and use the smoothed valued $s_{ij}(t)$ in order to have data less noisy than the sampled values x_{ijt}. The main features of dtw are as follows:

- It is a nonparametric procedure which does not require prior assumptions about the form of the warping functions (see, e.g., [9, 12] for the definition of parametric warping functions) or about the number and the timings of salient events (the landmarks).
- It relies on a minimization problem which can be solved efficiently by using dynamic programming.

- When applied to three-way data, the warping functions are estimated by considering the vector-valued time series $\mathbf{x}_{it} = [x_{11t}, \ldots, x_{1jt}, \ldots, x_{1pt}]$ and not by considering each univariate series x_{ijt} $(j = 1, \ldots, p)$ $(t = 1, \ldots, T)$ separately. Therefore, rather than estimating a univariate warping function w_{ij} for each variable j, the dtw estimates a p-variate warping function w_i.

These last two items differentiate the dtw algorithm used in this paper and the algorithm illustrated in [13, 14]. In Wang and Gasser the warping functions are univariate smooth continuous functions.

The paper is organized as follows. In Sect. 2 we illustrate the dtw algorithm used in paper. In Sect. 3 we focus on the application. We first describe the data and the study area and then we show how atmospheric data can be clustered and analyzed to achieve meaningful results.

2 The Warping Function and the Measure of Dissimilarity

The dtw algorithm was originally developed in engineering, for speech analysis and speech recognition, in order to align two sequences of values. Many enhancements of the method have been proposed in the data mining literature. Among other works, we refer to [1, 3–5]. In its original formulation, given the p-dimensional vector-valued series $\mathbf{x}_{1t}$ and $\mathbf{x}_{2t}$, where $\mathbf{x}_{it} = [x_{11t}, \ldots, x_{1jt}, \ldots, x_{1pt}]$, $i = 1, 2$ and $t = 1, \ldots, T$, the dtw first implies the construction of a $T \times T$ square lattice D, in which the element $d(r, c)$ $(r, c = 1, \ldots, T)$ is the distance $d(\mathbf{x}_{1r}, \mathbf{x}_{2c})$ between the values of series 1 at time r and the values of series 2 at time c. Any distance may be used in the construction of the square lattice D. However, before computing any Minkowski metric, the p variables should be standardized to take into account the different units of measurements and/or the different variability [6]. Each element $d(r, c)$ corresponds to the alignment between points $\mathbf{x}_{1r}$ and $\mathbf{x}_{2r}$ in the p-dimensional Euclidean space. The dtwc is defined as follows:

$$dtwc = min\sqrt{\frac{\sum_{k=1}^{K} d_k}{K}}, \tag{2}$$

where $T \leq K \leq (2T - 1)$, K is determined by the optimization process of the algorithm and the d_k are elements of D subject to the following constraints:

- Boundary condition: $d_1 = d(1, 1) = d(\mathbf{x}_{11}, \mathbf{x}_{21})$ and $d_K = d(T, T) = d(\mathbf{x}_{1T}, \mathbf{x}_{2T})$. This constraint requires that the first time and the last time in one series are aligned with the first time and the last time, respectively, in the other series. So, the first and the last time are not warped.
- Continuity constraint: given $d_k = d(\mathbf{x}_{1r}, \mathbf{x}_{2c})$ then $d_{k-1} = d(\mathbf{x}_{1r'}, \mathbf{x}_{2c'})$ where $(r - r') \leq 1$ and $(c - c') \leq 1$. This condition restricts two successive elements d_k to be adjacent (including diagonally) elements in D.

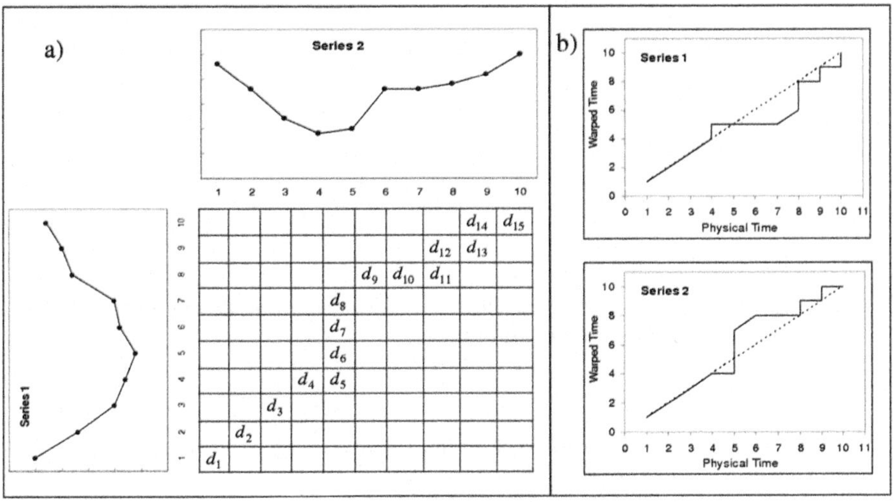

Fig. 2 (**a**) An example of the distances included in the dtwc; (**b**) the warping path

- Monotonicity constrain: given $d_k = d(\mathbf{x}_{1r}, \mathbf{x}_{2c})$ then $d_{k-1} = d(\mathbf{x}_{1r'}, \mathbf{x}_{2c'})$ where $(r - r') \geq 1$ and $(c - c') \geq 1$. This condition forces the couple of points for which the distance is taken into account in the dtwc to be monotonically spaced in time.

The dtw produces a relative shift between the two sampled curves. However, as shown in Fig. 2a, the algorithm defines a warping path and from this path we cannot draw two increasing warping functions to align $\mathbf{x}_1$ to $\mathbf{x}_2$ and to align $\mathbf{x}_2$ to $\mathbf{x}_1$, since a single point on one time series may map onto a large subsection of the other series (Fig. 2b). In order to find two monotonic—not strictly increasing—warping functions, one could eliminate the boundary condition $d_K = d(\mathbf{x}_{1T}, \mathbf{x}_{2T})$ and restrict the continuity constrain such that $(r - r') = 1$ for aligning $\mathbf{x}_1$ to $\mathbf{x}_2$ and such that $(c - c') = 1$ for aligning $\mathbf{x}_2$ to $\mathbf{x}_1$ (Fig. 3b). However, with this restriction, the dtwc becomes asymmetric and cannot be a dissimilarity measure: given two sequences i and i', and, to keep the notation simple, $\text{dtwc}(\mathbf{x}_i, \mathbf{x}_{i'}) = \text{dtwc}(ii')$, then $\text{dtwc}(ii') \neq \text{dtwc}(i'i)$. In order to define at the same time a dissimilarity measure and a warping function, we use a modified parameterized path. This path is characterized by a weaker continuity constraint, defined as follows:

Continuity constraint: given $d_k = d(\mathbf{x}_{1r}, \mathbf{x}_{2c})$ then $d_{k-1} = d(\mathbf{x}_{1r'}, \mathbf{x}_{2c'})$ where $(r - r') \leq 2$ & $(c - c') < 2$ or $(r - r') < 2$ & $(c - c') \leq 2$ (Fig. 3c).

With this continuity constraint, the classical boundary condition, and the monotonicity constraint, the dtw algorithm estimates a $wd_i(t)$ warping function, with the

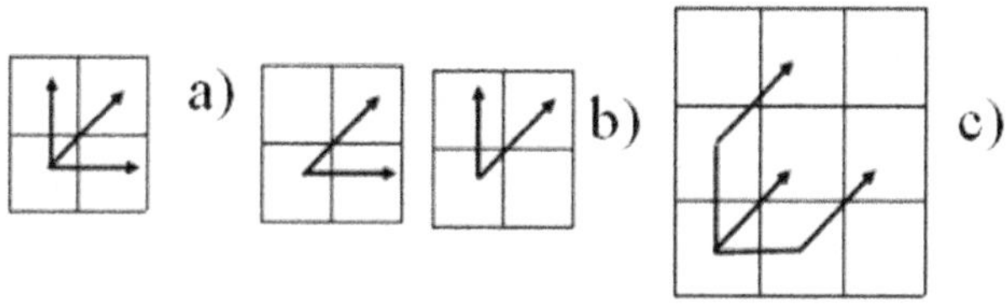

Fig. 3 Representation of the dtw step. (**a**) The classical dtw step; (**b**) step with restrictions on the continuity constraint; (**c**) step with a weaker continuity constraint

following properties:

1. $t_1 < t_2 \Rightarrow wd_i(t_1) \leq wd_i(t_2)$ (the function is monotonic increasing but not strictly increasing and it is not smooth)
2. $wd_i(0) = 0$
3. $wd_i(T) = T$

and a dtwc dissimilarity measure, satisfying the following conditions:

1. $\text{dtwc}(ii') \geq 0, i, i' = 1, \ldots, N$ (nonnegativity)
2. $\text{dtwc}(ii) = 0, i = 1, \ldots, N$ (this a condition weaker than the identity condition required for distance measures)
3. $\text{dtwc}(ii') = \text{dtwc}(i'i), i, i' = 1, \ldots, N$ (symmetry)

As outlined in the Introduction, wd_i is equal for every variable j ($j = 1, \ldots, p$), since it is estimated considering the vector-valued series $\mathbf{x}_{it} = [x_{11t}, \ldots, x_{1jt}, \ldots, x_{1pt}]$, $t = 1, \ldots, T$ and not by considering each univariate series x_{ijt} ($j = 1, \ldots, p$), ($t = 1, \ldots, T$) separately. Another feature that characterizes the warping function wd_i and that may be useful in applications with meteorological data, is the possibility to define the maximum number of time-lags between the physical time and the warped time. Indeed, considering for example daily series, only similar events that occur in nearby days are likely to be expression of the same feature (for example, a peak or an extreme value, in a certain period) and should be aligned. Salient events that occurs in days which are faraway, should be considered as two "different" features in the two series and should not be aligned. The maximum number of days between the timing of two events that are likely to be logically compared depends on the application and on the aim of the data analysis. In general, if u is the maximum number of lags for which we assume the same event may be timed differently in the different series, the simplest strategy is to introduce the following "windowing condition" in the dtw algorithm:

$$d_k = d(\mathbf{x}_{1r}, \mathbf{x}_{2c}) \text{ with } |r - c| \leq u$$

We refer to [11] for the definition of more refined constraints on the warping path, aimed at preventing unrealistic warping.

The dtwc dissimilarity matrix may be used for classifying time series with the following hierarchical methods: the average linkage, the complete and the single linkages. The centroid method is not appropriate, since the dendrogram obtained with this method with a dissimilarity measure is a non-monotonic cluster tree.

3 Classification of Meteorological Time Series

We perform a cluster analysis of atmospheric measurements gathered by a historical weather station in the urban area of the province of Modena, in the Emilia Romagna Region (Northern Italy). The station is the geophysical observatory of Modena. Even though the weather station does not conform to the W.M.O. regulations for the position of the instruments (which have been emitted many years after the construction of the geophysical observatory) it does permit the collection rainfall data, in the same location, from 1831. Information about the history of the geophysical observatory may be found in the web page http://www.ossgeo.unimo.it. Here we only report the main coordinates of the station:

- Boreal latitude: 44°38′50.76″
- East longitude from Greenwich: 10°55′45.50″
- Height of the barometric cockpit from the sea level: 64.2 m
- Height of the rain gauge from the ground: 41.9 m
- Height of the ground from the sea level: 34.6 m

The (cross-sectional) mean values and the maximum values of the total rainfall for the day (in mm) of the period 1831–2008 are reported in Fig. 4. The minimum daily value is always equal to 0. In average, the total amount of rain in a day is less than 4 mm and reaches the highest peaks in October and November and the minimum values in August. The pattern of the maximum values is different: salient peaks are present in quite all months. In some years, the total rain in a day has reached values higher than 75 mm. The series reported in Fig. 4 shows that the variable has a high variability between years and between days. There are many years presenting anomalous extreme values and it is clear that the cross-sectional mean underestimates both the value of the peaks and the order of magnitude of the phenomenon. We consider the available "three way" data set, with $p = 3$. The three variables are:

- X_1: minimum air temperature (in Celsius degree)
- X_2: maximum air temperature (in Celsius degree)
- X_3: total rainfall (in mm)

Air temperature is known only from the year 1861. We then cluster 148 sequences: the years 1861 to 2008. We perform a cluster analysis of the 148 years on the basis of the minimum temperature, the maximum temperature, and the total rainfall for the month. We will refer to these data, with $T = 12$, as monthly values of X_1, X_2, and X_3. Figure 5 reports time series of the minimum, the maximum, and the (cross-sectional) mean of the monthly values of X_1, X_2, and X_3. This figure shows that Modena experiences a "mediterranean" climate with mild wet winters and hot, less rainy, summers. While the temperature shows a clear seasonal pattern, and both the maximum and the minimum values follow the same average pattern, the amount of rainfall has a more irregular trend and the minimum and the maximum values show different patterns. Before computing the dtw dissimilarity measure, data are

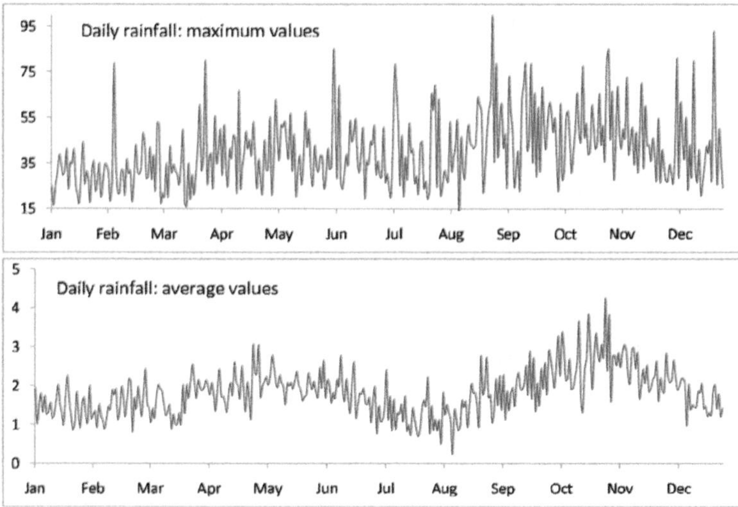

Fig. 4 Time series of the maximum (*top*) and the average (*bottom*) daily values of rain

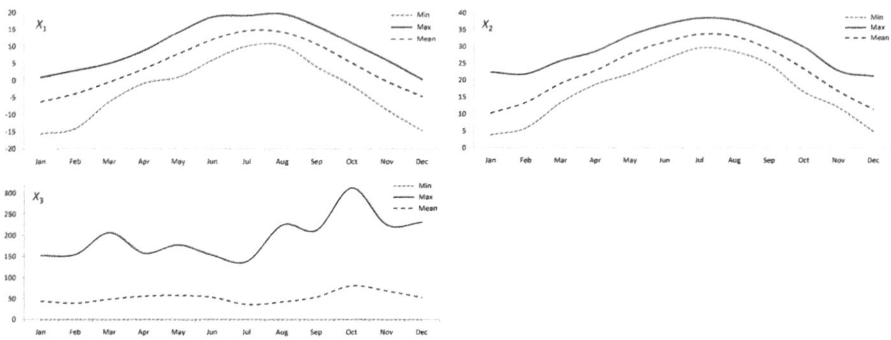

Fig. 5 Time series of the minimum, the maximum and the average monthly values of X_1, X_2, X_3

standardized so that in each t ($t = 1, \ldots, T$ with $T = 12$) each variable has 0 mean and unit variance.

In the warping function, we set $u = 2$, allowing for a maximum shift of 2 months. Figure 6 reports dendrograms obtained with the single, the complete, and the average linkages. The trees show that the single and the average linkages exhibit less ability to provide separation than the complete linkage. The single linkage is greatly affected by the "chain effect". The average linkage is less influenced by this effect but it still tends to aggregate single observations or very small groups in each stage of the hierarchy and many single observations remain isolated till the last stages. The complete linkage readily distinguishes clusters with more than one or two observations. On the basis of the ratio between the within variance and the total variance (which has a relatively high increase from partition in six clusters

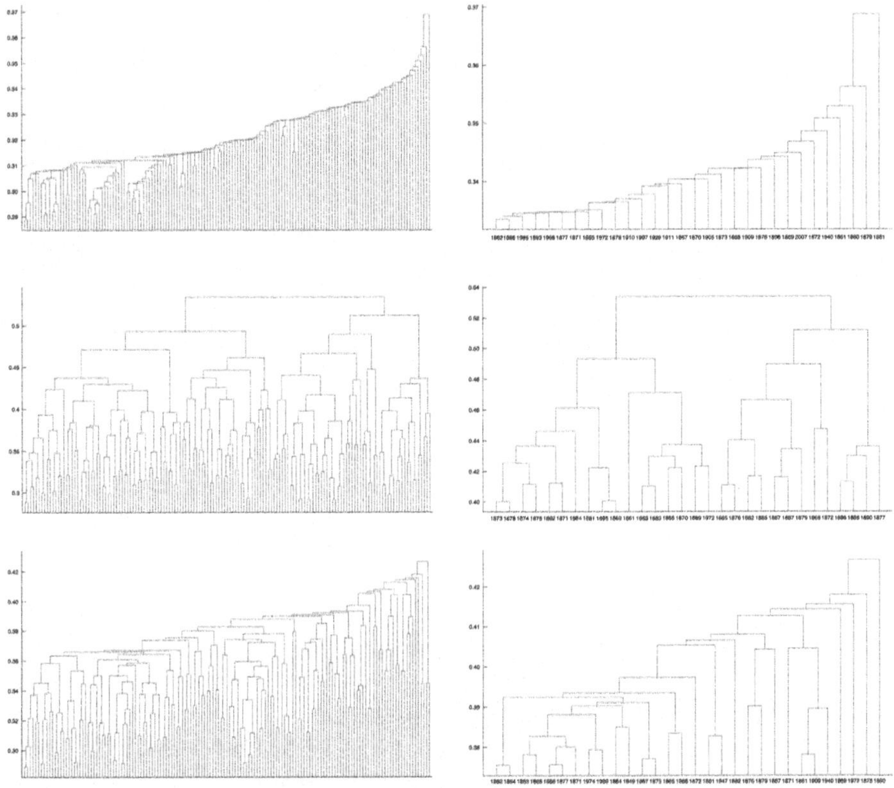

Fig. 6 Data set with $n = 148$, $T = 12$, and $p = 3$: full dendrogram (on the *left*) and dendrogram with 30 leaf nodes (on the *right*) resulting by collapsing lower branches of the full dendrogram, obtained with the single linkage (in the *top*), the complete linkage (in the *middle*), and the average linkage (in the *bottom*)

to partition in five clusters) we consider the classification in six groups. Analyzing cluster means, we see that partitions with less than six groups aggregate years with a very different behavior, while partitions with more than six groups lead to different clusters with similar average behavior. Groups, in the six-clusters partition, are as follows:

Cluster 1: {1861, 1917, 1941, 1942, 1947, 1953, 1980, 1985}
Cluster 2: {1863, 1866, 1883, 1889, 1892, 1898, 1900, 1902, 1904, 1905, 1910, 1912, 1914, 1915, 1919, 1920, 1923, 1924, 1925, 1926, 1927, 1928, 1930, 1933, 1934, 1936, 1937, 1939, 1943, 1944, 1951, 1954, 1955, 1956, 1964, 1965, 1969, 1970, 1971, 1972, 1973, 1974, 1977, 1978, 1982, 1984, 1986, 1991, 1996}
Cluster 3: {1868, 1929, 1938, 1963, 1993, 2002}
Cluster 4: {1865, 1867, 1869, 1870, 1872, 1876, 1879, 1882, 1885, 1887, 1890, 1896, 1897, 1911, 1913, 1916, 1921, 1931, 1948, 1952, 1957, 1958, 1961, 1967, 1976, 1979, 1981, 1987, 1988, 1990, 1992, 2006}

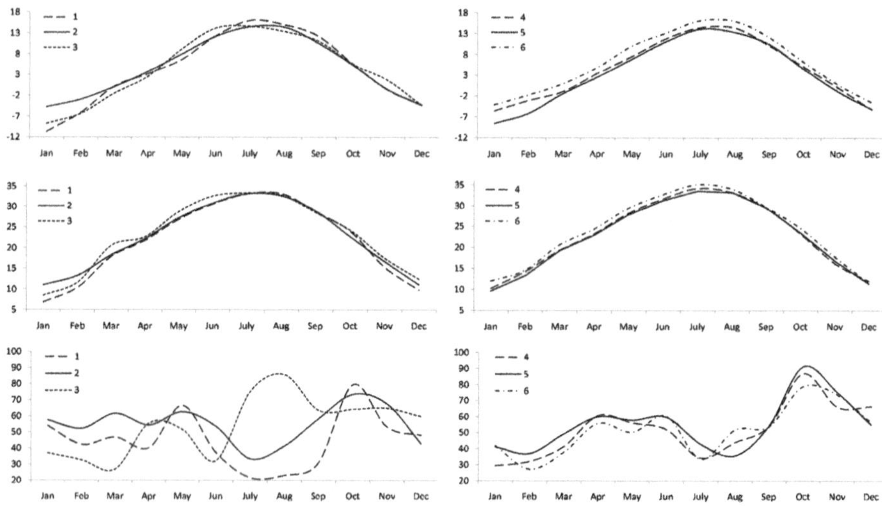

Fig. 7 Data set with $n = 148$, $T = 12$, and $p = 3$: group means in the partition in six clusters. Variable X_1 is reported in the *top*, variable X_2 in the *middle*, and variable X_3 in the *bottom*

Cluster 5: {1862, 1864, 1871, 1873, 1874, 1875, 1877, 1878, 1880, 1881, 1884, 1888, 1891, 1893, 1894, 1895, 1899, 1901, 1903, 1906, 1907, 1908, 1909, 1918, 1922, 1932, 1935, 1940, 1949, 1959, 1960, 1962, 1989}

Cluster 6: {1886, 1945, 1946, 1950, 1966, 1968, 1975, 1983, 1994, 1995, 1997, 1998, 1999, 2000, 2001, 2003, 2004, 2005, 2007, 2008}

Cluster means are reported in Fig. 7. Clusters 1 and 3 represent two small groups with anomalous years. Cluster 1 groups together former years (the most recent one is 1985), which are characterized by low maximum temperatures in quite all months, by a very dry summer season and dry months in the second part of autumn. This kind of climate is completely absent in the two last decades. A similar pattern characterizes group 5, in which are clustered several years from 1962 to 1989. In this group, the minimum temperatures are very low, the summer season is dry but the autumn months are extremely wet. Cluster 3 groups together 6 years (with the recent 2002) with a large amount of rain in the summer season and relatively dry spring months. The minimum and maximum temperatures in these years are in line with the average values. Cluster 6 groups many of the most recent years and the cluster means may be considered as representative of the actual climate situation. This group is characterized by high maximum and minimum temperatures and by a relatively large amount of rain in summer, in autumn, and in the beginning of the winter season. The time series of the group means (as long as the composition of the clusters) lead to the evidence that a climate change is present, at the beginning of the twentieth century. Both the minima and the maxima temperatures are higher, all over the years, and the seasonality in the rain is less evident, since the average amount of rain shows less variability across months.

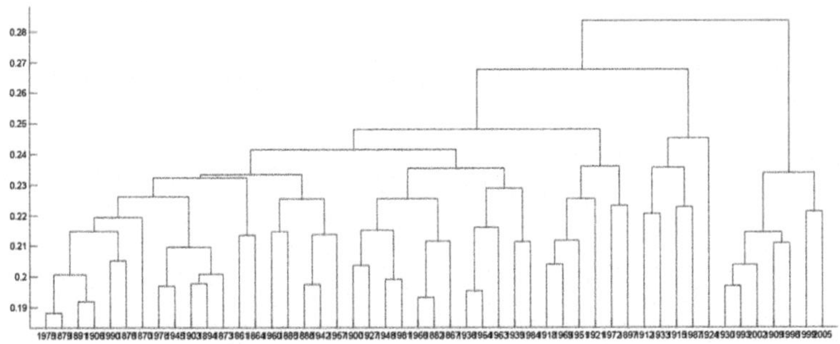

Fig. 8 Data with $n = 49$, $T = 12$, and $p = 3$: dendrogram obtained with the complete linkage

In order to gain insights into the climate change, we perform a second analysis, considering sequences of 3 years. Each series has 36 monthly values ($T = 36$) and $n = 49$. The first series is the triennium 1861–1863, the last series is the triennium 2005–2007. The label of each series is the second year (for example, for the first series the label is 1862 and for the last series the label is 2006). We consider triennium in order to allow a larger shift in the warping function and to allow the shift for the month of January (for the second and the third year) and for December (for the first and the second year). Indeed, considering series of 1 year, the warping in the winter months of January and December is not possible. We set $u = 3$ (the same length of a season). Figure 8 reports the dendrogram obtained with the complete linkage. Here again, the complete linkage seems less affected by the "chain effect" than the single and the average linkages and the tree shows the presence of well-separated clusters. We consider partitions in six and three groups. The cluster means of partition in six groups are shown in Fig. 9 and the group memberships are:

Cluster 1: {1861, 1900, 1912, 1915, 1918, 1921, 1924, 1930, 1933, 1936, 1939, 1951, 1954, 1960, 1963, 1969, 1972, 1975, 1978 }
Cluster 2: {1864, 1888, 1891, 1897, 1903, 1942, 1945, 1957, 1966, 1984, 1987, 1990}
Cluster 3: {1879}
Cluster 4: {1867, 1870, 1885, 1948}
Cluster 5: {1873, 1876, 1894, 1906, 1909, 1927}
Cluster 6: {1882, 1981, 1993, 1996, 1999, 2002, 2005}

This partition reveals the presence of an outlier, the triennium 1878–1880 in group 3, which is characterized by extreme (both very high and very low) values in the temperatures and in the rain. This group is merged with group 4 in partition in three clusters. Group 4 contains early years and is characterized by very low temperatures in winter and large amounts of rain in spring and autumn. Groups 1, 2, and 5, contain non-recent years and are merged together in partition in three clusters. The time series of the average values of these groups are smoother than the other series: the seasonality in the temperatures is more evident and the amount of rain

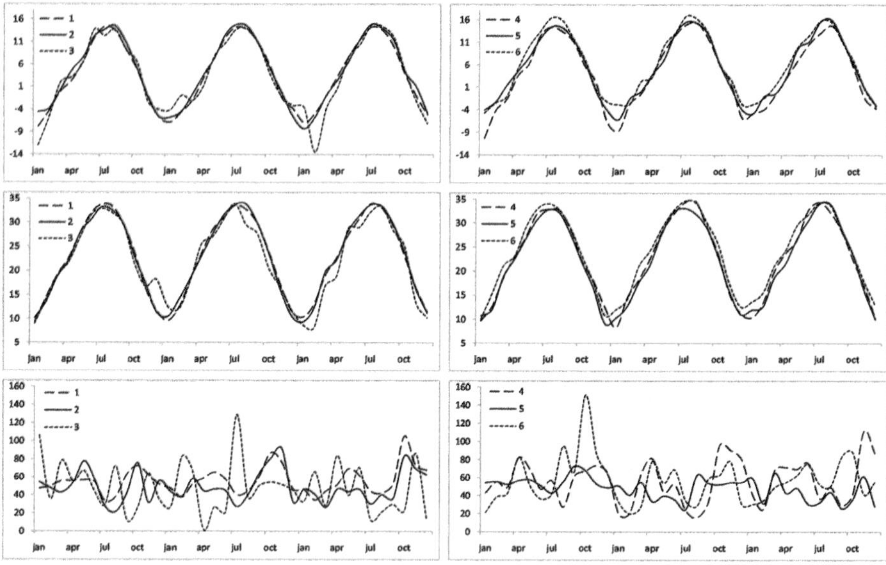

Fig. 9 Data set with $n = 49$, $T = 12$, and $p = 3$: group means in the partition in six clusters. Variable X_1 is reported in the *top*, variable X_2 in the *middle*, and variable X_3 in the *bottom*

across months presents less variability. Group 6 contains recent years. It remains a single group in partition in six clusters and it is not merged with other groups until the top level of the hierarchy. This feature gives evidence of the peculiarity of the years contained in the group. The average values of the temperatures (both the minima and the maxima) are higher than the values in the other groups. In particular, the minima temperatures are much higher than in the other groups. The amount or rain is greatly variable across months and shows anomalous peaks in the first year of the triennium. In general, the amount of rain is higher around April and October and the summer months are wetter than in other groups.

The climate change is more evident in this second analysis, since all recent years (after 1991) are clustered together. The group containing these years remains isolated until the last level of the dendrogram and the times series of the average values show peculiar patterns.

Acknowledgements This study was carried out under the research program PRIN 2010–2011 (grant 2010JHF437) funded by the Italian Ministry of Education, University, and Research.

References

1. Chu, S., Keogh, E., Hart, D., Pazzani, M.: Iterative deepening dynamic time warping. In: Second SIAM International Conference on Data Mining (2002)
2. Ferraty, F., Vieu, P.: Nonparametric Functional Data Analysis: Theory and Practice. Springer, New York (2006)

3. Keogh, E.: Exact indexing of dynamic time warping. In: 28th International Conference on Very Large Data Bases, Hong Kong, pp. 406–417 (2002)
4. Keogh, E., Pazzani, M.: Scaling up dynamic time warping for data mining applications. In: Proceedings of the 6th ACM SIGKDD International Conference on Knowledge Discovery & Data Mining, Boston, pp. 285–289, 20–23 August 2000
5. Keogh, E., Pazzani, M.: Dynamic time warping with higher order features. In: First SIAM International Conference on Data Mining (SDM'2001), Chicago (2001)
6. Milligan, G.W., Cooper, M.C.: A study of standardization of variables in cluster analysis. J. Classif. **5**, 181–204 (1988)
7. Ramsay, J.O.: Functional components of variation in handwriting. J. Am. Stat. Assoc. **95**, 9–15 (2000)
8. Ramsay, J.O., Dalzell, C.J.: Some tools for functional data analysis (with Discussion). J. Roy. Stat. Soc. Ser. B **53**, 539–572 (1991)
9. Ramsay, J.O., Li, X.: Curve registration. J. Roy. Stat. Soc. Ser. B **60**, 351–363 (1998)
10. Ramsay, J.O., Silverman, B.W.: Functional Data Analysis, 2nd edn. Springer, New York (2005)
11. Ratanamahatana, C.A., Keogh. E.: Making Time-series classification more accurate using learned constraints. In: Proceedings of SIAM International Conference on Data Mining (SDM '04), Lake Buena Vista, pp. 11–22, 22–24 April 2004
12. Silverman, B.: Incorporating parametric effects into functional principal components analysis. J. Roy. Stat. Soc. Ser. B **57**, 673–689 (1995)
13. Wang, K., Gasser, T.: Alignment of curves by dynamic time warping. Ann. Stat. **25**(3), 1251–1276 (1997)
14. Wang, K., Gasser, T.: Synchronizing sample curves nonparametrically. Ann. Stat. **27**(2), 439–460 (1999)

Asymmetric CLUster Analysis Based on SKEW-Symmetry: ACLUSKEW

Akinori Okada and Satoru Yokoyama

Abstract A procedure of cluster analysis to deal with asymmetric similarities is introduced, where the similarity from one object to the other object is not necessarily equal to the similarity from the latter to the former. The procedure analyzes one-mode two-way asymmetric similarities among objects to classify objects into clusters. Each cluster consists of a dominant (central) object and the other (noncentral) objects. The central object of a cluster represents the cluster and dominates the other objects in the cluster. In the present procedure, differences between two conjugate similarities (two times of skew-symmetries) are weighted by multiplying with the sum of the two corresponding similarities. Thus the larger the similarity between two objects is, the more prominently the difference is evaluated. The present procedure is applied to car switching data among car categories, and the result is compared with the result which was obtained by analyzing unweighted differences between two conjugate similarities. The comparison shows the weight is reasonable.

Keywords Asymmetry • Cluster analysis • Nonhierarchical • Similarity • Skew-symmetry

1 Introduction

Similarity relationships among objects are not always symmetric i.e., the similarity from an object to the other object is not always equal to the similarity from the latter to the former. While asymmetry sometimes can be important to understand

A. Okada (✉)
Graduate School of Management and Information Sciences, Tama University, 4-1-1 Hijirigaoka Tama-shi, Tokyo 206-0022, Japan
e-mail: okada@rikkyo.ac.jp

S. Yokoyama
Department of Business Administration Faculty of Economics, Teikyo University, 359 Otsuka Hachioji, Tokyo 192-0395, Japan
e-mail: satoru@main.teikyo-u.ac.jp

I. Morlini et al. (eds.), *Advances in Statistical Models for Data Analysis*, Studies in Classification, Data Analysis, and Knowledge Organization, DOI 10.1007/978-3-319-17377-1_20

similarity relationships among objects, most of researchers have ignored asymmetry in the analysis of asymmetric similarities. Some researchers have paid attention on asymmetry and have introduced several procedures to analyze and represent asymmetry in similarities.

Two sorts of procedures for analyzing asymmetric relationships have been developed. One is based on multidimensional scaling [2, Chap. 23] where asymmetric relationships among objects are represented geometrically in a multidimensional space. The other is based on cluster analysis where asymmetric relationships among objects are represented by a cluster structure. Most of procedures based on cluster analysis are agglomerative [4, 5, 11, 15] which have been extended from a seminal work of [7]. They focus the attention on which of two conjugate similarities is larger than the other or on which of the two skew-symmetries is positive, and which of them is negative. Akahori [1] introduced an agglomerative procedure for analyzing asymmetric similarities of time-series data. The procedure utilizes the same idea as stated above.

Olszewski [12, 13] presented nonhierarchical cluster analysis procedures for analyzing two-mode two-way data asymmetrically. Olszewski [12] introduced the asymmetric distance to represent the asymmetry, and Olszewski [13] introduced a coefficient to represent the asymmetry. Two procedures are based on the same principle in the clustering which is based on the distance from an object to a centroid of a cluster. Two conjugate similarities are not directly compared. Vicari [17] developed a nonhierarchical clustering procedure for analyzing one-mode two-way asymmetric similarities by using two different cluster structures (in the general model) to represent symmetric and skew-symmetric components of similarities, respectively. In the procedure, two conjugate similarities are not compared directly, but are approximated by the cluster structure.

A procedure to analyze asymmetric similarity named **A**symmetric **CLU**ster analysis based on **SKEW**-symmetry (ACLUSKEW) is introduced in the present study. ACLUSKEW focuses the attention on the difference between two conjugate similarities [17]. The similarity from a less dominant or salient object to a more dominant or salient object is larger than that from the more dominant or salient object to the less dominant or salient object [16]. An object is chosen as the dominant objects (central object) of a cluster. The dominant object represents the cluster to which it belongs. The dominant object of each cluster is chosen so that the sum of weighted differences between two conjugate similarities in the cluster (similarity from the non-dominant object to the dominant object) – (similarity from the dominant object to the non-dominant object) is maximized. The difference is weighted by multiplying with the sum of two corresponding similarities. While each cluster is represented by a dominant object in ACLUSKEW, the notion is comparable to that of [12, 13] where each cluster is represented by the centroid of the cluster.

2 The Procedure

ACLUSKEW can be regarded as an extension of k-means cluster analysis [8], and it deals with object $\times$ object or one-mode two-way similarities. Each cluster has its own dominant (central) object and the other nondominant (noncentral) objects which are dominated by the dominant object of the cluster [12, 13].

Let s_{ik} be the similarity from objects i to k, where s_{ik} is not necessarily equal to s_{ki}. Two differences between two conjugate similarities $(s_{ik} - s_{ki})$ and $(s_{ki} - s_{ik})$ represent two times of the skew-symmetries between objects i and k, respectively. The difference $(s_{ik} - s_{ki})$ is called the difference from objects i to k, and the difference $(s_{ki} - s_{ik})$ is called the difference from objects k to i hereafter. They have the same absolute value and have opposite signs. When $s_{ik} > s_{ki}$, object k dominates over object i, and when $s_{ik} < s_{ki}$, object i dominates over object k [16]. In the former case the difference from objects i to k is positive, and in the latter case the difference from objects i to k is negative. In the brand switching, this means that when two brands have asymmetries of the power of attracting consumers from the other brand each other, the more attractive brand can lure consumers from the less attractive brand.

Suppose that $(s_{ik} - s_{ki})$ is equal to $(s_{i\ell} - s_{\ell i})$, then objects k and ℓ equally dominate over object i. When $(s_{ik} + s_{ki})$ is larger than $(s_{i\ell} + s_{\ell i})$, the dominance of object k over object i can be more influential than the dominance of object ℓ can [3, 14]. This means that the larger the similarity between a more dominant or salient object and a less dominant or salient object is, the more significant the dominance relationship between two objects becomes. In ACLUSKEW the differences or two times of skew-symmetries between two conjugate similarities $(s_{ik} - s_{ki})$ or $(s_{ki} - s_{ik})$ are evaluated not by the values themselves but after multiplying with the sum of corresponding two similarities $(s_{ik} + s_{ki})$;

$$(s_{ik} - s_{ki}) \times (s_{ik} + s_{ki}).$$

Let N be the number of objects, and K be the number of clusters. The problem is to find K dominant objects and to classify each of the other $(N - K)$ nondominant objects into one of K clusters. A nondominant object i is assigned to the cluster represented by object k, which satisfies

$$\max_{k=1,\cdots,K}(s_{ik} - s_{ki}) \times (s_{ik} + s_{ki}).$$

The term $(s_{ik} - s_{ki})$ shows the difference from objects i to k. $(s_{ik} - s_{ki}) \times (s_{ik} + s_{ki})$ shows that the larger the $(s_{ik} + s_{ki})$ is, the more the difference $(s_{ik} - s_{ki})$ is weighted. This means that object k dominates over object i more than object ℓ does, when object i is more similar to object k than to object ℓ, even if $(s_{ik} - s_{ki}) = (s_{i\ell} - s_{\ell i})$.

The purpose of ACLUSKEW is to find K clusters which maximize the goodness of fit (GOF):

$$\text{GOF} = \sum_{k=1}^{K} \sum_{\substack{i \in cluster\, k \\ i \neq k}}^{N_k} signum(s_{ik} - s_{ki})[(s_{ik} - s_{ki})(s_{ik} + s_{ki})]^2, \tag{1}$$

for a given number of clusters K, where N_k is the number of objects in cluster k, $signum(s_{ik} - s_{ki}) = 1$ when $(s_{ik} - s_{ki}) > 0$, $signum(s_{ik} - s_{ki}) = 0$ when $(s_{ik} - s_{ki}) = 0$, and $signum(s_{ik} - s_{ki}) = -1$ when $(s_{ik} - s_{ki}) < 0$.

The method of finding K clusters is:

1. Determine the number of clusters K
2. Choose all combinations $({}_N C_K)$ of K objects from N objects as dominant objects
3. For each of ${}_N C_K$ combinations, assign each of $(N - K)$ objects to the cluster where $(s_{ik} - s_{ki}) \times (s_{ik} + s_{ki})$ is largest
4. Find clusters which give the largest GOF among ${}_N C_K$ results of K clusters
5. Repeat steps 1–4 using different values of K
6. Choose the solution (or determine the number of clusters) from the results obtained at step 4, where each of the results gives the largest GOF among the results of K clusters, based on the interpretation and GOFs of the results

3 An Application

ACLUSKEW is applied to car switching data among 16 car categories [6]. The data are represented by a 16×16 table. The (i, j) element of the table shows the number of cars corresponding to car category i which was traded in to purchase cars in car category j. The table was rescaled by multiplying with a rescaling constant to the row and the column so that the sum of row plus column elements is equal over all 16 sums [9, 10]. Namely, row i and column i are multiplied with a scaling constant c_i so that in the resulting (rescaled) table, sum of row i elements plus sum of column i elements is equal to the mean of the sum of row elements plus sum of column elements of the original (unscaled) table. The 16 car categories and the abbreviation of each car category are shown in Table 1. Table 2 shows rescaled data.

The rescaled car switching data were analyzed by ACLUSKEW for the number of clusters K=1 through 5. Obtained GOFs for K=1 through 5 are 276.6×10^{15}, 399.0×10^{15}, 425.0×10^{15}, 441.3×10^{15}, and 449.9×10^{15}, respectively. Results for the number of clusters K=2, 3, and 4 are represented in Table 3. The three-cluster result ($K = 3$) was adopted as the solution. The reason for choosing the three cluster

Table 1 Sixteen car categories

	Car category	Abbreviation	Domestic/captive import/import
1	Subcompact domestic	SUBD	Domestic
2	Subcompact captive imports	SUBC	Captive import
3	Subcompact imports	SUBI	Import
4	Small specialty domestic	SMAD	Domestic
5	Small specialty captive imports	SMAC	Captive import
6	Small specialty imports	SMAI	Import
7	Low price compact	COML	Domestic
8	Medium price compact	COMM	Domestic
9	Import compact	COMI	Import
10	Midsize domestic	MIDD	Domestic
11	Midsize imports	MIDI	Import
12	Midsize specialty	MIDS	Domestic
13	Low price standard	STDL	Domestic
14	Medium price standard	STDM	Domestic
15	Luxury domestic	LUXD	Domestic
16	Luxury imports	LUXI	Import

result as the solution is twofold; one is the clarity of the interpretation and the other is increment of GOF from K=2 to 3 and that from K=3 to 4. The dominant objects SUBD in cluster 1, SUBI in cluster 2, and MIDS in cluster 3 seem to be the smallest and the least expensive car categories in each cluster, suggesting that smaller or less expensive car categories are growing in each cluster. Cluster 1 consists of domestic car categories which seem smaller or less expensive than those in cluster 3. Cluster 2 consists mainly of import and captive import car categories, while one domestic car category (LUXD) is in Cluster 2. Cluster 3 consists of mainly domestic car categories, but has one import car category (SMAI). Three specialty car categories (SMAD, SMAI, and MIDS) out of four are in cluster 3.

4 Discussion

A procedure of nonhierarchical cluster analysis of one-mode two-way asymmetric similarities, named ACLUSKEW, was introduced and was applied to car switching data successfully. The three-cluster solution was obtained. From Table 3, when $K = 2$, SUBI and MIDS are dominant objects of clusters 1 and 2, respectively. Cluster 1 consists of a great variety of domestic, captive import, and import car categories ranging from subcompact to luxury car categories. Four of five import car categories and two captive imports (SUBC and SMAC) are included in cluster 1. Cluster 2 consists mainly of domestic car categories ranging from subcompact to standard car categories. It includes three (MIDS, SMAD, and SMAI) of four

Table 2 Rescaled data

To																
From	1	2	3	4	5	6	7	8	9	10	11	12	13	14	15	16
1 SUBD	14049	5644	6741	10639	4367	4589	8854	5453	1724	4646	1326	7365	2223	1178	708	465
2 SUBC	12350	26581	12163	9353	12888	6855	4323	7547	4436	3869	3864	5706	3043	1687	1153	852
3 SUBI	7282	4899	17736	5838	4454	11362	2487	1932	7593	2033	3830	4003	1283	851	849	1122
4 SUMD	6576	4197	6640	19974	5706	8959	4023	2909	2733	2904	3010	10001	1570	2039	2846	1394
5 SUMC	4189	3362	0	9675	52637	0	0	9714	0	5977	0	666	0	0	0	0
6 SMAI	3506	2700	7621	6340	4674	34045	2616	1378	7654	1968	9666	7299	579	1468	3202	5513
7 COML	13222	8411	9266	10136	6880	3821	23107	9859	3138	9140	1853	8219	6039	2131	806	740
8 COMM	13910	5850	8340	7934	7930	2685	9924	22308	3969	7875	2666	9763	4187	6077	1267	693
9 COMI	8267	9568	23486	5440	4671	10609	4904	4448	25470	5281	14516	2340	1244	2581	2122	2439
10 MIDD	12150	5375	8621	10116	5977	5058	9178	9951	4886	11893	3596	16871	8925	10275	4459	1577
11 MIDI	3563	1975	8334	3176	2034	6974	4932	2771	8161	6418	38485	4853	1406	2387	2965	9953
12 MIDS	5855	2782	3967	7913	6460	5092	1978	1754	2210	3177	2706	18882	4280	6939	6987	1582
13 STDL	14540	6211	7214	7757	2624	2267	11751	6785	3238	9027	2238	14600	30762	12428	4252	952
14 STDM	8594	4033	5941	5283	3008	3033	4563	6829	3355	6345	2378	11209	6591	33466	13488	1653
15 LUXD	2783	2016	3684	4091	661	1520	1463	1731	1860	1395	2822	4775	1914	5639	58489	7173
16 LUXI	385	921	3884	2050	0	4096	326	448	3380	337	9852	2075	501	2280	5585	69479

This table is reproduced from Table 2 of [9]

Table 3 Clusters when the number of clusters is 2, 3, and 4

Cluster	Dominant car category	Dominated car category
Number of clusters=2		
1	SUBI	SUBC, SMAC, COML, COMI, MIDI, LUXD, LUXI
2	MIDS	SUBD, SMAD, SMAI, COMM, MIDD, STDL, STDM
Number of clusters=3		
1	SUBD	SMAC, COML, COMM, STDL
2	SUBI	SUBC, COMI, MIDI, LUXD, LUXI
3	MIDS	SMAD, SMAI, MIDD, STDM
Number of clusters=4		
1	SUBD	COML, COMM, STDL
2	SUBI	SUBC, COMI, MIDI, LUXI
3	MIDS	SMAD, SMAI, MIDD
4	LUXD	SMAC, STDM

specialty car categories, while it does not include SMAC. When $K = 4$, SUBD, SUBI, MIDS, and LUXD are dominant objects of clusters 1 through 4, respectively. Cluster 1 consists of less expensive domestic car categories. Cluster 2 consists of all import car categories (but SMAI) and a captive import car category (SUBC). Cluster 3 consists of the three of four specialty car categories (but SMAC) and MIDD. Cluster 4 comprises of LUXD, SMAC, and STDM each of which belongs to clusters 2, 1, and 3, respectively, of the results obtained when $K = 3$. It seems that results of $K = 2$ and of $K = 4$ are more difficult to interpret the meaning of clusters than to interpret that of the solution ($K = 3$). This validates the three-cluster solution as well.

The solution seems to be compatible with earlier studies. Three dominant objects have the smallest, second, and fourth smallest radii obtained in the asymmetric multidimensional scaling analysis [10]. This suggests the three car categories are dominant in the car switching, because in the model of [10], the smaller the radius of an object is, the more dominant the object is. Comparing the present solution with the result of [10], it seems that (a) cluster 1 corresponds with Dimension 1, (b) cluster 2 corresponds with Dimension 2, and (c) cluster 3 corresponds with Dimension 3 of the analysis. Three clusters are compatible with the (unconstrained) configuration of [18] as well. SUBI and SUBD seem to be dominant categories in the configuration, because they are located in the lower right part of the configuration which the slide vector points.

In ACLUSKEW each cluster is represented by a dominant object, while in [12, 13] each cluster is represented by the centroid of the objects in the cluster. This is the same idea of *k*-means cluster analysis. Olszewski [12, 13] maximizes the sum of the squares of the distance between the centroid and the objects in the cluster, suggesting objects in a same cluster are similar to each other. On the other hand, ACLUSKEW maximizes GOF defined by Eq. (1). This means that ACLUSKEW pays attention to the differences from nondominant objects to a dominant object

Table 4 Three clusters of the solution obtained by analyzing unweighted data

Cluster	Dominant car category	Dominated car category
1	SUBD	SMAC, COMM, STDL, STDM
2	SUBI	SUBC, COML, COMI, MIDI, LUXD, LUXI
3	MIDS	SMAD, SMAI, MIDD

and pays no attention to similarities among nondominant clusters. Thus the feature of clusters derived by ACLUSKEW is different from that derived by Olszewski [12, 13]. Other procedures for analyzing asymmetric similarities or dissimilarities, referred to in Sect. 1, also give the clusters which consist of similar objects.

ACLUSKEW is characterized by that the difference of two conjugate similarities or two times of the skew-symmetry is multiplied with the sum of two corresponding similarities. The unweighted differences of conjugate similarities which are not multiplied with the sum of two similarities were analyzed by using the same algorithm of ACLUSKEW. The analysis resulted in GOFs 763.0, 862.2, and 918.6 for the number of clusters 2, 3, and 4, respectively. Obtained GOFs suggest to choose three-cluster result as the solution, which is shown in Table 4.

Comparing the solution ($K = 3$) shown in Table 3 and that shown in Table 4, we can see that both solutions have same dominant objects. But there are some disparities between the two; STDM in cluster 3 of Table 3 is in cluster 1 of Table 4, and COML in cluster 1 of Table 3 is in cluster 2 of Table 4. STDM seems more expensive than the other car categories in cluster 1 of Table 4, and it is not reasonable that STDM is in cluster 1 of Table 4. COML is neither an import nor a captive import car category, and it is not natural that COML is in cluster 2 of Table 4. This suggests that the three clusters of Table 3 are more reasonable than those of Table 4 are. It seems that multiplying the differences $(s_{ik} - s_{ki})$ and $(s_{ki} - s_{ik})$ with the sum of two corresponding similarities $(s_{ik} + s_{ki})$ is rational in finding clusters of the car switching data.

The GOF of Eq. (1) can be generalized as

$$\mathrm{GOF}_1 = \sum_{k=1}^{K} \sum_{\substack{i \in cluster\, k \\ i \neq k}}^{N_k} signum(s_{ik} - s_{ki})(s_{ik} - s_{ki})^p (s_{ik} + s_{ki})^q. \qquad (2)$$

When $p = q = 2$, Eq. (2) is equivalent to Eq. (1). When $p = 2$ and $q = 0$, $(s_{ik} - s_{ki})$ and $(s_{ki} - s_{ik})$ are not multiplied with $(s_{ik} + s_{ki})$ or unweighted at all.

Acknowledgements The authors would like to express their gratitude for suggestions on the brand switching given by Prof. Dr. Daniel Baier and by Prof. Dr. Hiroyuki Tsurumi. They also wish to appreciate an anonymous referee who gave them valuable and constructive comments to the earlier version of the manuscript.

References

1. Akahori, K.: Jugyou no kategori bunseki ni okeru keiretsu no chushutsu [Abstraction of behavior sequence on categorical analysis of instruction]. Kodo Keiryogaku [in Japanese] **15**(2), 1–8 (1988)
2. Borg, I., Groenen, P.J.F.: Modern Multidimensional Scaling: Theory and Applications, 2nd edn. Springer, New York (2005)
3. Feinberg, F.M., Kahn, B.E., McAlister, L.: Market share response when customers seek variety. J. Mar. Res. **29**, 227–237 (1992)
4. Fujiwara, H.: Hitaisho sokudo to toshitsuseikeisu o michiita kurasuta bunsekiho [Methods for cluster analysis using asymmetric measures and homogeneity coefficient]. Kodo Keiryogaku [in Japanese] **7**(2), 12–21 (1980)
5. Fujiwara, H., Ozaki, H.: Hitaisho ruijido ni yoru kurasuta bunseki [Cluster analysis using asymmetric similarity measures]. Denshi Tsushin Gakkai Giho [in Japanese] **ET79-9**, 1–6 (1979)
6. Harshman, R.A., Green, P.E., Wind, Y., Lundy, M.E. : A model for the analysis of asymmetric data in marketing research. Market. Sci. **1**, 205–242 (1982)
7. Hubert, L.: Min and max hierarchical clustering using asymmetric similarity. Psychometrika **38**, 63–72 (1973)
8. MacQueen, J.B.: Some methods for classification and analysis of multivariate observations. In: Proceeding of the 5th Berkeley Symposium on Mathematical Statistics and Probability, vol.2, pp. 281–297. University of California Press, Berkeley (1967)
9. Okada, A.: Asymmetric multidimensional scaling of car switching data. In: Gaul, W., Schader, M.B. (eds.) Data, Expert Knowledge and Decisions, pp. 279–220. Springer, Berlin (1988)
10. Okada, A., Imaizumi, T.: Nonmetric multidimensional scaling of symmetric proximities. Behaviormetrika **21**, 81–96 (1987)
11. Okada, A., Iwamoto, T.: University enrollment flow among the Japanese prefectures: a comparison before and after the joint first stage achievement test by asymmetric cluster analysis. Behaviormetrika **23**, 169–185 (1996)
12. Olszewski, D.: Asymmetric *k*-means algorithm. In: Dobnikar, A., Lotrič, U., Ster, B. (eds.) International Conference on Adaptive and Natural Computing Algorithms (ICANNGA 2011), Part II. Lecture Notes in Computer Science, vol. 6594, pp. 1–10. Springer, Heidelberg (2011)
13. Olszewski, D.: *k*-means clustering of asymmetric data. In: Corchado, E., Snášel, V., Abraham, A., Woźniak, M., Graña, M., Cho, S.-B. (eds.) Hybrid Artificial Intelligent Systems, Part II. Lecture Notes in Computer Science, vol. 7208, pp. 243–254. Springer, Heidelberg (2012)
14. Patterson, P.G., Smith, T.: A cross-cultural study of switching barriers and propensity to stay with service providers. J. Retail. **79**, 107–120 (2003)
15. Takeuchi, A., Saito, T., Yadohisa, H.: Asymmetric agglomerative hierarchical clustering algorithms and evaluations. J. Classif. **24**, 123–143 (2007)
16. Tversky, A.: Features of similarity. Psychol. Rev. **84**, 327–352 (1997)
17. Vicari, D.: Partitioning asymmetric dissimilarity data. In: Book of Short Papers of jcs-cladag 2012, Analysis of Modeling of Complex Data in Behavioural and Social Sciences. Retrieved 10 May 2013, from www.jcs-cladag12.tk (2012)
18. Zielman, B., Heiser, W.J.: Analysis of asymmetry by a slide-vector. Psychometrika **58**, 101–114 (1993)

Parsimonious Generalized Linear Gaussian Cluster-Weighted Models

Antonio Punzo and Salvatore Ingrassia

Abstract Mixtures with random covariates are statistical models which can be applied for clustering and for density estimation of a random vector composed by a response variable and a set of covariates. In this class, the generalized linear Gaussian cluster-weighted model (GLGCWM) assumes, in each mixture component, an exponential family distribution for the response variable and a multivariate Gaussian distribution for the vector of real-valued covariates. For parsimony sake, a family of fourteen models is here introduced by applying some constraints on the eigen-decomposed covariance matrices of the Gaussian distribution. The EM algorithm is described to find maximum likelihood estimates of the parameters for these models. This novel family of models is finally applied to a real data set where a good classification performance is obtained, especially when compared with other well-established mixture-based approaches.

Keywords Cluster-weighted models • Eigen decomposition • Generalized linear models • Model-based clustering • Parsimonious mixtures

1 Introduction

Let $\left(Y, \boldsymbol{X}'\right)'$ be a random vector where Y is the response variable and $\boldsymbol{X}$ is the p-variate vector of real-valued random covariates. Moreover, let $p\left(y, \boldsymbol{x}\right)$ be the joint density of $\left(Y, \boldsymbol{X}'\right)'$. A flexible framework for density estimation and for clustering of data $\left\{\left(y_i, \boldsymbol{x}_i'\right)'\right\}_{i=1}^{n}$ from $\left(Y, \boldsymbol{X}'\right)'$ is represented by the family of mixture models with random covariates. With respect to classical mixture models with fixed covariates, they have the advantage to allow the covariates to affect the cluster structure (the so-called *assignment dependence* property; see [10], for details).

An eminent member of the family of mixtures with random covariates is the cluster-weighted model (CWM; [7]), also called saturated mixture regression

---

A. Punzo (✉) • S. Ingrassia
Department of Economics and Business, University of Catania, 95129 Catania, Italy
e-mail: antonio.punzo@unict.it; s.ingrassia@unict.it

I. Morlini et al. (eds.), *Advances in Statistical Models for Data Analysis*,
Studies in Classification, Data Analysis, and Knowledge Organization,
DOI 10.1007/978-3-319-17377-1_21

model in [24]. The CWM principle consists in factorizing $p(y, \boldsymbol{x})$, in each mixture component, into the product between the conditional density of $Y|\boldsymbol{X} = \boldsymbol{x}$ and the marginal density of $\boldsymbol{X}$ by assuming a (parametric) functional relation for the expectation of Y on $\boldsymbol{x}$. Some recent literature about this model can be found in [11–13, 15–18, 22, 23]. In particular, as a special case of their model, Ingrassia et al. [13] propose the linear Gaussian CWM which adopts a generalized linear model for the relationship of Y on $\boldsymbol{x}$ in each mixture component. This implies the possibility to model, for example, a count response via a Poisson distribution and a dichotomous response by a Bernoulli distribution.

However, when p increases, the number of parameters to be estimated in this model increases too, especially due to the contribute of the covariance matrices of the component Gaussian distributions. To make the approach parsimonious, in line with [4], a family of fourteen models is introduced in Sect. 2 by applying some constraints on the eigen-decomposed component covariance matrices. The EM algorithm is illustrated in Sect. 3 for maximum likelihood parameter estimation for the members of the family. An application to real data is presented in Sect. 4 where a good classification performance is obtained, especially when compared with well-established mixture-based approaches.

2 Parsimonious Generalized Linear Gaussian CWMs

The generalized linear Gaussian cluster-weighted model (GLGCWM; [13]) is a finite mixture model, with k components, of equation

$$p(y, \boldsymbol{x}; \boldsymbol{\Psi}) = \sum_{j=1}^{k} p\left(y|\boldsymbol{x}; \theta_j, \zeta_j\right) \phi\left(\boldsymbol{x}; \boldsymbol{\mu}_j, \boldsymbol{\Sigma}_j\right) \pi_j, \tag{1}$$

where π_j is the weight of the jth component, with $\pi_j > 0$ and $\sum_{j=1}^{k} \pi_j = 1$,

$$\phi\left(\boldsymbol{x}; \boldsymbol{\mu}_j, \boldsymbol{\Sigma}_j\right) = (2\pi)^{-\frac{p}{2}} \left|\boldsymbol{\Sigma}_j\right|^{-\frac{1}{2}} \exp\left\{-\frac{1}{2}\left(\boldsymbol{x} - \boldsymbol{\mu}_j\right)' \boldsymbol{\Sigma}_j^{-1} \left(\boldsymbol{x} - \boldsymbol{\mu}_j\right)\right\}$$

is the density of a p-variate Gaussian random vector with mean $\boldsymbol{\mu}_j$ and covariance matrix $\boldsymbol{\Sigma}_j$

$$p\left(y|\boldsymbol{x}; \theta_j, \zeta_j\right) = \exp\left\{\frac{y\theta_j - b\left(\theta_j\right)}{a\left(\zeta_j\right)} + c\left(y; \zeta_j\right)\right\}, \tag{2}$$

for specific functions $a(\cdot)$, $b(\cdot)$, and $c(\cdot)$, is an exponential family distribution with parameter θ_j (and, possibly, ζ_j). In (1), $\boldsymbol{\Psi}$ contains all of the parameters of the mixture. It is well known that the exponential family model in (2) is strictly related to the generalized linear models. Here, a monotone and differentiable link function $g(\cdot)$ is introduced which relates the expected value μ_j, of the response $Y|j$, to the covariates $\boldsymbol{X}$ through the relation

$$g(\mu_j) = \eta_j = \beta_{0j} + \boldsymbol{\beta}'_{1j}\boldsymbol{x}.$$

Since the interest is now in the parameters $\left(\beta_{0j}, \boldsymbol{\beta}'_{1j}\right)' = \boldsymbol{\beta}_j$, the distribution of $Y|\boldsymbol{x}, j$ will be denoted by $p\left(y|\boldsymbol{x}; \boldsymbol{\beta}_j, \zeta_j\right)$.

Because there are $p(p+1)/2$ free parameters for each $\boldsymbol{\Sigma}_j$, it is usually necessary to introduce parsimony into the general model (1) for real applications. To this end, we consider the eigen decomposition

$$\boldsymbol{\Sigma}_j = \lambda_j \boldsymbol{\Gamma}_j \boldsymbol{\Delta}_j \boldsymbol{\Gamma}'_j, \qquad j = 1, \ldots, k, \tag{3}$$

where $\lambda_j = \left|\boldsymbol{\Sigma}_j\right|^{1/p}$, $\boldsymbol{\Delta}_j$ is the scaled ($\left|\boldsymbol{\Delta}_j\right| = 1$) diagonal matrix of the eigenvalues of $\boldsymbol{\Sigma}_j$ sorted in decreasing order, and $\boldsymbol{\Gamma}_j$ is a $p \times p$ orthogonal matrix whose columns are the normalized eigenvectors of $\boldsymbol{\Sigma}_j$, ordered according to their eigenvalues (see [4]). Each component in the right side of (3) has also a different geometric interpretation: λ_j determines the volume of the cluster, $\boldsymbol{\Delta}_j$ its shape, and $\boldsymbol{\Gamma}_j$ its orientation. The constraints we pose on the three components of (3) generate the family of fourteen parsimonious GLGCWMs models summarized in Table 1.

3 Maximum Likelihood Estimation: The EM Algorithm

The EM algorithm [5] can be used to find maximum likelihood (ML) estimates for the parameters. Once k is assigned, it basically takes into account the complete-data log-likelihood

$$l_c(\boldsymbol{\Psi}) = \sum_{i=1}^{n}\sum_{j=1}^{k} z_{ij} \ln\left[p\left(y_i|\boldsymbol{x}_i; \boldsymbol{\beta}_j, \zeta_j\right)\right] + \sum_{i=1}^{n}\sum_{j=1}^{k} z_{ij} \ln\left[\phi\left(\boldsymbol{x}_i; \boldsymbol{\mu}_j, \boldsymbol{\Sigma}_j\right)\right] + \sum_{i=1}^{n}\sum_{j=1}^{k} z_{ij} \ln\left(\pi_j\right), \tag{4}$$

where $z_{ij} = 1$ if $\left(y_i, \boldsymbol{x}'_i\right)'$ comes from component j and $z_{ij} = 0$ otherwise. For the most general VVV-GLGCWM, E and M steps can be detailed as follows.

The E-step, on the $(q+1)$th iteration, $q = 0, 1, \ldots$, simply requires the calculation of the current conditional expectation of Z_{ij} given the observed sample, where Z_{ij} is the random variable corresponding to z_{ij}. In particular, for $i = 1, \ldots, n$

Table 1 Nomenclature, covariance structure, type of Maximum Likelihood (ML) solution in the M-step of the EM algorithm (CF="closed form" and IP="iterative procedure"), and number of free covariance parameters for each parsimonious GLGCWM

Family	Model	Volume	Shape	Orientation	$\boldsymbol{\Sigma}_j$	ML	Free covariance parameters
Spherical	EII	Equal	Spherical	–	$\lambda \boldsymbol{I}$	CF	1
	VII	Variable	Spherical	–	$\lambda_j \boldsymbol{I}$	CF	k
Diagonal	EEI	Equal	Equal	Axis-aligned	$\lambda \boldsymbol{\Delta}$	CF	p
	VEI	Variable	Equal	Axis-aligned	$\lambda_j \boldsymbol{\Delta}$	IP	$k+p-1$
	EVI	Equal	Variable	Axis-aligned	$\lambda \boldsymbol{\Delta}_j$	CF	$1+k(p-1)$
	VVI	Variable	Variable	Axis-aligned	$\lambda_j \boldsymbol{\Delta}_j$	CF	kp
General	EEE	Equal	Equal	Equal	$\lambda \boldsymbol{\Gamma} \boldsymbol{\Delta} \boldsymbol{\Gamma}'$	CF	$p(p+1)/2$
	VEE	Variable	Equal	Equal	$\lambda_j \boldsymbol{\Gamma} \boldsymbol{\Delta} \boldsymbol{\Gamma}'$	IP	$k+p-1+p(p-1)/2$
	EVE	Equal	Variable	Equal	$\lambda \boldsymbol{\Gamma} \boldsymbol{\Delta}_j \boldsymbol{\Gamma}'$	IP	$1+k(p-1)+p(p-1)/2$
	EEV	Equal	Equal	Variable	$\lambda \boldsymbol{\Gamma}_j \boldsymbol{\Delta} \boldsymbol{\Gamma}_j'$	CF	$p+kp(p-1)/2$
	VVE	Variable	Variable	Equal	$\lambda_j \boldsymbol{\Gamma} \boldsymbol{\Delta}_j \boldsymbol{\Gamma}'$	IP	$kp+p(p-1)/2$
	VEV	Variable	Equal	Variable	$\lambda_j \boldsymbol{\Gamma}_j \boldsymbol{\Delta} \boldsymbol{\Gamma}_j'$	IP	$k+p-1+kp(p-1)/2$
	EVV	Equal	Variable	Variable	$\lambda \boldsymbol{\Gamma}_j \boldsymbol{\Delta}_j \boldsymbol{\Gamma}_j'$	CF	$1+k(p-1)+kp(p-1)/2$
	VVV	Variable	Variable	Variable	$\lambda_j \boldsymbol{\Gamma}_j \boldsymbol{\Delta}_j \boldsymbol{\Gamma}_j'$	CF	$kp(p+1)/2$

and $j = 1, \ldots, k$, it follows that

$$\tau_{ij}^{(q+1)} := E_{\boldsymbol{\Psi}^{(q)}}\left[Z_{ij}\left|\left(y_i, \boldsymbol{x}_i'\right)'\right.\right] = \frac{\pi_j^{(q)} p\left(y_i|\boldsymbol{x}_i; \boldsymbol{\beta}_j^{(q)}, \zeta_j^{(q)}\right) \phi\left(\boldsymbol{x}_i; \boldsymbol{\mu}_j^{(q)}, \boldsymbol{\Sigma}_j^{(q)}\right)}{p\left(y_i, \boldsymbol{x}_i; \boldsymbol{\Psi}^{(q)}\right)}, \tag{5}$$

which is the posterior probability that the unlabeled observation $\left(y_i, \boldsymbol{x}_i'\right)'$ belongs to the jth component, using the current fit $\boldsymbol{\Psi}^{(q)}$ for $\boldsymbol{\Psi}$.

In the M-step, on the $(q+1)$th iteration, $q = 0, 1, \ldots$, the values z_{ij} in (4) are simply replaced by their current expectations $\tau_{ij}^{(q+1)}$ obtained in (5). This leads to

$$E\left[l_c\left(\boldsymbol{\Psi}\right)\right] = \sum_{i=1}^{n}\sum_{j=1}^{k} \tau_{ij}^{(q+1)} \ln\left[p\left(y_i|\boldsymbol{x}_i; \boldsymbol{\beta}_j, \zeta_j\right)\right]$$
$$+ \sum_{i=1}^{n}\sum_{j=1}^{k} \tau_{ij}^{(q+1)} \ln\left[\phi\left(\boldsymbol{x}_i; \boldsymbol{\mu}_j, \boldsymbol{\Sigma}_j\right)\right] + \sum_{i=1}^{n}\sum_{j=1}^{k} \tau_{ij}^{(q+1)} \ln\left(\pi_j\right). \tag{6}$$

Because the three terms on the right have zero cross-derivatives, they can be maximized separately. Let us set $\boldsymbol{\pi} = \left\{\pi_j\right\}_{j=1}^{k}$, $\boldsymbol{\beta} = \left\{\boldsymbol{\beta}_j\right\}_{j=1}^{k}$, and $\boldsymbol{\zeta} = \left\{\zeta_j\right\}_{j=1}^{k}$.

The maximum of Eq. (6) with respect to $\boldsymbol{\pi}$, subject to the constraints on those parameters, yields

$$\pi_j^{(q+1)} = \sum_{i=1}^{n} \tau_{ij}^{(q+1)} \Big/ n.$$

Maximization of (6) with respect to $\boldsymbol{\beta}$ (and possibly to $\boldsymbol{\zeta}$) is equivalent to independently maximize each of the k expressions

$$l_c^{(Y,j)} = \sum_{i=1}^{n} \tau_{ij}^{(q+1)} \ln\left[p\left(y_i|\boldsymbol{x}_i; \boldsymbol{\beta}_j, \zeta_j\right)\right]. \tag{7}$$

The maximization of (7) is equivalent to the maximization problem of the generalized linear model (for the complete data), except that each observation $\left(y_i, \boldsymbol{x}_i'\right)'$ contributes to the log-likelihood for each component with a known weight $\tau_{ij}^{(q+1)}$, which is obtained in the preceding E-step. Maximization of (7), with respect to $\boldsymbol{\beta}_j$ (and, possibly, to ζ_j), can be carried out numerically; details can be found in [25, pp. 120–124].

For model VVV, maximization of (6) with respect to $\boldsymbol{\mu}_j$ and $\boldsymbol{\Sigma}_j, j = 1, \ldots, k$, is equivalent to independently maximize each of the k expressions

$$l_c^{(X,j)} = \sum_{i=1}^{n} \tau_{ij}^{(q+1)} \ln \left[\phi \left(\boldsymbol{x}_i; \boldsymbol{\mu}_j, \boldsymbol{\Sigma}_j \right) \right].$$

In particular we obtain

$$\boldsymbol{\mu}_j^{(q+1)} = \frac{1}{n_j^{(q+1)}} \sum_{i=1}^{n} \tau_{ij}^{(q+1)} \boldsymbol{x}_i$$

and

$$\boldsymbol{\Sigma}_j^{(q+1)} = \frac{1}{n_j^{(q+1)}} \sum_{i=1}^{n} \tau_{ij}^{(q+1)} \left(\boldsymbol{x}_i - \boldsymbol{\mu}_j^{(q+1)} \right) \left(\boldsymbol{x}_i - \boldsymbol{\mu}_j^{(q+1)} \right)',$$

where $n_j^{(q+1)} = \sum_{i=1}^{n} \tau_{ij}^{(q+1)}$.

The EM algorithm for the other parsimonious GLGCWMs changes only in the way the terms of the decomposition of $\boldsymbol{\Sigma}_j$ are obtained in the M-step. These updates are analogous to those given in [4], and we defer the reader to this work for details.

To fit the proposed models, we adopt the **flexCWM** package [14] for the R computing environment [20]. Among the possible initialization strategies (see, e.g. [3]), a random (hard) initialization of $z_i = (z_{i1}, \ldots, z_{ik})'$, $i = 1, \ldots, n$, was considered. With regard to the stopping rule of the algorithm, the well-known Aitken acceleration method [2] was used.

4 Real Data Analysis: The `f.voles` Data Set

The `f.voles` data set, detailed in [6, Table 5.3.7] and available in the **Flury** package for R, consists of measurements of female voles from two species, *Microtus californicus* and *Microtus ochrogaster*. The data consist of 86 observations for which we have a binary variable Species denoting the species (45 *M. ochrogaster* and 41 *M. californicus*), a variable Age measured in days, and further six variables related to skull measurements. The names of the variables are the same as in the original analysis of this data set by Airoldi, Hoffmann [1]: L_2 = condylo-incisive length, L_9 = length of incisive foramen, L_7 = alveolar length of upper molar tooth row, B_3 = zygomatic width, B_4 = interorbital width, and H_1 = skull height. All of the variables related to the skull are measured in units of 0.1 mm.

The purpose of [1] was to study age variation in *M. californicus* and *M. ochrogaster* and to predict age on the basis of the skull measurements. For our purpose, we assume that data are unlabelled with respect to Species. The interest is in evaluating

Table 2 Clustering of the `f.voles` data using three different approaches

(a) VEE-GLGCWM			(b) MLGR						(c) MLGRC		
	Est.			Est.						Est.	
True	1	2	True	1	2	3	4	5	True	1	2
M. ochrogaster	43	2	*M. ochrogaster*	14	5	6	15	5	*M. ochrogaster*	15	30
M. californicus	–	41	*M. californicus*	10	3	8	9	11	*M. californicus*	3	38

The best model is selected by the BIC

clustering using the GLGCWMs as well as comparing the algorithm with some well-established mixture-based techniques. Therefore, Age can be considered the natural Y variable and the $p = 6$ skull measurements can be considered as the vector of covariates $\boldsymbol{X}$.

By considering a Gaussian distribution for Y in each mixture component, all fourteen GLGCWMs were fitted, assuming no known group membership, for $k \in \{1, \ldots, 5\}$, for a total of $14 \times 5 = 70$ models. The model with the largest BIC value (-3863.451) was VEE with $k = 2$; the use of the BIC [21] for this class of models was justified by some results, based on simulated data, reported in [13, 16]. Table 2(a) displays the clustering results from this model (group memberships are individuated by *maximum a posteriori* probabilities). Table 2 also shows clustering results for mixtures of linear Gaussian regressions (MLGRs) and mixtures of linear Gaussian regressions with concomitants (MLGRCs), using the covariates as concomitants. They are estimated via the `stepFlexmix()` function of the R-package **flexmix** [9] by using the range $k \in \{1, \ldots, 5\}$ and by selecting the best number of mixture components with the BIC. The best results were clearly obtained for the VEE-GLGCWM where the number of groups was correctly selected and only two *M. ochrogaster* observations were misclassified as *M. californicus*. On these data, the other two approaches did not show a good clustering performance. The better results for the VEE-GLGCWM are motivated by the fact that the distribution of the covariates is different in the two groups induced by the dichotomous variable Species, considered here as the group variable; in all these cases of assignment dependence (cf. Sect. 1), the CWM could represent a reference approach of analysis.

5 Discussion

Recently, in the class of mixture models with random covariates, the GLGCWM was presented. The model has the drawback to be potentially overparameterized when the number of covariates and/or the number of mixture components increases. To overcome this problem, we introduced a family of fourteen parsimonious GLGCWMs; some of these models substantially reduce the number of parameters with respect to the unconstrained GLGCWM. An EM algorithm was adopted for maximum likelihood parameter estimation. We illustrated our approach on real data where our method gave impressive superior clustering results when compared to other more famous mixture approaches. Although we showed the usefulness of our models for model-based clustering, note that they can also be applied for classification and discriminant analysis. Finally, in the fashion of [8, 16, 19], future work will investigate the use of likelihood-ratio tests to "select" the best parsimonious model in the proposed family.

Acknowledgements The authors acknowledge the financial support from the grant "Finite mixture and latent variable models for causal inference and analysis of socio-economic data" (FIRB 2012-Futuro in ricerca) funded by the Italian Government (RBFR12SHVV).

References

1. Airoldi, J., Hoffmann, R.: Age variation in voles (Microtus californicus, M. ochrogaster) and its significance for systematic studies. In: Occasional Papers of the Museum of Natural History, vol. 111. University of Kansas, Lawrence (1984)
2. Aitken, A.: On Bernoulli's numerical solution of algebraic equations. In: Proceedings of the Royal Society of Edinburgh, vol. 46, pp. 289–305 (1926)
3. Bagnato, L., Punzo, A.: Finite mixtures of unimodal beta and gamma densities and the k-bumps algorithm. Comput. Stat. **28**(4), 1571–1597 (2013)
4. Celeux, G., Govaert, G.: Gaussian parsimonious clustering models. Pattern Recogn. **28**(5), 781–793 (1995)
5. Dempster, A., Laird, N., Rubin, D.: Maximum likelihood from incomplete data via the EM algorithm. J. R. Stat. Soc. Ser. B Methodol. **39**(1), 1–38 (1977)
6. Flury, B.: A First Course in Multivariate Statistics. Springer, New York (1997)
7. Gershenfeld, N.: Nonlinear inference and cluster-weighted modeling. Ann. N. Y. Acad. Sci. **808**(1), 18–24 (1997)
8. Greselin, F., Punzo, A.: Closed likelihood ratio testing procedures to assess similarity of covariance matrices. Am. Stat. **67**(3), 117–128 (2013)
9. Grün, B., Leisch, F.: **FlexMix** version 2: Finite mixtures with concomitant variables and varying and constant parameters. J. Stat. Softw. **28**(4), 1–35 (2008)
10. Hennig, C.: Identifiablity of models for clusterwise linear regression. J. Classif. **17**(2), 273–296 (2000)
11. Ingrassia, S., Minotti, S.C., Vittadini, G.: Local statistical modeling via the cluster-weighted approach with elliptical distributions. J.Classif. **29**(3), 363–401 (2012)
12. Ingrassia, S., Minotti, S.C., Punzo, A.: Model-based clustering via linear cluster-weighted models. Comput. Stat. Data Anal. **71**, 159–182 (2014)
13. Ingrassia, S., Punzo, A., Vittadini, G., Minotti, S.C.: The generalized linear mixed cluster-weighted model. J. Classif. **32**(1), 85–113 (2015)
14. Mazza, A., Punzo, A., Ingrassia, S.: **flexCWM**: Flexible Cluster-Weighted Modeling. Available at http://cran.r-project.org/web/packages/flexCWM/index.html (2014)
15. Punzo, A.: Flexible mixture modeling with the polynomial Gaussian cluster-weighted model. Stat. Model. **14**(3), 257–291 (2014)
16. Punzo, A., Ingrassia, S.: On the use of the generalized linear exponential cluster-weighted model to asses local linear independence in bivariate data. QdS J. Methodol. Appl. Stat. **15**, 131–144 (2013)
17. Punzo, A., Ingrassia, S.: Clustering bivariate mixed-type data via the cluster-weighted model. Comput. Stat. (2015)

18. Punzo, A., McNicholas, P.D.: Robust clustering in regression analysis via the contaminated Gaussian cluster-weighted model. Available at: http://arxiv.org/abs/1409.6019 (2014) [arXiv.org e-print 1409.6019]
19. Punzo, A., Browne, R.P., McNicholas, P.D.: Hypothesis testing for parsimonious Gaussian mixture models. Available at: http://arxiv.org/abs/1405.0377 (2014) [arXiv.org e-print 1405.0377]
20. R Core Team: R: A Language and Environment for Statistical Computing. R Foundation for Statistical Computing, Vienna (2013)
21. Schwarz, G.: Estimating the dimension of a model. Ann. Stat. **6**(2), 461–464 (1978)
22. Subedi, S., Punzo, A., Ingrassia, S., McNicholas, P.D.: Clustering and classification via cluster-weighted factor analyzers. Adv. Data Anal. Classif. **7**(1), 5–40 (2013)
23. Subedi, S., Punzo, A., Ingrassia, S., McNicholas, P.D.: Cluster-weighted *t*-factor analyzers for robust model-based clustering and dimension reduction. Stat. Methods Appl. **24** (2015)
24. Wedel, M.: Concomitant variables in finite mixture models. Statistica Neerlandica **56**(3), 362–375 (2002)
25. Wedel, M., Kamakura, W.: Market Segmentation: Conceptual and Methodological Foundations, 2nd edn. Kluwer Academic, Boston (2001)

New Perspectives for the *MDC* Index in Social Research Fields

Emanuela Raffinetti and Pier Alda Ferrari

Abstract The great interest in quantitative social research has led to the development of specific statistical techniques suitable in dealing with dependence between variables also in the presence of ordinal data. A specific index, hereafter called monotonic dependence coefficient (*MDC*), was provided as a monotonic dependence measure. Due to its properties and specific features, *MDC* overcomes the Pearson's correlation coefficient, since it captures not only linear dependence relationships but also any general monotonic one. The *MDC* adequacy is validated by a simulation study assessing its performance with respect to the traditional Pearson's correlation coefficient. Finally, a real application of *MDC* to real data is also illustrated.

Keywords Dependence relationship • Monte Carlo simulations • Ordinal data

1 Background

The quantification of dependency is an interesting and relevant topic to researchers dealing with the study of social issues, particularly because it is easy to misinterpret the traditional correlation measures when variables are expressed according to different measurement scales.

The purpose of this paper is presenting a specific monotonic dependence measure and illustrating its main properties. This measure is based on two previous proposals. In fact, it was firstly employed as an index of "equity in a taxation process" (see, e.g., [2]) and subsequently used as a concordance measure for a multiple linear regression model (see, e.g., [4]). A new version of such index was studied and developed by Raffinetti and Ferrari [3], in order to provide both an extension of its applicability to any real-valued variable and an interpretation in terms of monotonic dependence relationship. For this reason, hereafter the index

E. Raffinetti (✉) • P.A. Ferrari
Department of Economics, Management and Quantitative Methods, Università degli Studi di Milano, Via Conservatorio 7, 20122 Milano, Italy
e-mail: emanuela.raffinetti@unimi.it; pieralda.ferrari@unimi.it

I. Morlini et al. (eds.), *Advances in Statistical Models for Data Analysis*, Studies in Classification, Data Analysis, and Knowledge Organization,
DOI 10.1007/978-3-319-17377-1_22

will be called monotonic dependence coefficient (*MDC*) and denoted with the acronym *MDC*.

Some recalls to *MDC* characteristics are needed, such as for instance its expression and its properties. Let Y and X be two variables (Y numerical, X numerical/ordinal) and let us consider a simple linear regression model between a response variable Y and a covariate X. Specifically, in the case X is ordinal the simple linear regression model is performed by using it on the rank scale. Let us denote with $y_{(i)}$ the ordered (in nondecreasing sense) Y values and with y_i^* the same Y values reordered according to the ranks of the Y linear estimated values, obtaining in such a way n pairs $(y_{(i)}, y_i^*)$ $(i = 1, \ldots, n)$. A general *MDC* formula is given by

$$MDC = \frac{2\sum_{i=1}^{n} i(y_i^* - y_0) - n(n+1)(M_Y - y_0)}{2\sum_{i=1}^{n} i(y_{(i)} - y_0) - n(n+1)(M_Y - y_0)}, \quad i = 1, \ldots, n, \tag{1}$$

where $y_0 = min(0, y^-)$, with y^- representing the minimum response variable Y value, if negative, and M_Y is the Y mean value.

Since the proposed measure ranges between -1 and $+1$, reaching value zero in case of independence, and it is suitable to assess monotonic dependence of Y from X, it seems of interest investigating about its properties[1]. Here, this problem was faced and discussed. Additional studies were carried out to detect similarities and dissimilarities of *MDC* with respect to Pearson's (r) correlation coefficient. For such a purpose, a Monte Carlo simulation study was run and the related findings are shown and commented in Sect. 2. In particular, *MDC* performs better than r, since it captures linear dependence relationship as well as any monotonic dependence one. Due to such a role, *MDC* has also the capability of catching the existing dependence relationship of a variable from another one, preserving it also when pieces of information are lost. This feature is highlighted in Sect. 3, where a real application of our proposal is illustrated by comparing the *MDC* performance with the one of r. Finally, Sect. 4 aims at summarizing the main *MDC* features and properties.

2 A Comparison of *MDC* and *r* Through Monte Carlo Simulations

Given the similarities of *MDC* with r (especially when the involved variables are expressed according to an interval or ratio scale), an attractive research issue concerns the analysis of their performance in different situations of monotonic dependence relationship. For this purpose, a Monte Carlo simulation study was carried out and implemented in `R`. The idea focuses on generating samples from

[1]Note that, if the least squares estimate of the regression coefficient β is smaller than 0, the sign of the linear estimated values has to be changed to obtain a negative MDC value.

two different families of bivariate distributions. Firstly, a sampling scheme from a multivariate exponential power (MEP) distributions family was taken into account. MEP distributions represent a generalization of Normal distributions involving a specific "non-normality" parameter κ which detects for symmetrical distribution the departure from normality. Subsequently, a sampling scheme based on non-normal distributions family was considered, with the aim of evaluating the effect of skewed distributions on the two indices behavior. Results regarding both the scenarios are discussed in Sects. 2.1 and 2.2.

2.1 Sampling From Bivariate MEP Distributions

The first study is provided by running a Monte Carlo simulation based on generating samples from the family of bivariate MEP distributions, this family being one of the possible generalizations of normal distributions in terms of ellipsoidal departures. For more details about MEP distributions, see, e.g., [5]. However, what is basic to point out is that MEP distributions depend on a specific parameter, denoted with κ and expressing the "non-normality" condition. For $\kappa < 2$ and $\kappa > 2$, respectively, leptokurtic and platikurtic distributions are obtained, while for $\kappa = 2$ normal distributions are derived. In such paper the sampling scheme is based on choosing values $\kappa = \{1, 2, 8\}$ for, respectively, describing leptokurtic, normal, and platikurtic bivariate distributions. Through the illustrated procedure, the sampling distribution of *MDC* and r for variables generated from MEP distributions was obtained under different experimental conditions. For each value of κ, a variance–covariance matrix was built according to the following pairwise correlation coefficients $\rho = \{0.2, 0.4, 0.6, 0.8\}$. Under each of such scenarios and referring to the R code provided by Solaro [5], we drew samples of size 100, 500, and 1,000 and we iterated these steps 10,000 times. For the sake of shortness, here we report only results corresponding to the scenario characterized by a pairwise correlation coefficient

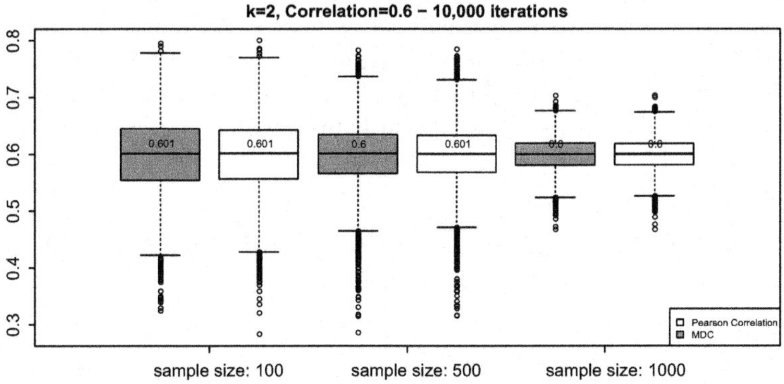

Fig. 1 Pairwise correlation $\rho = 0.6$ and $\kappa = 2$

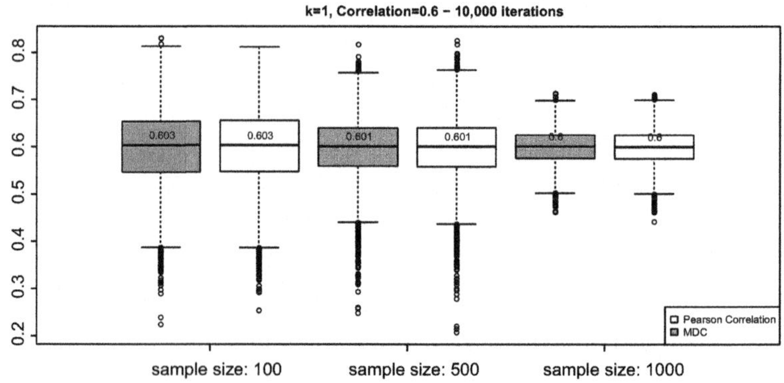

Fig. 2 Pairwise correlation $\rho = 0.6$ and $\kappa = 1$

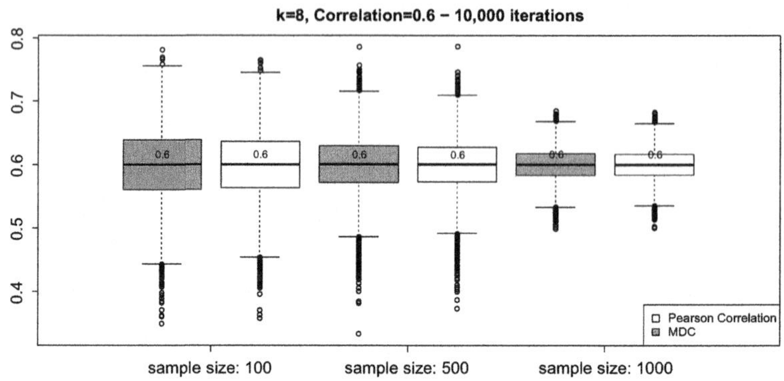

Fig. 3 Pairwise correlation $\rho = 0.6$ and $\kappa = 8$

ρ equal to 0.6, stressing that similar findings can be reached also with respect to the remained pairwise correlation levels. Simulation results are shown by boxplots in Figs. 1, 2 and 3: above each boxplot the median value is specified.

In case of bivariate MEP distributions with $\kappa = 2$, monotonic dependence coincides with the linear one. Results shown in Fig. 1 confirm this issue: in fact *MDC* and r achieve the same median value (but also mean value). Similar conclusions arise also in case of MEP distributions with $\kappa = 1$ and $\kappa = 8$.

2.2 Sampling From Bivariate Non-Normal Distributions

In this section, we consider a family of bivariate non-normal distributions. The first contribution in generating non-normal variables is due to [1] who defines, in the univariate case, a non-normal variable as a linear combination of the first three powers of a standard normal variable. Vale and Maurelli [6] developed the same method for generating multivariate non-normal distributions with specified inter-correlations and marginal means, variances, skewness, and kurtosis. According to the procedure in [6], the two parameters that affect the normality condition are skewness and kurtosis. This family of distributions is particularly interesting since it allows us to vary the skewness and kurtosis parameters in order to assess their impact on *MDC* and r behavior when the normality condition is violated. We specify that in this case the simulation study was carried out by resorting to the R code made available by Zopluoglu [7]. Since simulation results coming from the sampling scheme built on bivariate MEP distributions highlighted that the kurtosis parameter seems not to affect the *MDC* behavior with respect to that of r, now we let vary only the skewness parameter, denoted with γ, by fixing the kurtosis parameter, pointed out as κ, equal for both the variables. Such a value was chosen in order to fulfill two goals. First, to consider an intermediate value for κ lying between 2 and 8 (used for simulations based on MEP distributions) and secondly to define a possible combination of the kurtosis and skewness parameters well performing according to the provided R code. In our case, a running combination was thus obtained by $\kappa = (5.5, 5.5)$ and γ equal for both the variables and fixed to $\gamma = (1, 1)$ and $\gamma = (2, 2)$. Analogously to conditions defined for simulations based on MEP distributions, we chose the same four levels of pairwise correlation coefficients $\rho = \{0.2, 0.4, 0.6, 0.8\}$ and the same sample size ($n = 100, 500, 1{,}000$). Since once again the findings are very similar for all the selected pairwise correlation coefficients, we restrict the discussion only on $\rho = 0.6$.

With data being generated from a non-normal bivariate distribution, existing dependence relationships do not coincide with the linear ones and thus we expect that *MDC* provides greater values than r. Boxplots represented in Figs. 4 and 5 satisfy such expectation: the median value (but also the mean value) of the sampling distribution of *MDC* is always higher than that of Pearson's correlation coefficient, highlighting its capability in catching any monotonic dependence relationship.

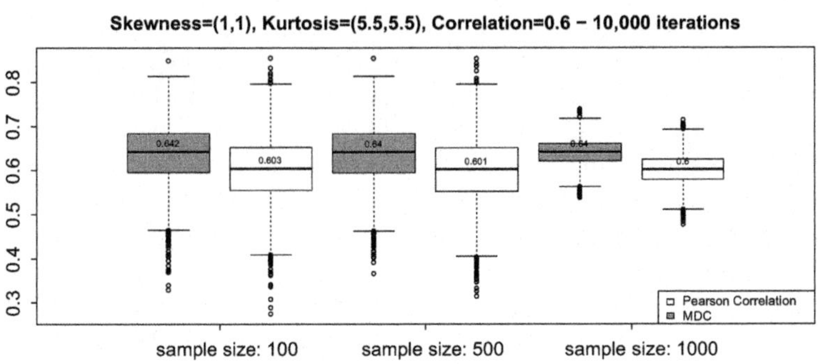

Fig. 4 Pairwise correlation $\rho = 0.6$, $\gamma = (1, 1)$, and $\kappa = (5.5, 5.5)$

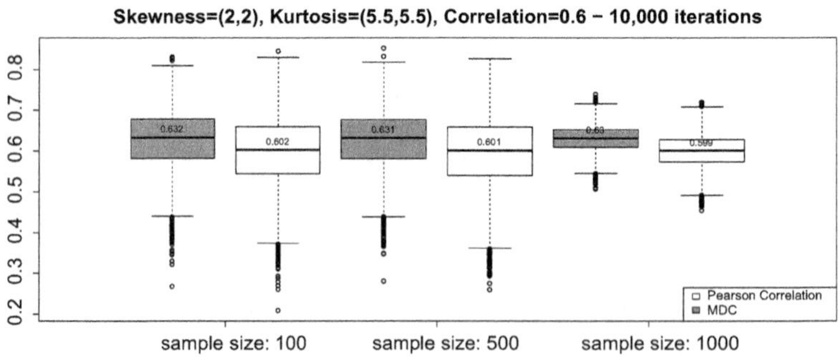

Fig. 5 Pairwise correlation $\rho = 0.6$, $\gamma = (2, 2)$, and $\kappa = (5.5, 5.5)$

3 Further Investigations in Application Contexts

In this section we introduce an application of our proposed measure in order to make a comparison between its behavior and that associated to Pearson's r correlation coefficient. We remind that the aim of such example is not addressed to provide a detailed analysis of all data involved and described in the considered dataset. Here, we simply show how the *MDC* index can lead to more robust results with respect to Pearson's r correlation coefficient.

Some information about the used dataset are due. The employed dataset is available as an SPSS Data file and it is called "Employee Data.sav." The file contains data extracted from a bank's employee records in an investigation into discrimination in 1987 and it is built on 473 statistical units.

As illustrated in [3], the *MDC* index can be usefully applied both in a quantitative context, when studying the dependence relationship between two quantitative variables, and in a context when the dependent variable is quantitative and the independent one is ordinal or discrete. With regard to this issue, here the focus is based on dependence of a variable Y, representing the beginning salary (in dollars), on another variable X, representing the individual education years. This variable takes values according to the frequency distribution represented in Table 1.

The purpose of this application is threefold. More precisely, we aim at comparing the *MDC* performance with respect to that of r under three specific scenarios. Firstly, we take into account both the variables Y and X as directly provided by the dataset and representing the real situation of analysis. This in order to assess the existence of a dependence relationship between the beginning salary (continuous variable) and education years (discrete variable). Secondly, the comparison between *MDC* and r is carried out as if the original data, concerning the explanatory variable X, are classified into five groups whose frequency distribution is provided in Table 2. More in detail, the education years variable is re-expressed according to the previous groups and a rank scale such that employees included in group 1, 2, 3, 4, and 5 are characterized by an education level encoded, respectively, by 1, 2, 3, 4, and 5. Finally, the study is focused on considering the average value of education years

Table 1 Frequency distribution of variable *X* education year

X: education years	8	12	14	15	16	17	18	19	20	21
Absolute frequencies	53	190	6	116	58	11	9	27	2	1

Table 2 Frequency distribution of groups and average of education years within each group

Groups	1	2	3	4	5
Absolute frequencies	243	122	69	36	3
Education year ranges	[8, 14)	[14, 16)	[16, 18)	[18, 20)	Over 20
Average of education years	11.128	14.951	16.159	18.750	20.333

Table 3 *MDC* and *r* values to evaluate dependence of beginning salary on education years

	X-education years expressed as		
	Original data (a)	Five ordered categories (b)	Average at group (c)
MDC	0.787	0.781	0.781
r	0.633	0.746	0.688

within each group as available data. To be more thorough the education year ranges within each group are also reported.

By serving in this way, the effect associated to transformation of the *X* variable scale on the capability of both the indices in catching the dependence relationship is evaluated.

Results regarding the two analyses are presented in Table 3.

With regard to original data (a), the value of *MDC* (0.787) is higher than *r* (0.633) and such difference is well marked. This is because of the original discrete dependent variable nature. As well known, Pearson's *r* correlation coefficient is sensitive to variable nature: typically, it shrinks with variables which are not continuous. *MDC* is not affected by the variable nature since it is based only on reordering the response variable values according to the corresponding linear estimates.

In the second case (b), when data about the explanatory variable are encoded into five ordered categories, if on one hand *MDC* reaches almost the same value (0.781), on the other hand *r* raises considerably (0.746), highlighting its sensitivity to scale transformation.

The third case (c) reports the results based on data available only in terms of average years of education within each group. If on one hand *r* gets worse (0.688) with respect to case (b), our proposed *MDC* remains unchanged in its value (0.781), highlighting once again the unreliability of Pearson's correlation coefficient.

The described real example confirms the *MDC* robustness in catching the dependence relationships even when one of the variable is expressed according to different measurement scales. Such issue supports the *MDC* adequacy in dependence relationship investigation with respect to Pearson's correlation coefficient which can lead, as previously shown, to misleading results.

4 Conclusions

In this contribution we analyze the properties of a dependence measure able to catch any monotonic relationship between a real-valued response variable and a numerical or ordinal independent variable (even tied). Our measure is also invariant with respect to the quantification of categories for the ordinal variable. This finding is a direct implication of the *MDC* construction, *MDC* being built by comparing the original values of the dependent variable with the same values reordered according to the ranks of their corresponding linear estimated values.

The behavior of *MDC* was also investigated through a Monte Carlo simulation study in which the *MDC* performance was compared to that of Pearson's (r) correlation coefficient, by firstly sampling from bivariate MEP distributions characterized by specific non-normality parameter κ values which were let vary for defining leptokurtic ($\kappa < 2$), normal ($\kappa = 2$) and platikurtic ($\kappa > 2$) distributions. Findings coming from this study highlight that *MDC* and r achieve a similar performance for this family. In order to stress the *MDC* role in catching any monotonic dependence relationship, a second Monte Carlo simulation study was also carried out by sampling from a non-normal bivariate distribution characterized by different values of skewness. When distributions are non-normal, monotonic dependence and linear dependence might be very different relationships and as a consequence Pearson's correlation coefficient could not well capture monotonic dependence relationship far from the linear one. In such situations, *MDC* is shown to perform better than Pearson's correlation coefficient.

Finally, an application to real data was provided to confirm the capability of the index to catch the dependence relationship between the variables, preserving it also in case of the ordinal/discrete independent variable nature.

Acknowledgements The authors wish to acknowledge financial support from the European Social Fund Grant (Lombardy Region, Italy) and the anonymous reviewers for their helpful comments.

References

1. Fleishman, A.I.: A method for simulating nonnormal distributions. Psychometrika **48**, 273–279 (1978)
2. Muliere, P.: Some Notes About the Horizontal Equity of a Taxation (in Italian), in Honor of Francesco Brambilla, vol. 2. Bocconi Communication, Milan (1986)
3. Raffinetti E., Ferrari P.A.: The *RCI* as a measure of monotonic dependence. In: Vicari, D., Okada, A., Ragozini, G., Weihs, C. (eds.) Analysis and Modeling of Complex Data

in Behavioural and Social Sciences. Series: Studies in Classification, Data Analysis, and Knowledge Organization. Springer, Heidelberg (2014)
4. Raffinetti, E., Giudici, P.: Multivariate ranks-based concordance indexes. In: Advanced Statistical Methods for the Analysis of Large Data-Sets, pp. 465–473. Springer, Heidelberg (2012)
5. Solaro, N.: Random variate generation from multivariate exponential power distribution. Stat. Appl. **II**(2), 25–44ă(2004)
6. Vale, C., Maurelli, V.: Simulating multivariate nonnormal distributions. Psychometrika **48**(3), 465–471 (1983)
7. Zopluoglu, C.: Application in R: Generating Multivariate Non-normal Variables, University of Minnesota. Available on line at link http://www.academia.edu/1744752/R-Routine-for-generating-multivariate-non-normal-data (2011)

Clustering Methods for Ordinal Data: A Comparison Between Standard and New Approaches

Monia Ranalli and Roberto Rocci

Abstract The literature on cluster analysis has a long and rich history in several different fields. In this paper, we provide an overview of the more well-known clustering methods frequently used to analyse ordinal data. We summarize and compare their main features discussing some key issues. Finally, an example of application to real data is illustrated comparing and discussing clustering performances of different methods.

Keywords EM algorithm • Finite mixture models • k-means • Ordinal data • Pairwise likelihood

1 Introduction

The aim of cluster analysis is discovering the natural groups of a set of objects, such that clusters differ considerably from each other. The literature on clustering is rich and wide, even if it has mainly been developed for continuous data. Only in the last decades, there has been an increasing interest in clustering categorical data; however, the amount of work done is still relatively small. Categorical variables are encountered in many fields, such as in behavioural, social and health sciences. These variables, frequently of ordinal type, measure attitudes, abilities or opinions. However, due to the lack of metric properties, modelling properly this kind of variables could be challenging. For this reason, it is still common to analyse ordinal data following a naive approach whereby their nature is ignored. Ranks are treated as interval-scaled, and thus clustering techniques developed for continuous data are applied. A way to circumvent the problem is to apply a two-step procedure, named tandem analysis, where, first, the ordinal variables are reduced into continuous

M. Ranalli (✉)
Department of Statistics, The Pennsylvania State University, USA
e-mail: monia.ranalli@psu.edu

R. Rocci
Dipartimento IGF, Università di Roma Tor Vergata, Roma, Italy
e-mail: roberto.rocci@uniroma2.it

I. Morlini et al. (eds.), *Advances in Statistical Models for Data Analysis*, Studies in Classification, Data Analysis, and Knowledge Organization,
DOI 10.1007/978-3-319-17377-1_23

factors, then, objects are grouped on the basis of their factor scores. An alternative to tandem analysis is to use a distance-based clustering technique, where distances are computed taking into account the measurement scale of the data. The main drawback of these approaches is that they do not model the data generation process, and so standard statistical tools cannot be used. A model-based approach solves the aforementioned problems but opens new issues, especially from a computational point of view.

The aim of this paper is to review and compare the four approaches: naive, tandem, distance-based and model-based, illustrating their weaknesses and strengths. The plan of the paper is as follows. In Sect. 2, we deal with the naive approach reviewing the main optimization and model-based clustering techniques for continuous data. The tandem and distance-based approaches are described in Sect. 3, while the model-based approach is illustrated in Sect. 4. In Sect. 5, some clustering techniques, representative of the approaches presented, are applied to a real dataset and a discussion regarding the clustering performance is provided. In the last section, some concluding remarks are pointed out.

2 The Naive Approach: Clustering Techniques for Continuous Data

The naive approach consists in using a clustering method developed for continuous data to analyse ordinal data, treating the ranks as interval scaled. For the sake of brevity and comparability, here, we recall only the partitioning techniques most used in practice (see [1, 15] and references therein, for non-partitioning techniques). Within this framework, we distinguish between optimization and model-based clustering techniques. The most well-known optimization algorithm is the k-means [18]. Letting $\mathbf{X} = \{\mathbf{x}_n : n = 1, \ldots, N\}$ be the sample of P-dimensional observations, it is based on the minimization of the loss

$$\ell_{km}(\boldsymbol{\psi}, \mathbf{Z}; \mathbf{X}) = \sum_{n=1}^{N} \sum_{g=1}^{G} z_{ng} \left(\mathbf{x}_n - \boldsymbol{\mu}_g\right)' \left(\mathbf{x}_n - \boldsymbol{\mu}_g\right), \tag{1}$$

where $\mathbf{Z} = [z_{ng}]$ is a binary membership matrix, with rows that sum to 1, such that $z_{ng} = 1$ if observation n belongs to cluster g and 0 otherwise, and $\boldsymbol{\psi} = \{\boldsymbol{\mu}_1, \ldots, \boldsymbol{\mu}_G\}$ is the set of cluster centroids. In order to be performed, k-means needs two pre-specified inputs: number of components G and a cluster initialization. Different initializations can lead to different final partitions, since k-means only converges to local minima, even if it has been shown that it could converge to the global optimum if the clusters are well separated [20]. Some possible extensions of the k-means are the following: fuzzy c-means [2], according to which the observations are not assigned exclusively to one cluster (the so called soft assignment), and k-medoids [15], where the mean is substituted for a data point.

The model-based analogue of k-means is the finite mixture of Gaussians (FMG). They are considered a powerful tool for clustering [7, 19] and effectively capturing sample heterogeneity, since it is assumed that a population is a convex combination of a finite number of densities, each of which may represent a cluster [11]. An FMG assumes that $\mathbf{x}_n$ has a density function defined as

$$f(\mathbf{x}_n; \boldsymbol{\psi}) = \sum_{g=1}^{G} p_g \phi\left(\mathbf{x}_n; \boldsymbol{\mu}_g, \boldsymbol{\Sigma}_g\right), \tag{2}$$

where the p_g's are the mixing weights, $\phi\left(\mathbf{x}_n; \boldsymbol{\mu}_g, \boldsymbol{\Sigma}_g\right)$ is the density of a P-variate normal distribution with mean vector $\boldsymbol{\mu}_g$ and covariance matrix $\boldsymbol{\Sigma}_g$, while $\boldsymbol{\psi} = \{p_1, \ldots, p_G, \boldsymbol{\mu}_1, \ldots, \boldsymbol{\mu}_G, \boldsymbol{\Sigma}_1, \ldots, \boldsymbol{\Sigma}_G\}$ is the set of model parameters. The maximum likelihood estimates of model parameters are usually computed through the Expectation-Maximization (EM, [3]). When the observations are i.i.d. this can be seen as the maximization by a coordinate ascend algorithm of the fuzzy loss [10]

$$\ell_{FMG}(\boldsymbol{\psi}, \mathbf{Z}; \mathbf{X}) = \sum_{n=1}^{N} \sum_{g=1}^{G} z_{ng} \log\left[p_g \phi(\mathbf{x}_n; \boldsymbol{\mu}_g, \boldsymbol{\Sigma}_g)\right] - \sum_{n=1}^{N} \sum_{g=1}^{G} z_{ng} \log(z_{ng}), \tag{3}$$

where $\mathbf{Z} = [z_{ng}]$ is the fuzzy membership matrix with non-negative elements that sum to 1 by row and express how much an observation belongs to a cluster. Cluster g has an ellipsoidal shape described by $\boldsymbol{\Sigma}_g$ and centred at $\boldsymbol{\mu}_g$. A hard classification can be obtained by assigning the observations to the component with the maximum a posteriori probability, i.e. with the maximum degree of membership z_{ng}. It is interesting to note that k-means loss can be seen as a particular case of (3). To be precise, (3) becomes equivalent to (1) by setting $z_{ng} \in \{0, 1\}$ (groups do not overlap), $p_g = 1/G$ (groups have the same size) and $\boldsymbol{\Sigma}_g = \sigma^2 \mathbf{I}$ (variables are of the same variances and uncorrelated within the groups, i.e. locally independent). It follows that k-means works better than FMG when the constraints are true and worse when they are false. However, both have a common drawback when applied to ordinal variables: they do not take properly into account the measurement scale of the variables.

3 Tandem and Distance-Based Approaches

The aforementioned problem can be circumvented by reducing the ordinal variables into continuous factors before applying the clustering algorithm. This produces a two-step procedure named tandem analysis [1]. In the first step, principal component analysis for qualitative data (PRINQUAL, [25]) or multiple correspondence analysis (MCA, [9]) is performed to summarize the association between a set of variables by a small number of dimensions; then, to discover the cluster structure, k-means on the reduced data is applied using the scores of PRINQUAL or MCA. The main problem with the tandem approach is that there is no guarantee that the reduced data obtained in step one is optimal for recovering the cluster structure in the second step [1]. This may hide or even distort the true cluster structure underlying the data. As a solution to the problem, data reduction and cluster analysis should be combined into the same loss, such as in [13, 14, 22]. In this way, the latent factors are identified to highlight the cluster structure rather than, in some cases, to obscure it. This procedure, named simultaneous tandem analysis, is effective when there are "noisy" dimensions in the data, i.e. latent or manifest variables that do not have information about the cluster structure. In all other cases it is better to consider a distance-based technique. In other words, it is an optimization method where the dissimilarities between objects are computed by taking into account the ordinal nature of the data (e.g. see [24]). This could also be obtained by running a simultaneous tandem where in the first step the number of dimensions is not reduced. A drawback of these approaches is that they are not model-based. As a consequence, they do not model the data generation process and standard statistical tools cannot be used to make decisions about model parameters.

4 Model-Based Approach

In the model-based approach, the most frequently used clustering technique for categorical data is latent class analysis [8] and some constrained versions that have been provided for ordinal data (see, e.g., [4]). These models are based on the local independence assumption. They consider the cluster membership as a nominal latent factor and assume that the manifest variables are independent given that factor. Of course, this model is inadequate every time that there are dependencies among the manifest variables within the clusters. A way to overcome this limitation is to consider an FMG that allows dependencies within clusters to be modelled by means of the covariance matrices. Following the Underlying Response Variable (URV) approach used in latent variables models, the FMG model can be adapted to ordinal data by assuming that the observed variables are a discretization of underlying non-observable continuous variables distributed as an FMG. In what follows, we will discuss three different proposals in this direction.

We start by describing the key figures for the proposal of [21]. This aims at capturing the cluster structure underlying the data. Since the local independence assumption is not required, in comparison with the latent class models, it is possible to obtain a simpler and more realistic solution with a smaller number of groups. Let $x_1, x_2, \ldots, x_P$ be ordinal variables and $c_i = 1, \ldots, C_i$ the associated categories for $i = 1, 2, \ldots, P$. There are $R = \prod_{i=1}^{P} C_i$ possible response patterns $\mathbf{x}_r = (x_1 = c_1, x_2 = c_2, \ldots, x_P = c_P)$, with $r = 1, \ldots, R$. The ordinal variables are generated by thresholding $\mathbf{y}$: a multivariate continuous random variable distributed as an FMG.

The link between $\mathbf{x}$ and $\mathbf{y}$ is expressed by a threshold model defined as

$$x_i = c_i \Leftrightarrow \gamma^{(i)}_{c_i-1} \leq y_i < \gamma^{(i)}_{c_i}. \tag{4}$$

Let $\boldsymbol{\psi} = \{p_1, \dots, p_G, \boldsymbol{\mu}_1, \dots, \boldsymbol{\mu}_G, \boldsymbol{\Sigma}_1, \dots, \boldsymbol{\Sigma}_G, \boldsymbol{\gamma}\}$ be the set of model parameters. The probability of response pattern $\mathbf{x}_r$ is given by

$$Pr(\mathbf{x}_r; \boldsymbol{\psi}) = \sum_{g=1}^{G} p_g \int_{\gamma^{(1)}_{c_1-1}}^{\gamma^{(1)}_{c_1}} \cdots \int_{\gamma^{(P)}_{c_P-1}}^{\gamma^{(P)}_{c_P}} \phi(\mathbf{y}; \boldsymbol{\mu}_g, \boldsymbol{\Sigma}_g) d\mathbf{y} = \sum_{g=1}^{G} p_g \pi_r(\boldsymbol{\mu}_g, \boldsymbol{\Sigma}_g, \boldsymbol{\gamma}), \tag{5}$$

where $\pi_r(\boldsymbol{\mu}_g, \boldsymbol{\Sigma}_g, \boldsymbol{\gamma})$ is the probability of response pattern $\mathbf{x}_r$ in cluster g. Thus, for a random i.i.d. sample of size N the log-likelihood is

$$\ell(\boldsymbol{\psi}; \mathbf{X}) = \sum_{r=1}^{R} n_r \log \left[\sum_{g=1}^{G} p_g \pi_r \left(\boldsymbol{\mu}_g, \boldsymbol{\Sigma}_g, \boldsymbol{\gamma} \right) \right], \tag{6}$$

where n_r is the observed sample frequency of response pattern $\mathbf{x}_r$ and $\sum_{r=1}^{R} n_r = N$. A similar proposal has been done by Everitt [5, 6], who introduces a mixture model for mixed data. The joint distribution of the variables is a homoscedastic FMG where some variables are observed as ordinal. In particular, the ordinal variables are seen as generated by thresholding some marginals of the joint FMG with different thresholds in each component. The model proposed by Lubke and Neale [17] is specified for ordinal variables that are generated by thresholding an heteroscedastic mixture of Gaussians, whose covariance matrices are reparametrized as a factor analysis model. In all models, estimation is carried out by full maximum likelihood. It implies the numerical computation of multidimensional integrals, which is time consuming and becomes infeasible when more than 4 or 5 variables are involved. For this reason, in [21], the authors propose to estimate the model within the EM framework maximizing the pairwise log-likelihood, i.e. the sum of all possible log-likelihoods based on the bivariate marginals. The obtained estimators have been proven to be consistent, asymptotically unbiased and normally distributed. In general, they are less efficient than the full maximum likelihood estimators, but in many cases the loss in efficiency is very small or almost null [16, 23]. In formulas, the pairwise log-likelihood is of the form

$$\begin{aligned} p\ell(\boldsymbol{\psi}; \mathbf{X}) &= \sum_{i=1}^{P-1} \sum_{j=i+1}^{P} \ell(\boldsymbol{\psi}; (x_i, x_j)) \\ &= \sum_{i=1}^{P-1} \sum_{j=i+1}^{P} \sum_{c_i=1}^{C_i} \sum_{c_j=1}^{C_j} n^{(ij)}_{c_i c_j} \log \left[\sum_{g=1}^{G} p_g \pi^{(ij)}_{c_i c_j} (\boldsymbol{\mu}_g, \boldsymbol{\Sigma}_g, \boldsymbol{\gamma}) \right], \end{aligned} \tag{7}$$

where $n^{(ij)}_{c_ic_j}$ is the observed frequency of a response in categories c_i and c_j for variables x_i and x_j, respectively, while $\pi^{(ij)}_{c_ic_j}(\boldsymbol{\mu}_g, \boldsymbol{\Sigma}_g, \boldsymbol{\gamma})$ is the probability obtained by integrating the corresponding bivariate marginal of the normal distribution with parameters $(\boldsymbol{\mu}_g, \boldsymbol{\Sigma}_g)$ between the corresponding threshold parameters. It is clear that the pairwise approach is feasible as it requires only the evaluation of integrals on bivariate normal distributions, regardless of the number of observed or latent variables $\mathbf{y}$. Nevertheless the estimation of all parameters is carried out simultaneously. As regards the classification, in [20] it has been suggested to use an iterative proportional fitting based on the pairwise posterior probabilities obtained as output of the pairwise EM algorithm in order to approximate the joint posterior probabilities. For identification reasons, a component is fixed as a reference group; thus, its mean vector is set to $\mathbf{0}$ and its variances to 1. On one hand, the model proposed in [21] can be seen as a particular case of Everitt's proposal [5]. In fact, here only categorical ordinal variables are considered and the identifiability constraint is reformulated such that means and covariance matrices can be computed. On the other hand, it is more flexible, since each component has its own mean vector and covariance matrix and it is computationally more efficient to be estimated, since a pairwise likelihood approach is suggested. Indeed in Everitt, the means and the variances of the latent variables are fixed to zero and one, respectively; the correlations are invariant and only the thresholds are free to change over the components. In comparison with the proposal of [17, 21] it is computationally feasible regardless of the number of variables involved. Moreover, it is able to recover the true partition and the true parameters (even if the accuracy depends on the sample size). For more details see [21].

5 A Comparison Between the Different Approaches

Fisher's Iris data is a well-known dataset in multivariate analysis. These data consist of 150 four-dimensional observations of three different species of Iris: Iris setosa, Iris versicolour and Iris virginica. For each plant, four continuous measurements have been observed: sepal length, sepal width, petal length and petal width. In order to analyse these data as ordinal data, the variables have been categorized. First of all, the variables have been normalized by the means and the standard deviations of the first group (Iris setosa). Then, the threshold parameters have been chosen equidistant and such that the cluster structure has not been completely destroyed by the categorization. We have compared the partitions obtained treating the ordinal variables as metric with those obtained treating the variables as they are, i.e. ordinal. In the first case we have applied k-means and the FMG; in the second case, we have performed tandem analysis (MCA followed by k-means) and the latent Gaussian mixture model for ordinal data proposed in [21]. The performance in recovering the true clustering structure has been evaluated through the adjusted rand index (ARI, [12]). In the following table we present the ARIs corresponding

Table 1 Adjusted rand indices—clustering performances for the fitted models

Ordinal variables as metric		Ordinal variables as ordinal	
k-means	0.6615	MCA & *k*-means (3 fact.)	0.5676
HomFMG(D)	0.6634	MCA & *k*-means (2 fact.)	0.7874
HomFMG(F)	0.5153	LMG (TV)	**0.9222**
HetFMG(F)	0.4128	LMG (RV)	**0.8005**

to different approaches and models: *k*-means; homoscedastic FMG with diagonal, HomFMG(D), and full, HomFMG(F), covariance matrix; heteroscedastic FMG with full, HetFMG(F), covariance matrices; MCA followed by *k*-means (with two or three factors) and latent mixture of Gaussians, LMG, for ordinal data. In the latter case we have considered different strategies to initialize the pairwise EM algorithm. Here, we have reported the results corresponding to two different initializations: starting from the true sample value of the parameters (TV) and starting for random values (RV), considering 1000 different starting points. In all other cases, the algorithms have been initialized randomly with 1000 different starting points. From the results, we can conclude that there is a reasonable difference in clustering performances comparing the two columns of Table 1. The poorest performance is given by the heteroscedastic Gaussian mixture with an ARI equal to 0.4128. As expected, the performances of *k*-means and HomFMG(D) are almost the same (0.6615 and 0.6634, respectively). As regards HomFMG(F), the ARI is lower (0.5153). Treating the ordinal variables as they are, the poorest performance is given by the MCA with three factors followed by *k*-means (0.5676). On the other hand, MCA with two factors followed by *k*-means yields a satisfactory ARI (0.7874). In classification terms, the best partition is obtained under the LMG with ARI equal to 0.9222, starting from the empirical true values. It results in an efficient way to cluster ordinal data; even if we initialize the pairwise EM algorithm randomly, the corresponding ARI is high enough (0.8005) to be the best clustering method. However, an open issue is finding an efficient strategy to initialize the pairwise EM algorithm especially when, as in this case, the sample size is small.

6 Concluding Remarks

In this paper, we briefly surveyed the more well-known clustering methods used to analyse ordinal data. Due to the lack of metric properties, clustering categorical data is more challenging and their graphical representation is more difficult in comparison with continuous data. Here, we have outlined some approaches used to cluster ordinal data, discussing their strengths and weaknesses: naive, tandem, distance-based and model-based. Then the differences in clustering performance have been illustrated through an application to real data. The theoretical comparison and the ARI values obtained in the application have shown that the ordinal variables have to be analysed taking their nature into account properly. In particular, the use of a model-based approach extending the URV methodology to cluster analysis seems to be the most appropriate strategy to cluster ordinal data. However, it suffers computational problems due to the numerical complexity implied by the computation of multidimensional integrals. As indicated in [21], a possible solution could be to estimate model parameters by maximizing the pairwise likelihood. Although there still remains some work to do, for example to find good starting points for the EM algorithm, the method seems to be promising.

References

1. Arabie, P., Hubert, L.: Cluster analysis in marketing research. In: Bagozzi, R. (ed.) Advanced Methods of Marketing Research. Oxford, Blackwell (1994)
2. Bezdek, J.C.: Pattern Recognition with Fuzzy Objective Function Algorithms. Academic Publishers, Dordrecht (1981)
3. Dempster, A.P., Laird, N.M., Rubin, D.B.: Maximum likelihood from incomplete data via the em algorithm. J. R. Stat. Soc. Ser. B (Methodol.) **39**(1), 1–38 (1977)
4. DeSantis, S.M., Houseman, E.A., Coull, B.A., Stemmer-Rachamimov, A., Betensky, R.A.: A penalized latent class model for ordinal data. Biostatistics **9**(2), 249–262 (2008)
5. Everitt, B.: A finite mixture model for the clustering of mixed-mode data. Stat. Probab. Lett. **6**(5), 305–309 (1988)
6. Everitt, B., Merette, C.: The clustering of mixed-mode data: a comparison of possible approaches. J. Appl. Stat. **17**(3), 283–297 (1990)
7. Fraley, C., Raftery, A.E.: Model-based clustering, discriminant analysis, and density estimation. J. Am. Stat. Assoc. **97**, 611–631 (2002)
8. Goodman, L.A.: Exploratory latent structure analysis using both identifiable and unidentifiable models. Biometrika **61**(2), 215–231 (1974)
9. Greenacre, M.: Theory and Applications of Correspondence Analysis. Academic Press, New York (1984)
10. Hathaway, R.J.: Another interpretation of the EM algorithm for mixture distributions. Stat. Probab. Lett. **4**(2), 53–56 (1986)
11. Hennig, C.: Methods for merging gaussian mixture components. Adv. Data Anal. Classif. **4**(1), 3–34 (2010)
12. Hubert, L., Arabie, P.: Comparing partitions. J. Classif. **2**(1), 193–218 (1985)
13. Hwang, H., Montréal, H., Dillon, W., Takane, Y.: An extension of multiple correspondence analysis for identifying heterogeneous subgroups of respondents. Psychometrika **71**(1), 161–171 (2006)
14. Iodice D'Enza, A., Palumbo, F.: Iterative factor clustering of binary data. Comput. Stat. **28**(2), 789–807 (2013)
15. Kaufman, L., Rousseeuw, P.J.: Finding Groups in Data: An Introduction to Cluster Analysis. Wiley Series in Probability and Statistics, 1st edn. Wiley, New York (2005)
16. Lindsay, B.G.: Composite likelihood methods. Contemp. Math. **80**, 221–239 (1988)
17. Lubke, G., Neale, M.: Distinguishing between latent classes and continuous factors with categorical outcomes: class invariance of parameters of factor mixture models. Multivar. Behav. Res. **43**(4), 592–620 (2008)
18. MacQueen, J.B.: Some methods for classification and analysis of multivariate observations. In: Proceedings of the Fifth Berkeley Symposium on Mathematical Statistics and Probability, vol. 1, pp. 281–297. University of California Press, California (1967)
19. McLachlan, G., Peel, D.: Finite Mixture Models. Wiley, New York (2000)

20. Meila, M.: The uniqueness of a good optimum for k-means. In: Proceedings of the 23rd International Conference on Machine Learning, pp. 625–632 (2006)
21. Ranalli, M., Rocci, R.: Mixture models for ordinal data: a pairwise likelihood approach. Stat. Comput. (2014). doi:10.1007/s11222-014-9543-4
22. Van Buuren, S., Heiser, W.J.: Clustering objects into k groups under optimal scaling of variables. Psychometrika **54**(4), 699–706 (1989)
23. Varin, C., Reid, N., Firth, D.: An overview of composite likelihood methods. Statistica Sinica **21**(1), 1–41 (2011)
24. Walesiak, M., Dudek, A.: Finding groups in ordinal data: an examination of some clustering procedures. In: Locarek-Junge, H., Weihs, C. (eds.) Classification as a Tool for Research. Studies in Classification, Data Analysis, and Knowledge Organization, pp. 185–192. Springer, Berlin (2010)
25. Young, F., Takane, Y., Leeuw, J.: The principal components of mixed measurement level multivariate data: an alternating least squares method with optimal scaling features. Psychometrika **43**(2), 279–281 (1978)

Novelty Detection with One-Class Support Vector Machines

John Shawe-Taylor and Blaž Žličar

Abstract In this paper we apply one-class support vector machine (OC-SVM) to identify potential anomalies in financial time series. We view anomalies as deviations from a prevalent distribution which is the main source behind the original signal. We are interested in detecting changes in the distribution and the timing of the occurrence of the anomalous behaviour in financial time series. The algorithm is applied to synthetic and empirical data. We find that our approach detects changes in anomalous behaviour in synthetic data sets and in several empirical data sets. However, it requires further work to ensure a satisfactory level of consistency and theoretical rigour.

Keywords Financial time series • Novelty detection • One-class SVM

1 Introduction

We apply one-class support vector machine (OC-SVM) to synthetic and empirical data and test its ability to detect anomalous behaviour in a time series. Anomalous behaviour in this case is a combination of consecutive data points in a time series that do not belong to a distribution identified by the algorithm. We first briefly introduce the theory behind the OC-SVM. Then we present its application to novelty detection in a time series by using lagged returns as inputs. Results, main conclusions and recommendations for further research are outlined at the end.

J. Shawe-Taylor • B. Žličar (✉)
Department of CS, University College London, London, UK
e-mail: j.shawe-taylor@cs.ucl.ac.uk; b.zlicar@cs.ucl.ac.uk

I. Morlini et al. (eds.), *Advances in Statistical Models for Data Analysis*,
Studies in Classification, Data Analysis, and Knowledge Organization,
DOI 10.1007/978-3-319-17377-1_24

2 Background: Novelty Detection and One-Class SVM

We begin by quoting a result from [1] that bounds the likelihood that data generated according to the same distribution used to train OC-SVM will generate a false alarm.

Theorem 1 *Fix $\gamma > 0$ and $\delta \in (0, 1)$. Let $(\mathbf{c}, r)$ be the centre and radius of a hypersphere in a feature space determined by a kernel $\kappa(\mathbf{x}, \mathbf{x}') = \langle \phi(\mathbf{x}), \phi(\mathbf{x}') \rangle$ from a training sample $S = \{\mathbf{x}_1, \ldots, \mathbf{x}_\ell\}$ drawn randomly according to a probability distribution $\mathscr{D}$. Let $g(\mathbf{x})$ be the function defined by*

$$g(\mathbf{x}) = \begin{cases} 0, & \text{if } \|\mathbf{c} - \boldsymbol{\phi}(\mathbf{x})\| \leq r; \\ \left(\|\mathbf{c} - \boldsymbol{\phi}(\mathbf{x})\|^2 - r^2\right)/\gamma, & \text{if } r^2 \leq \|\mathbf{c} - \boldsymbol{\phi}(\mathbf{x})\|^2 \leq r^2 + \gamma; \\ 1, & \text{otherwise}. \end{cases}$$

Then with probability at least $1 - \delta$ over samples of size ℓ we have

$$\mathbb{E}_{\mathscr{D}}[g(\mathbf{x})] \leq \frac{1}{\ell} \sum_{i=1}^{\ell} g(\mathbf{x}_i) + \frac{6R^2}{\gamma\sqrt{\ell}} + 3\sqrt{\frac{\ln(2/\delta)}{2\ell}},$$

where R is the radius of a ball in feature space centred at the origin containing the support of the distribution.

Hence, the support of the distribution outside the sphere of radius $r^2 + \gamma$ centred at $\mathbf{c}$ is bounded by the same quantity, since $g(\mathbf{x}) = 1$ for such inputs and is less than 1 elsewhere. Note moreover that the function $g(\mathbf{x})$ can be evaluated in kernel form if the optimisation is solved using its dual.

The theorem provides the theoretical basis for the application of the OC-SVM and it is perhaps worth dwelling for a moment on some implications for time-series analysis.

- Firstly, the assumption made by the theorem about the distribution $\mathscr{D}$ generating the training and test data has strong and weak elements:
  - It is strong in the sense that there are no assumptions made about the form of the input distribution $\mathscr{D}$. It therefore applies equally to long-tailed distributions as it does to multivariate Gaussians. We will perform some experiments with real-world data in which any assumptions about the form of the generating distribution would be difficult to justify.
  - It is weak in the sense that it assumes the training data are generated independently and identically (i.i.d.), something that will not to be strictly true for time series. This assumption becomes more reasonable when the training data are drawn from separate parts of the time series.

- The theorem is one sided in that it bounds the probability that data that arose from the training distribution are mistaken for novel outliers. It does not, however, say anything about the likelihood that novel data are not detected.

In connection with the final point, Vert and Vert [3] provide an interesting analysis showing that if we generate negative data from an artificial background measure μ and train as a 2-class SVM, in the limit of large data the SVM will identify the level sets of the pdf of the training distribution relative to μ. This suggests that it finds the minimal density with respect to μ consistent with capturing a given fraction of the input distribution. Hence, in this case, we can make assertions about the efficiency with which outliers are detected.

3 Problem: Novelty Detection in Financial Time Series

In order to apply OC-SVM to a single time series we follow the approach proposed by Ma and Perkins [2]. We extend this approach by adding an exponential decay parameter so that the more recent lags carry more weight than the older lags, since we are interested in detecting anomalies in the very short window before the occurrence of the extreme volatility, the underlying hypothesis being that the behaviour of the market changes before the occurrence of the spike in the time series.

3.1 Data Preprocessing

A data matrix is constructed in such a way so that the first column represents the original time series and every next column is a lag of the previous column. More specifically, for a time series variable x composed of observations $x(t)$ where $t = 1, \ldots, T$ (T being the number of time points, observations) we perform a vector-to-matrix transformation so that the dimensionality of the original column vector $\mathbf{x}$ changes from $Tx1$ to $(T\text{-}d\text{-}1)$ x d forming a data matrix X. Here d represents our choice of the dimensionality expansion, i.e. the number of all columns in the newly formed matrix X in effect reflecting the number of lags we chose to include in the analysis. Alternatively, we can think of this in terms of extending the dimensionality of a data point $x(t)$ to a row vector $\mathbf{x}$(t) so that

$$\mathbf{x}(t) = [x(t) \ldots x(t-d+2) \;\; x(t-d+1)] \tag{1}$$

Then the newly formed data matrix in terms of row vectors becomes

$$X = [\mathbf{x}(d), \ldots, \mathbf{x}(T)]^{\mathsf{T}} \tag{2}$$

with dimensions *(T-d-1)* x *d* (as suggested by Ma and Perkins [2]). We then take a step further and add a decay parameter c so that the weight of each next column falls exponentially with each lag. If we denote a row vector $\mathbf{d} = [1, \ldots, d]$ then we define $c^{\mathbf{d}}$ to be the row vector

$$\mathbf{c} = c^{\mathbf{d}} = [c^1, \ldots, c^d] \tag{3}$$

where c is an arbitrarily chosen decay parameter $0 < c < 1$. The new data matrix taking into account the exponential time decay is then

$$X_c = X \odot D \tag{4}$$

where D is a matrix of decay factors $D = \mathbf{1}^\mathsf{T} * \mathbf{c}$ and multiplication between X and D is element by element multiplication. Alternatively, if we denote the number of columns in X as $j = 1, \ldots, d$, then we can simply define the matrix D as having entries $D_{ij} = c^j$. X_c is then centred row-wise in a standard manner using a centering matrix C

$$C = I_d - \frac{1}{d} O_d \tag{5}$$

so that the final centred data matrix with a time decay is:

$$X_c^c = X_c\, C \tag{6}$$

3.2 Novelty Detection Algorithm

In this section we present a step-by-step pseudo-algorithm of OC-SVM based novelty detection in time-series analysis.

Input: a time series *x(t)* of length T. **Output:** points in time identified as novelties.

---

(1) Vector-matrix transformation: Calculate X_c^c using a range of different lags $d = [2 : 20]$ to obtain 19 matrices of different dimensions $XE = [X_2^c \ldots X_{20}^c]$. The value of the decay parameter is set arbitrarily at $c = 0.97$.

(2) Data sets: Each X_E^c is split in a train set (*X-train*) of length $\frac{2}{3}T$ and the remaining third of observations comprise a test set (*X-test*). Further split *X-train* in half, that is into a *sub-X-train* and a *val-X-train* set.

(3) Train OC-SVM: Apply OC-SVM to *sub-X-train* so as to obtain α_i for each X_c^c in the array of matrices $XE = [X_2^c \ldots X_{20}^c]$ individually and for all values of

$\gamma = 2^i$ (where $i = [-10 : 10]$)[1] and $\nu = 2^j$ where ($j = [-15 : -1]$). Then find pseudo-optimal d_o, γ_o and sensitivity parameter ν_o by locating the OC-SVM that achieved the highest accuracy on *val-X-train*.[2]

(4) Test optimal OC-SVM(d_o,γ_o,ν_o): Use pseudo-optimal values determined in (3) to train OC-SVM(d_o,γ_o,ν_o) on *X-train*. Test the model on *X-test* and obtain the novelty signal for the test set.

---

4 Experiments

Firstly, we describe the construction of synthetic time series and present the empirical data sets (three stock market indices). Next, we comment on the results and outline the challenges.

4.1 Data

Both synthetic and empirical data sets are of about the same length ($T \approxeq 5800$). Synthetic time series are comprised of an original signal in the training set while in the test sets we add anomalies (i.e. time intervals where the original time series is corrupted by an anomalous signal) on the intervals $T = [5000{:}5050, 5300{:}5350, 5600{:}5650]$. We train the OC-SVM algorithm on a data set comprised solely of original (non-anomalous) time series and then test the optimal specification of the model on a test set that includes pre-defined intervals with anomalies. Synthetic time series are constructed as follows:

$$x(t) = \begin{cases} x_a(t)\ for\ t \in [5000, 5050] \wedge [5300, 5350] \wedge [5600, 5650] \\ x_o(t)\ for\ t \notin [5000, 5050] \wedge [5300, 5350] \wedge [5600, 5650] \end{cases} \tag{7}$$

Here $x_o(t)$ denotes the original time series and $x_a(t)$ denotes the anomalous time series. Synthetic data sets are then the following three time series types:

1. **Synthetic 1** time series is a sinusoid with a small standardised random noise in the training set, but with increased standard deviation of the error process on

---

[1] We use radial basis kernel (RBF) so that $k(\mathbf{x}, \mathbf{y}) = exp(-\gamma||\mathbf{x} - \mathbf{y}||^2)$.

[2] We use a term pseudo-optimal since we simply choose a specification that is able to contain all training data and consequently label *sub-X-train* data sample as novelty-free. Clearly, this is not necessarily the optimal solution nor is it the only solution and presents one of the main challenges related to novelty detection with OC-SVM.

specific intervals within the test set. The *synthetic 1* signal follows

$$x_1(t) = sin(\frac{40\pi}{N}) + \epsilon(t) \tag{8}$$

where

$$\epsilon(t) = \sigma\ z(t) \tag{9}$$

and $z(t) \sim N(0, 1)$ with sigma term in the error process equal to $\sigma_o = 0.1$ for the original signal x_o and a slightly higher sigma term $\sigma_a = 0.2$ for the anomalous signal x_a.

2. **Synthetic 2** time series is constructed as a random process taking into account the empirical (sample) mean and standard deviation of an empirical time series (in our case the S&P500 stock market index). The original time series is constructed by adding an error term sampled from a standardised normal distribution and the anomalous signal is obtained by adding an error noise sampled from a student-t distribution. We write the original time series as

$$x_2(t) = \mu + \epsilon(t) \tag{10}$$

where

$$\epsilon_o(t) = \sigma_{SP500}\ z(t) \tag{11}$$

and $z(t) \sim N(0, 1)$.

The anomalous signal follows the same process only with its noise term sampled from student-t distribution with six degrees of freedom

$$\epsilon_a(t) = \sigma_{SP500}\ z(t) \tag{12}$$

and $z(t) \sim t_6$.

3. **Synthetic 3** time series is obtained by subtracting the mean from the *synthetic 2* signal and then taking the absolute value of the obtained time series. Such absolute returns are often used as a proxy for a volatility process in financial research. In other words, the *synthetic 3* signal is equal to absolute error term in Eq. (10).

$$x_3(t) = |\epsilon(t)| = |\sigma_{SP500}\ z(t)| \tag{13}$$

with $z(t) \sim N(0, 1)$ for the original time series and $z(t) \sim t_6$ for the anomalous time series.

Empirical data sets are time series of three stock market indices: S&P500, DAX30 and NIKKEI225. We work with adjusted daily closing prices obtained from *Yahoo.Finance* for a period of approximately 13 years, where we perform the following two transformations of the original series:

1. **Log returns** are calculated by obtaining the difference between natural logarithms of a price at time t and price at time $t - 1$:

$$r(t) = \ln(\frac{p_t}{p_{t-1}}) \tag{14}$$

   Then we subtract the mean of the return time series:

$$r_{\text{dm}}(t) = r(t) - \mu \tag{15}$$

2. **Absolute returns** are obtained by simply taking the absolute value of the log returns. Absolute returns are a frequently used proxy for a volatility measure in financial research

$$r_{\text{abs}}(t) = |r(t)| \tag{16}$$

4.2 Results

In this section we describe the results for OC-SVM algorithm applied to synthetic and empirical data sets without and with the exponential decay parameter in the preprocessing phase. The two algorithms are denoted with OC-SVM-ND (no decay) and OC-SVM-ED (exponential decay), respectively. Optimal model specification is indicated by adding optimal parameter values in brackets so that OC-SVM(d_o,γ_o,ν_o) denotes the specification of OC-SVM with optimal index values for the dimension (indicating number of lags), γ in RBF kernel and ν parameter in OC-SVM.[3] Please note that this is a naive pseudo-optimisation simply assuming that the best OC-SVM is the one that is able to put a bound around the data in a training set. We use LIBSVM support vector machine toolbox. Finally, figures are moved to Appendix to prevent cluttering.

[3]Where the numbers refer to the index not the value itself. For example, OC-SVM-ND (1,2,3) would denote the optimal specification of OC-SVM without the decay parameter, where the optimal lag dimension is the first dimension in the dimension array $d = [2 : 20]$, i.e. $d_o = 2$, the optimal ν refers to the second position in the j array $j = [-15 : -1]$, i.e. $\nu_o = 2^{-14}$, and the optimal γ refers to the third position in the i array $i = [-10 : 10]$, i.e. $\gamma_o = 2^{-8}$.

4.2.1 Synthetic Data

Synthetic time series are constructed so that we can investigate the performance and the ability of OC-SVM to detect novelties that were artificially inserted in the testing part of the various types of data sets. Ideally, no novelties would be detected in the valuation part of the training set. In the test set the best performance is the one detecting novelties in the previously determined anomalous intervals.

Figure 1 displays the results for valuation part of the training set on the left side (with *synthetic* 1 at the top and OC-SVM-ND novelty signal at the bottom) and the test set on the right side. It shows that the model has the ability to learn the original signal in the training set since it correctly detects no novelties (novelty signal is equal to 1 at all times). When the same model is applied to the test set we see that it correctly identifies areas where anomalous data have been added to the original signal (grey areas). However, it also falsely returns novelty signal where no novelties have been added to the original synthetic signal, indicating that this particular OC-SVM specification is perhaps still too sensitive to outliers. Figure 2 shows results of the OC-SVM-ED (with exponential decay) applied to the same *synthetic 1* time series. This model is also successful in identifying the anomalous areas in the test set with slightly lower number of false positives. The results for other two synthetic signals are displayed in Figs. 3, 4, 5, and 6. Both algorithms, without and with decay parameter, are able to detect the anomalous areas with a small number of false positives. Only Fig. 3 stands out as it displays a poor performance of OC-SVM-ND in the test set of a *synthetic 2* time series.

4.2.2 Empirical Data

Our empirical experiments focus on whether or not the algorithm detects anomalies slightly before volatility spikes. Figures 7 and 8 display results of OC-SVM applied to absolute returns of the empirical time series without and with decay, respectively. Figure 9 shows the valuation part of a training sample on the left and test sample on the right side for the S&P500 stock market index. Top row displays the time series of S&P500, middle row the return time series and bottom row the novelty signal for OC-SVM-ND. In this case our algorithm detects two intervals as anomalous (around time points 600 and 1300). Figure 9 displays identical figure twice with the only difference being that the two charts in fifth and sixth row on the right side are magnified around the novelty point so as to show the case of early novelty detection (around point 600). However, the volatility spike around time point 1300 is not detected in advance. The same results in both of these volatility cases are obtained when OC-SVM is applied using the exponential decay parameter (Fig. 10). Also, the results are very similar when both types of algorithms (with and without the decay parameter) are applied to S&P500 absolute returns time series. In case of the DAX30 index none of the two algorithms detect sudden increase in volatility in advance when applied to time series of returns (Figs. 11 and 12). When applied to absolute returns (Figs. 13 and 14) both algorithms (OC-SVM-ND and OC-SVM-ED) detect the second volatility spike in advance (around the time point 550) but their first detections (around the time point 350) temporally coincide with the volatility spike. In case of the NIKKEI225 index both OC-SVM-ND and OC-SVM-ED detect the biggest volatility increase in advance (around the time point 600) when applied to returns and absolute returns. However, both specifications also fail to detect in advance the second spike around the time point 1200, again when applied to either returns or absolute returns (Figs. 15, 16, 17 and 18).

5 Conclusions and Further Research

In this paper we investigate the application of the OC-SVM to novelty detection in financial time series. We add an exponential decay parameter when preprocessing the data to account for the reduced dependency related to older data points. We test the OC-SVM on synthetic data sets and find that our algorithm manages to consistently identify anomalous areas inserted artificially in our test sets. Building on these results we then apply the algorithm to empirical data, namely financial time series of three stock market indices: S&P500, DAX30 and NIKKEI225. The idea is that the projection of the market data into the feature space effectively allows for an inspection of market patterns we would normally not detect in the input space. This means that in cases when anomalous behaviour in the markets (reflected in the change of the distribution of the time series) has preceded the spike, our algorithm might be able to detect these anomalies. However, when the spike in volatility is the result of an unexpected exogenous event, OC-SVM will not be able to alert the user in advance since the time series is not reflecting the impending risk.[4] Our experiments to some extent support this reasoning as the results show instances where OC-SVM, with and without a decay parameter, detects novelties occurring before volatility spikes, but such results are by no means conclusive. The unsupervised nature of OC-SVM allows for the detection of previously unseen observation, however it simultaneously prevents us from targeting a type of process (event). This makes it useful for novelty detection in synthetic data sets (where novelty points are known in advance) while making it problematic for the application to empirical data sets.[5] Optimisation of the hyperparameter ν in OC-SVM is a challenge in itself and when applied to financial time-series analysis this problem becomes even more difficult. Future research could perhaps investigate a possible connection between ν and the level of randomness of the underlying time series. Also, it would be interesting to investigate the usefulness of one-class SVM for novelty detection in multivariate data.

[4]Note that for the synthetic data this does not arise since the volatility takes immediate effect.

[5]Especially when applied to extremely noisy data such as financial time series.

Appendix

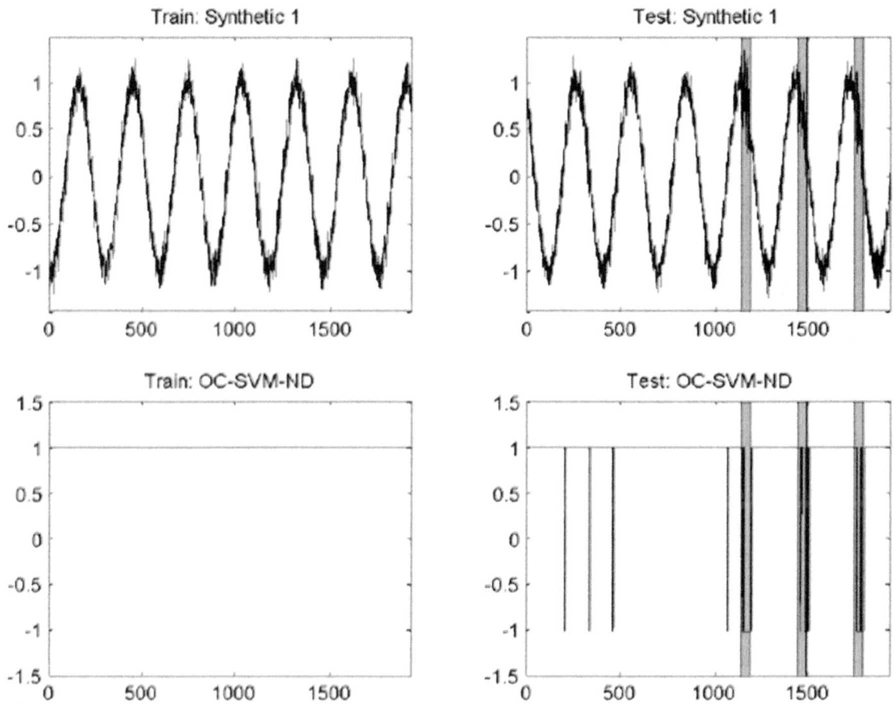

Fig. 1 Synthetic 1: OC-SVM-ND (19,7,1)

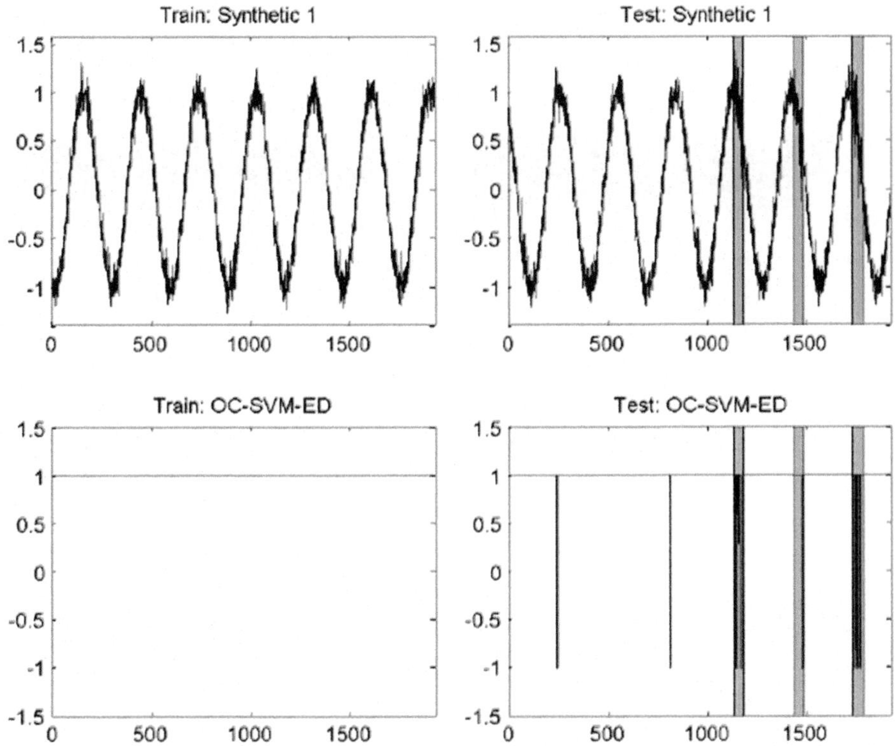

Fig. 2 Synthetic 1: OC-SVM-ED (12,7,1)

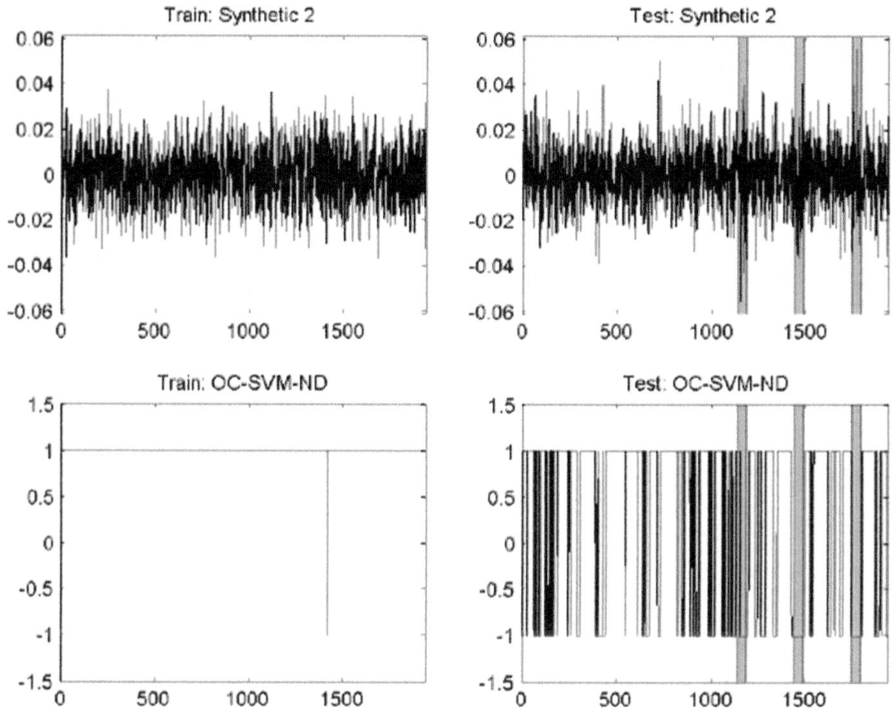

Fig. 3 Synthetic 2: OC-SVM-ND (15,9,5)

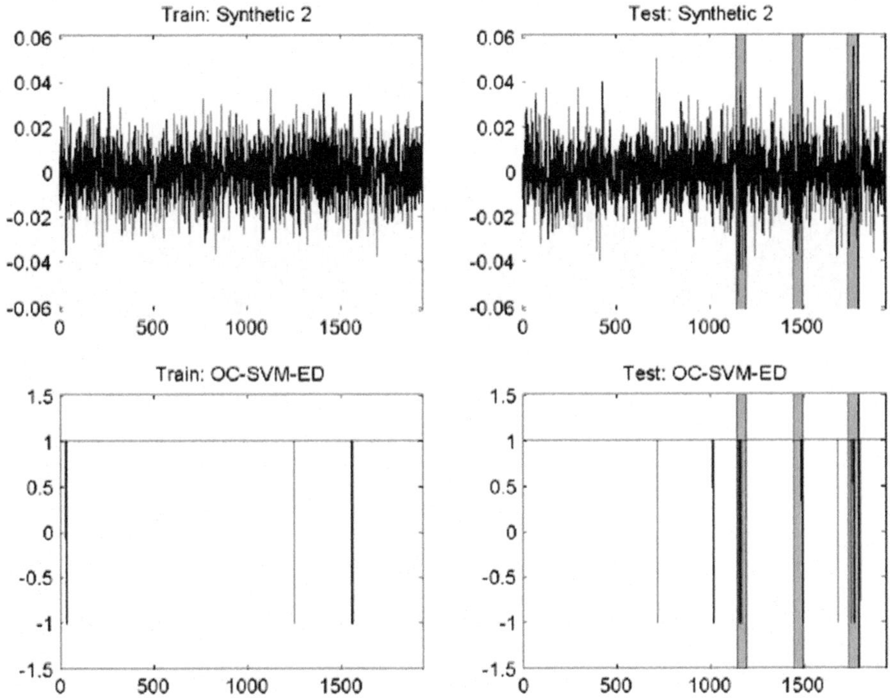

Fig. 4 Synthetic 2: OC-SVM-ED (1,7,9)

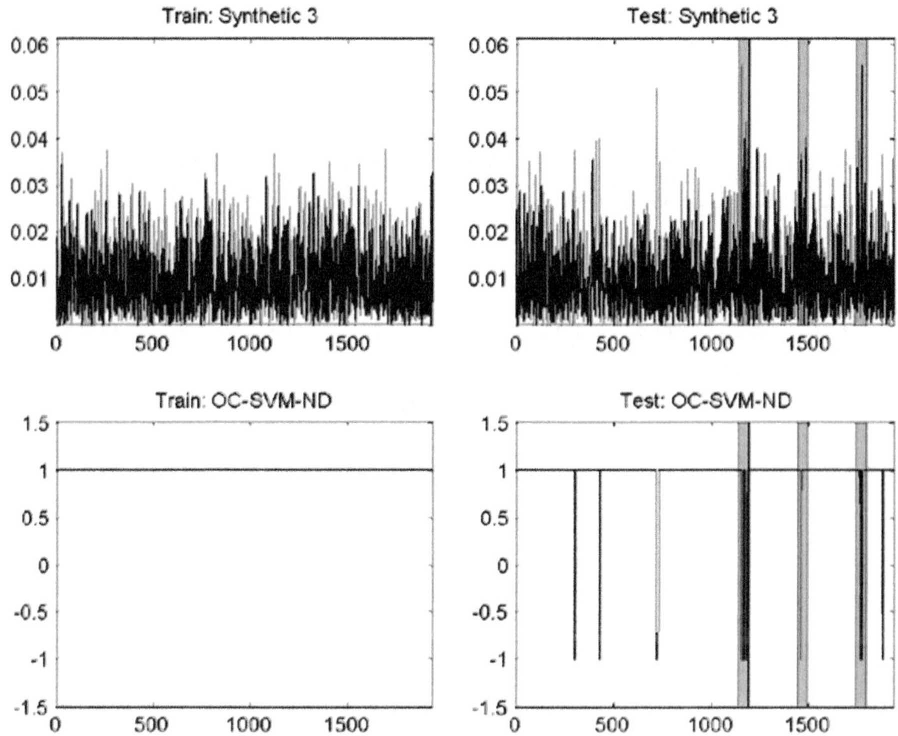

Fig. 5 Synthetic 3: OC-SVM-ND (8,8,8)

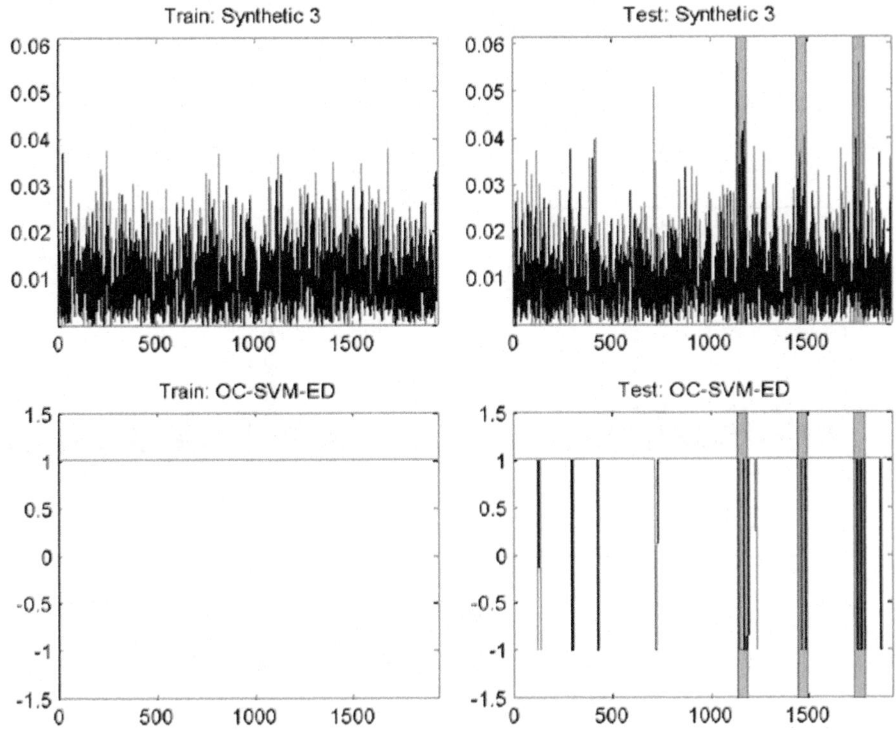

Fig. 6 Synthetic 3: OC-SVM-ED (13,9,7)

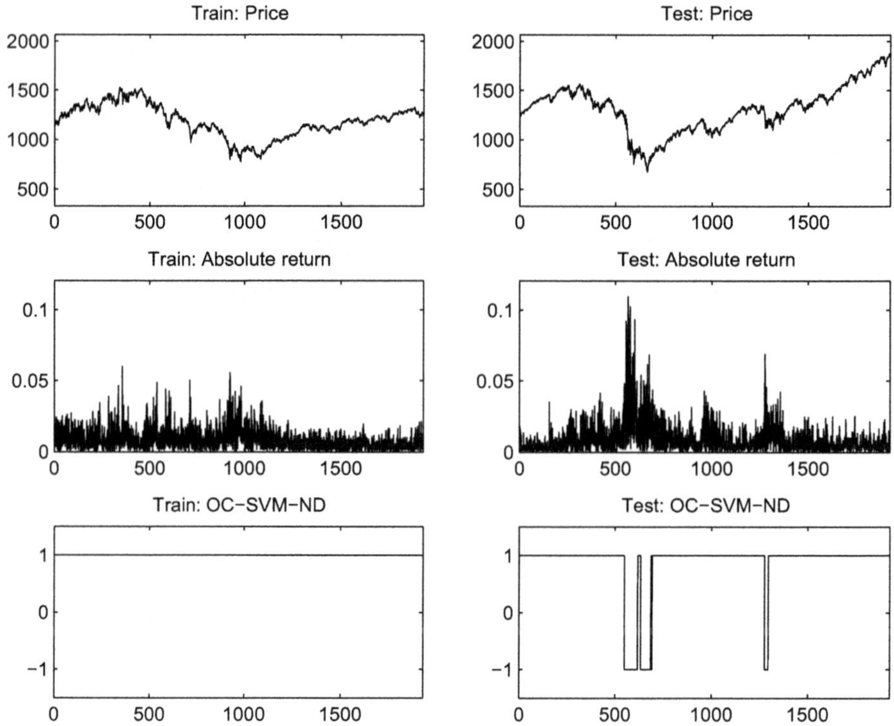

Fig. 7 S&P500 absolute returns: OC-SVM-ND (17,10,4)

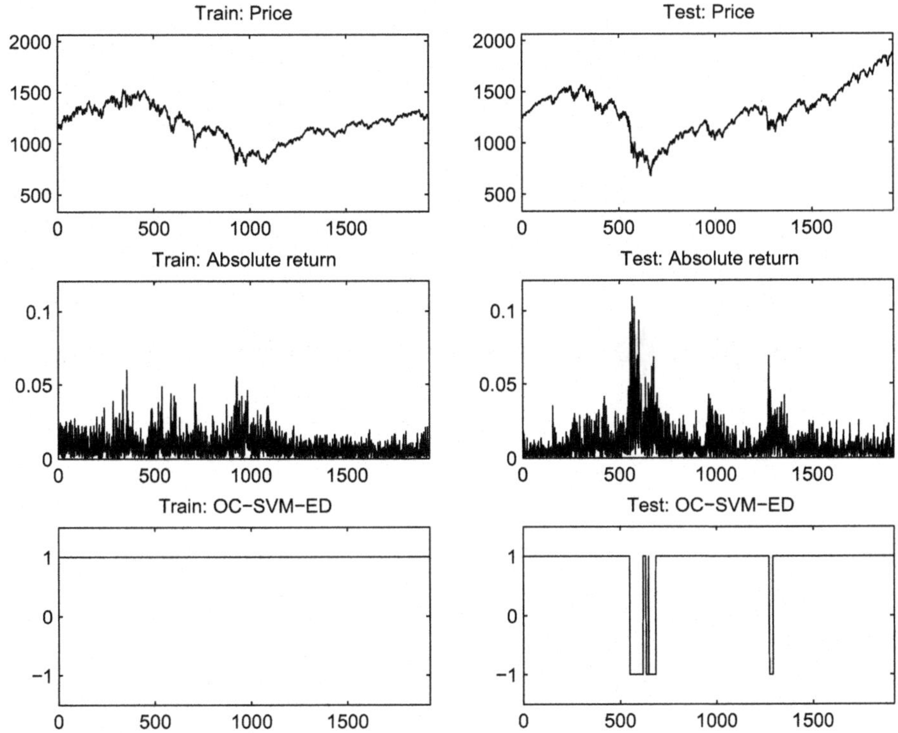

Fig. 8 S&P500 absolute returns: OC-SVM-ED (17,10,5)

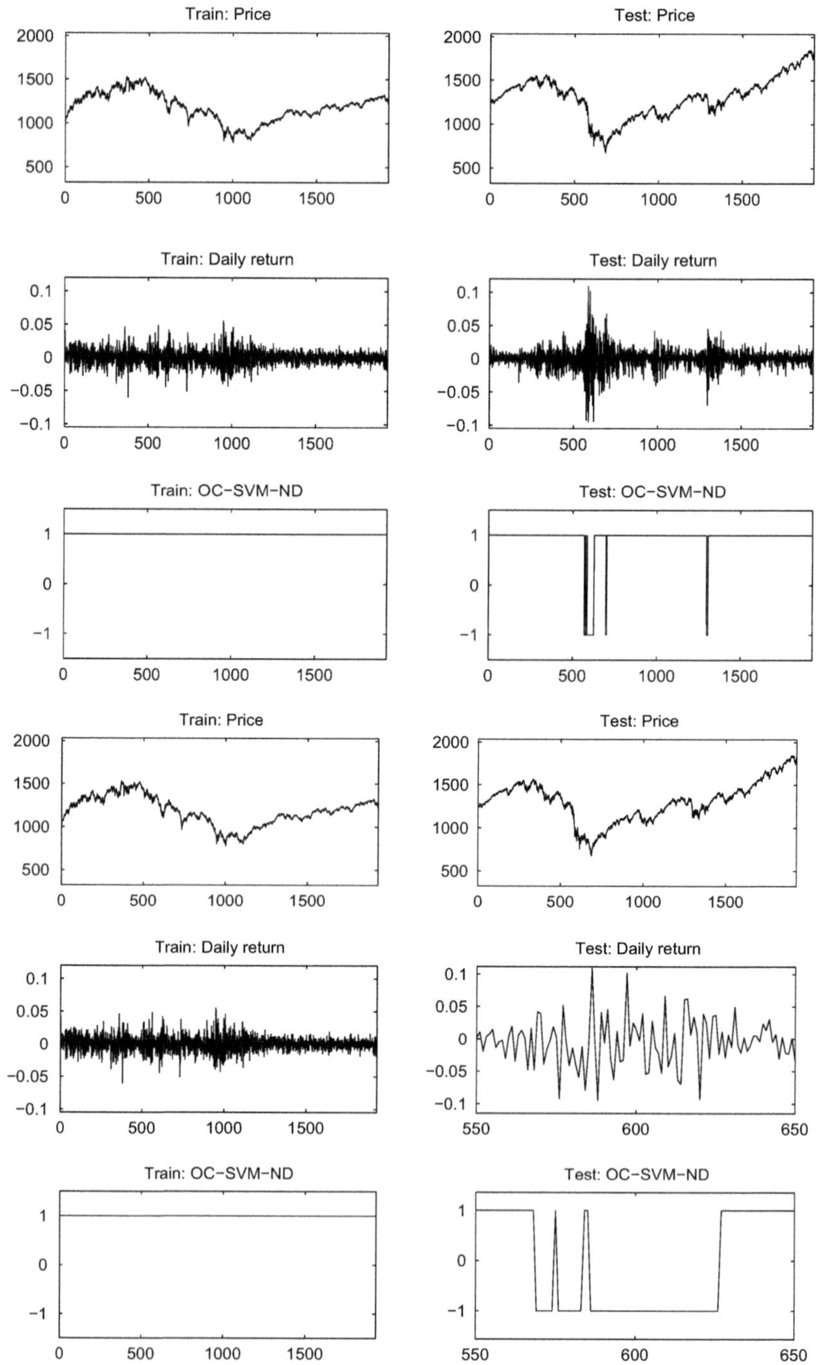

Fig. 9 S&P500 returns: OC-SVM-ND (6,8,5); *bottom figure* is identical to the *top* one with magnified *bottom charts* (*Test: Daily return* and *Test: OC-SVM-ND*) to demonstrate the early novelty detection by the algorithm

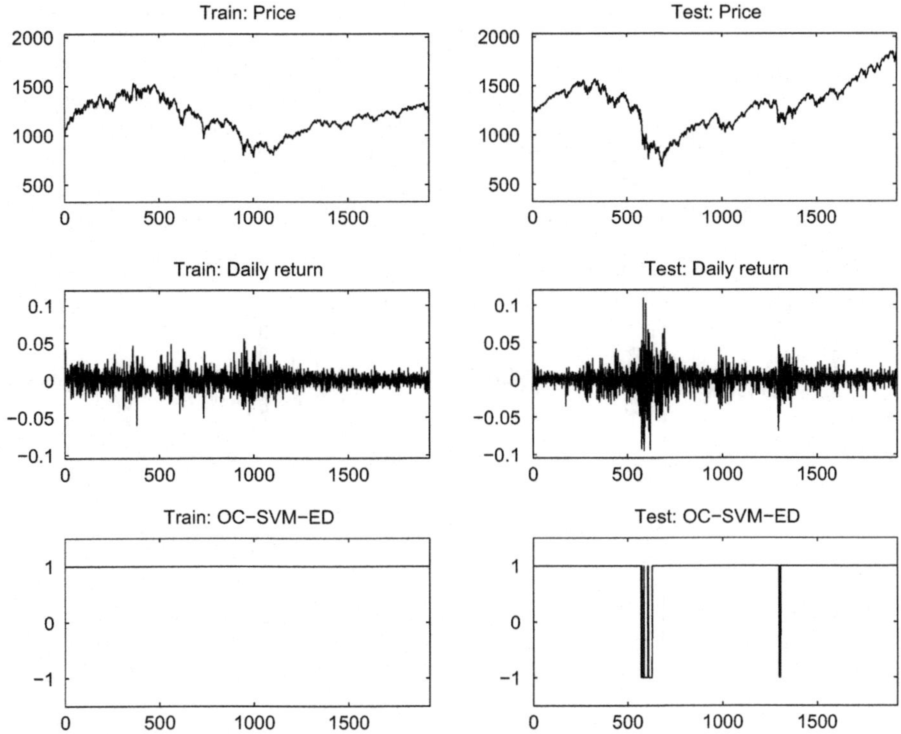

Fig. 10 S&P500 returns: OC-SVM-ED (7,7,7)

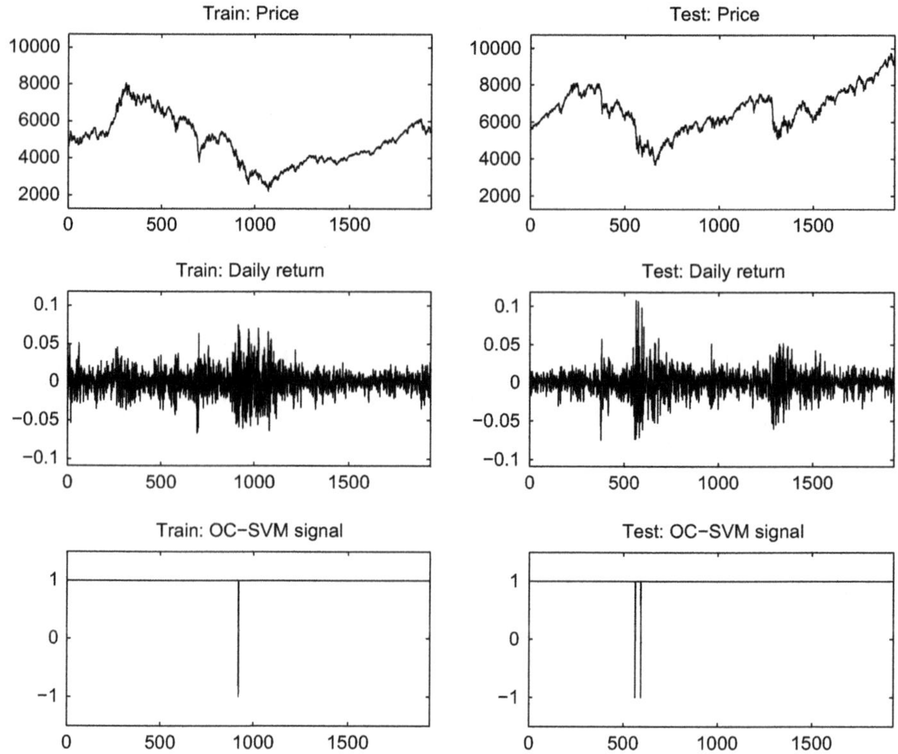

Fig. 11 DAX returns: OC-SVM-ND (1,6,8)

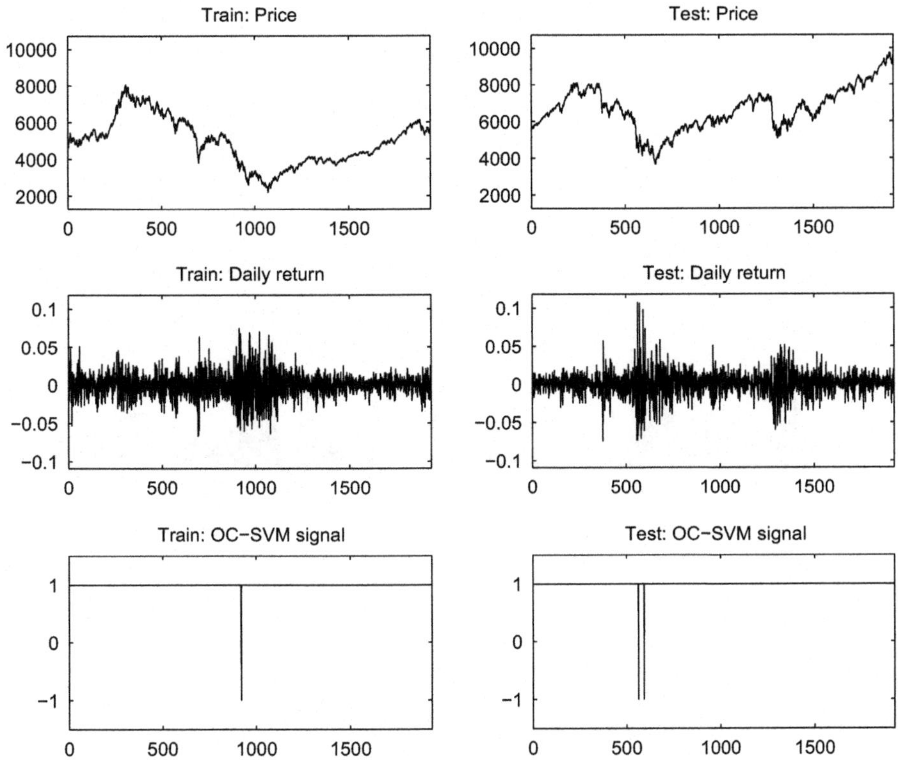

Fig. 12 DAX returns: OC-SVM-ED (1,6,8)

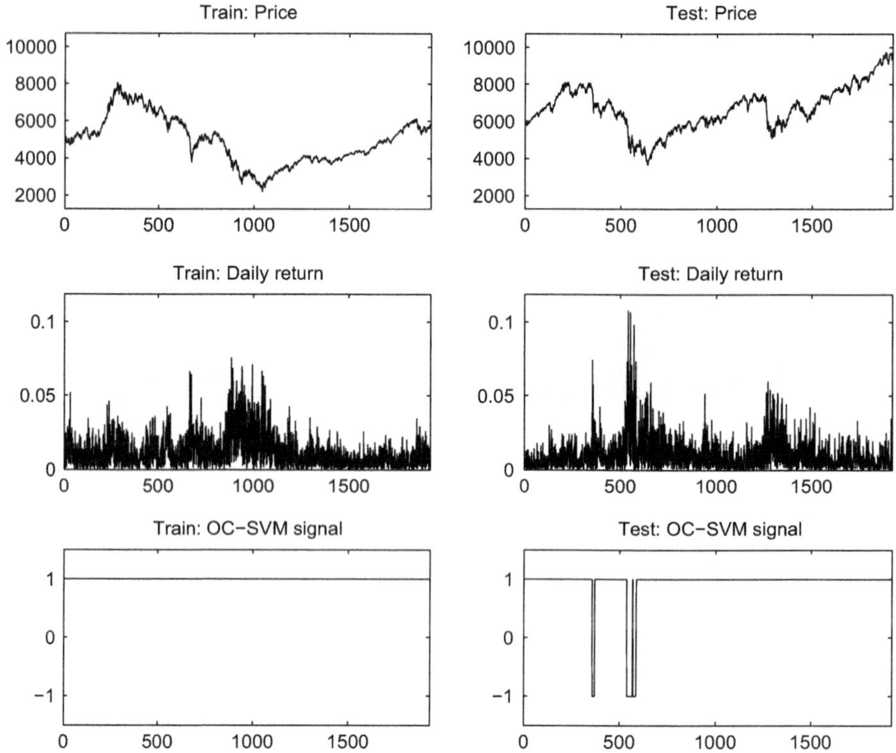

Fig. 13 DAX absolute returns: OC-SVM-ND (15,9,4)

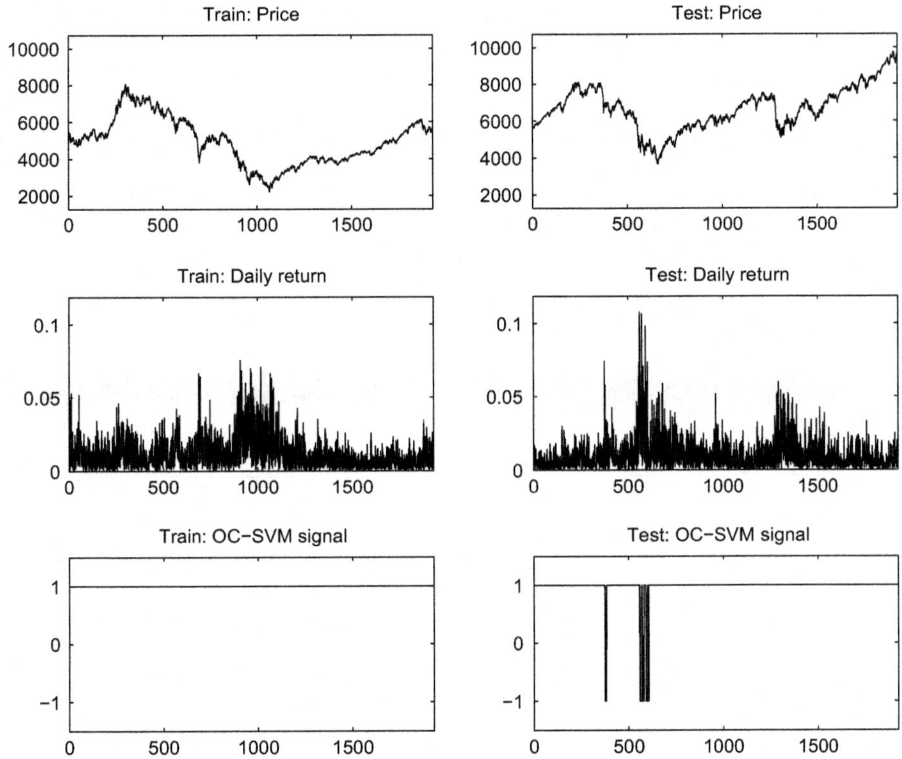

Fig. 14 DAX absolute returns: OC-SVM-ED (7,8,6)

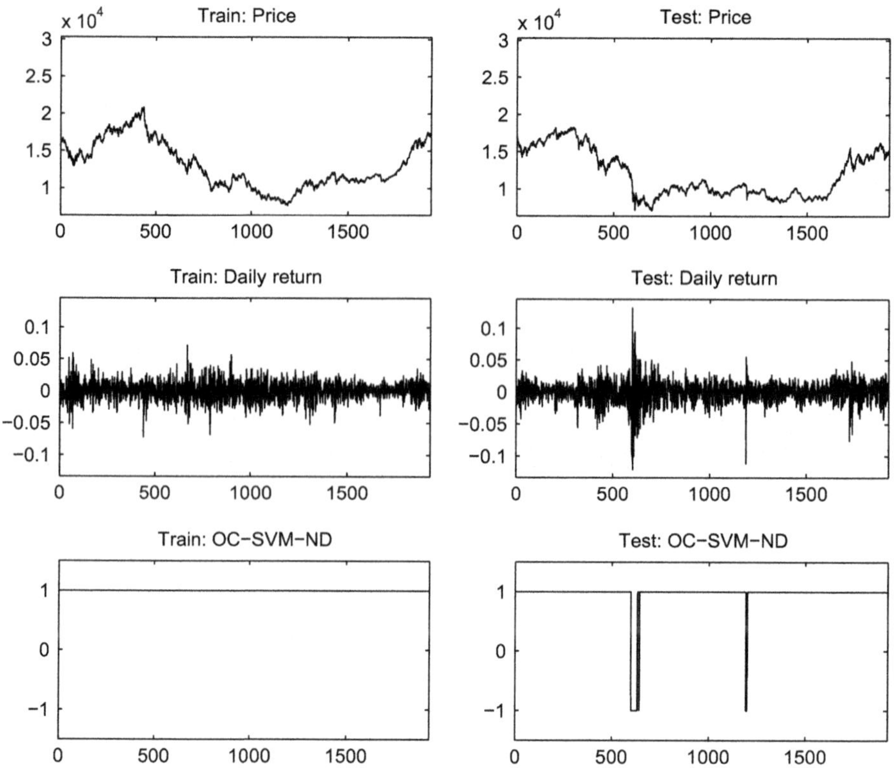

Fig. 15 NIKKEI225 returns: OC-SVM-ND (16,9,1)

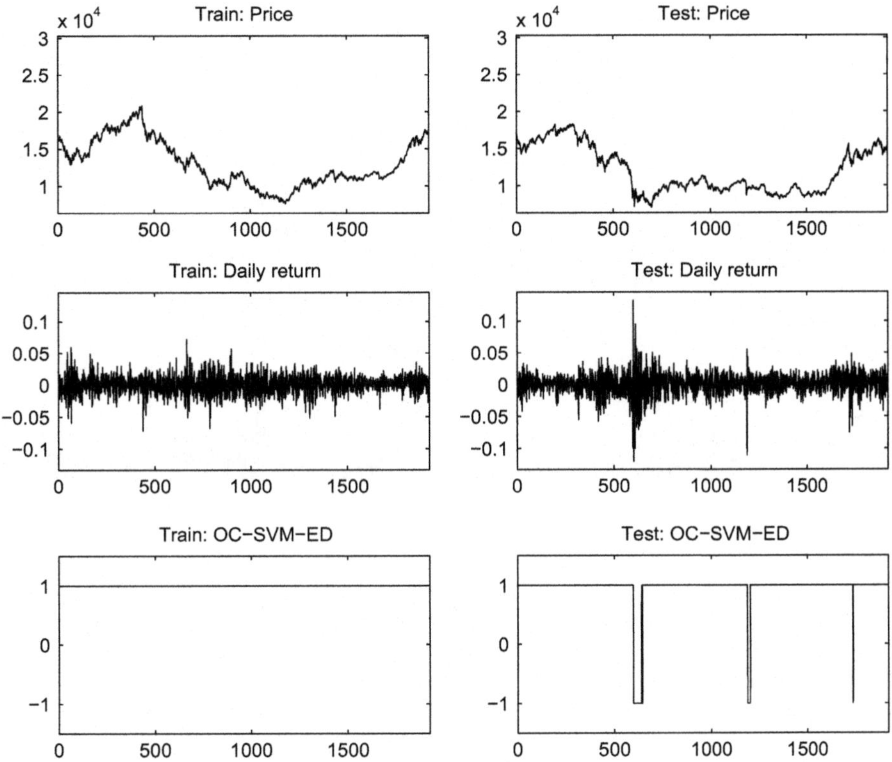

Fig. 16 NIKKEI225 returns: OC-SVM-ED (16,9,1)

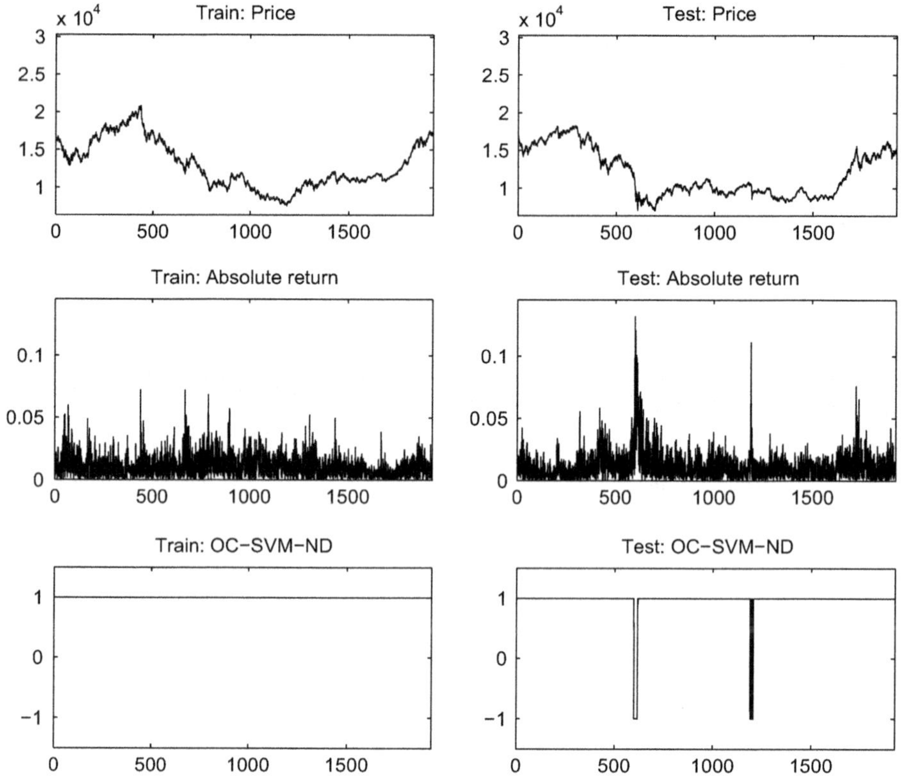

Fig. 17 NIKKEI225 absolute returns: OC-SVM-ND (16,9,1)

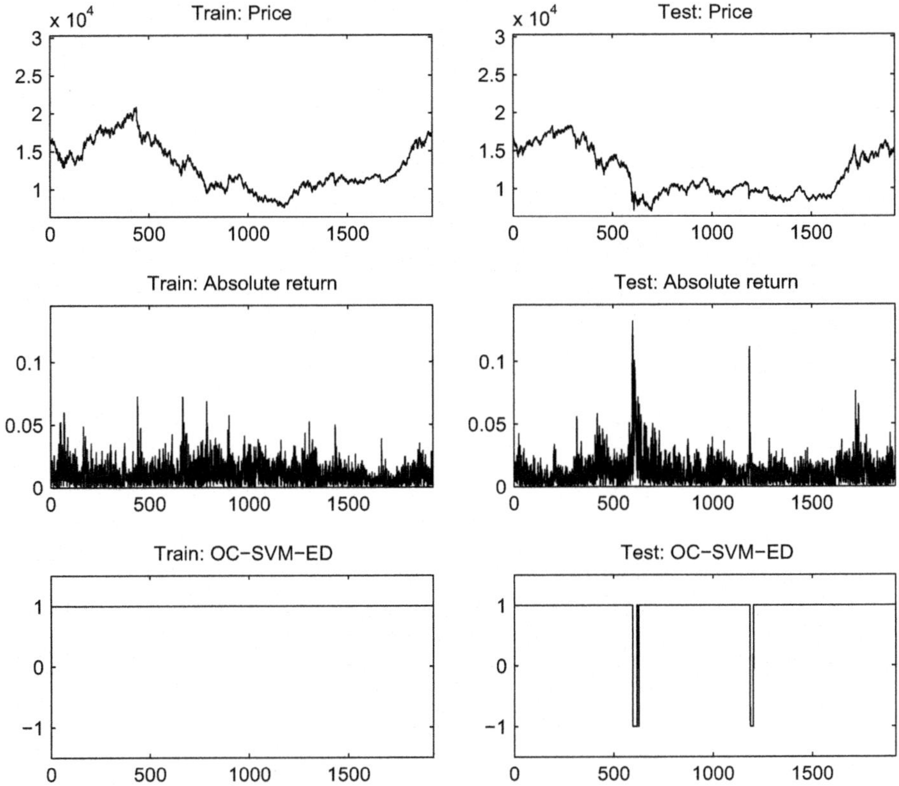

Fig. 18 NIKKEI225 absolute returns: OC-SVM-ED (16,9,1)

References

1. Cristianini, N., Shawe-Taylor, J.: Kernel Methods for Pattern Analysis. Cambridge University Press, Cambridge (2004)
2. Ma, J., Perkins, S.: Time-series novelty detection using one-class support vector machines. In: Proceedings of the International Joint Conference on Neural Networks (2003)
3. Vert, R., Vert, J-P.: Consistency and convergence rates of one-class SVMs and related algorithms. J. Mach. Learn. Res. **7**, 815–854 (2006)

Using Discrete-Time Multistate Models to Analyze Students' University Pathways

Isabella Sulis, Francesca Giambona, and Nicola Tedesco

Abstract The methodologies adopted in the last decades to analyze students' university careers using cohort studies focus mainly on the risk to observe one of the possible competing states, specifically dropout or graduation, after several years of follow-up. In this perspective all the other event types that may prevent the occurrence of the target event are treated as censored observations. A broader analysis of students' university careers from undergraduate to postgraduate status reveals that several competing and noncompeting events may occur, some of which have been denoted as absorbing while others as intermediate. In this study we propose to use multistate models to analyze the complexity of students' careers and to assess how the risk to experience different states varies along the time for students' with different profiles. An application is provided to show the usefulness of this approach.

Keywords Cohort studies • Multistate models • Risk factors • Students' careers

1 Introduction

The higher education system in Italy has been widely criticized for its ineffectiveness since it is characterized by lower university graduation rates comparing to the European countries. The efficiency of the formative processes, mainly measured in terms of regularity of students'careers, is one of the dimensions on the basis of which the central government assesses the quality of the institutions and allocates the financial support to them. For this reason the last governmental reforms had as priorities the increasing of the retention rate, the reduction in time to degree and the lowering of the average age of graduates. The building up of a system of efficiency/effectiveness indicators has been a good starting point to assess strengths and weaknesses of the institutions and to make comparative assessments across them [1, 3, 9]. At the same time, the information provided by the joint use of ad

I. Sulis (✉) • F. Giambona • N. Tedesco
Dipartimento di Scienze Sociali e delle Istituzioni, Università di Cagliari, Cigliari, Italy
e-mail: isulis@unica.it

I. Morlini et al. (eds.), *Advances in Statistical Models for Data Analysis*,
Studies in Classification, Data Analysis, and Knowledge Organization,
DOI 10.1007/978-3-319-17377-1_25

hoc surveys and administrative archives has allowed policy makers to shed some light on the main determinants of students' risk of accumulating inefficiencies or failures along the university studies [1, 13].

The researches carried out in the last two decades mainly focus on two main directions: (i) the refinement of suitable methodological approaches to identify risk factors of non-regularity in university students' careers; (ii) the following up of students' careers using cohort studies [11, 12]. Furthermore, most authors investigated the relationship between the probability of having a successful or unsuccessful career (e.g., graduation vs dropout) and students' socioeconomic educational and cultural characteristics, without considering the complexity of the overall university pathway[8, 10, 13]. A broader analysis of students' university careers from undergraduate to postgraduate status reveals that after the enrolment (initial state) several sequences of events can occur, some of which have been denoted as absorbing while others as intermediate. The absorbing states are those from which it is not possible to make transitions, whereas the intermediate states are those that students can reach after having experienced some previous states and from which they can transit to others [5, 7]. Furthermore, when the occurrence of a state may prevent the occurrence of others they need to be treated in a modelling approach as competing states. Multistate models [2, 5–7] combine peculiarities of competing risk models but allow also to deal with intermediate and recurrent events.

The main aim of this paper is to explore the usefulness of multistate models as methodological approach for analyzing the complexity of university students' careers from undergraduate to postgraduate. Specifically, the main potential of multistate models is that they allow us to: (i) consider the sequence of events experienced by students during their careers (first level graduation, second level graduation, dropout at different levels, postgraduate studies, etc.); (ii) analyze the role played by students' socio-anagraphic characteristics in shaping students' academic careers; (iii) assess how the risk to experience different states varies along the time [2, 11].

2 Methodological Approach to Analyze Students' Careers

Standard survival analysis methodologies have been widely adopted in the analysis of students' careers [8, 13]. However, classical approaches drastically simplify the complex structure of university students' careers bounding the analysis to a single level of the university studies (first level/second level/postgraduate) and mainly focusing on the risk to observe one or more of the possible competing states (dropout/graduation) during the follow-up and treating all the other event types, which may prevent the occurrence of the target events, as censored observations [8, 10, 11]. Multistate models are systems of multivariate survival equations which allow us to assess the risk of experiencing several types of competing and noncompeting events and to move through a series of concatenate states following certain paths of possible transitions [2, 5–7]. In this sense, they are considered an extension of the class of the competing risk models which from an initial state allow just the transition to several mutually exclusive absorbing states. Multistate models generalize the potential of competing risk models allowing the presence of intermediate events. To summarize we can say that multistate models depart from classical event-history since they allow researchers to jointly deal with (i) several absorbing (end-points) events, (ii) several intermediate events, and (iii) several types of censored events [7].

By denoting with T the time of reaching state j from state i, the hazard rate of the transition $i \rightarrow j$ is defined by

$$\lambda_{ij}(t) = \lim_{\Delta_t \downarrow 0} \frac{\text{Prob}(t \leq T < t + \Delta t | T \geq t)}{\Delta t}. \tag{1}$$

The risk of transition across the states of students' careers can be described by a set of hazard ratios (HRs), which vary as function of students or degree programmes, covariates (Z) [2, 6, 7]. In this first explorative analysis we adopted a Cox's proportional-hazards model [4]:

$$\lambda_{ij}(t|Z) = \lambda_{ij,0}(t) \exp(\beta_{ij}^T Z) \tag{2}$$

where $\lambda_{ij,0}(t)$ is the baseline hazard of transition at time t from $i \rightarrow j$ for students with baseline covariates pattern and $\lambda_{ij(t|Z)}$ is the hazard of transition in the same states for individuals with a certain pattern of covariates Z. β_{ij} is the vector of regression coefficients that describe the effect of the Z covariates on the risk of transition. The cumulative transition hazard is $\Lambda_{ij}(t|Z) = \int_0^t \lambda_{ij}(u|z)du$, where $\Lambda(t)$ defines a squared matrix of possible transitions. Consequently each transition probability $P_{ij}(s,t) = P(t = j|s = i)$ (which expresses the probability of transition from i to j) in the interval (s,t) is provided by the Aalen-Johansen estimator [5]. The model has been estimated using the `mstate` package implemented by de Wreede et al.[5] in R and the Breslow method has been specified to handle ties [7, 11]. The model allows us to specify different covariate effects for the different transitions and different baseline transition hazards for each transition. The likelihood ratio test is used for model selection [5, 7].

3 An Application of Multistate Models to the Italian context

3.1 Definition of States

Longitudinal cohort data on students' careers have been provided from the administrative archive of an Italian University. For the sake of this application we consider the 4336 students who enrolled for the first time in the 2006/2007 academic year (a.y.) in the first level degree programmes. The last information on students' status recorded in the dataset refers to March 2013. All the students start the university career at the same time and they are exposed in 7 years to the same adverse (different kinds of dropouts) and positive (different level of graduations and enrollment to postgraduate studies) events. Any transition made by the students after their enrollment to the first level degree programme is recorded in the dataset.

After 7 academic years from the enrolment (S_1) the possible states in which students can be observed are: dropout during the first level degree (S_2); formal dropout during the first level degree (S_3); graduated in the first level degree (S_4); enrolled at the second level degree (S_5); graduated in the second level degree (S_6); formal dropout during the second level degree (S_7); enrolled at one of the postgraduate programmes (S_8). We define implicit dropouts as those students who have not being paid their academic fees by more than two academic years (last year of observation in the archive 2010–2011 a.y.) whereas formal dropouts are those who leave the university study with an administrative act. Thus, implicit dropouts can be observed just in the first level degree programmes because we observe the cohort of students for just 7 years. The possible paths of students' careers are depicted in Fig. 1. Different shapes are used to identify different conditions (enrolment=circle, dropout=square, graduation=rhombus, triangle=censor) and the arrows link states from which students can transit.

After 7 years of follow-up the students could have experienced one or more states (absorbing or intermediate) before to enter in the last state recorded in the archive.

We have restricted the analysis to the main states' observable in students' careers by eliminating from their pathways all events which rarely occur (e.g., enrollment to a single exam after S_4 or S_5, enrollment to the first level master degree after the first level graduation, formal dropout more than one time during the first level degree programme if at the end the student did not get a degree) and as postgraduate programmes we considered just second level masters and Ph.D. studies.

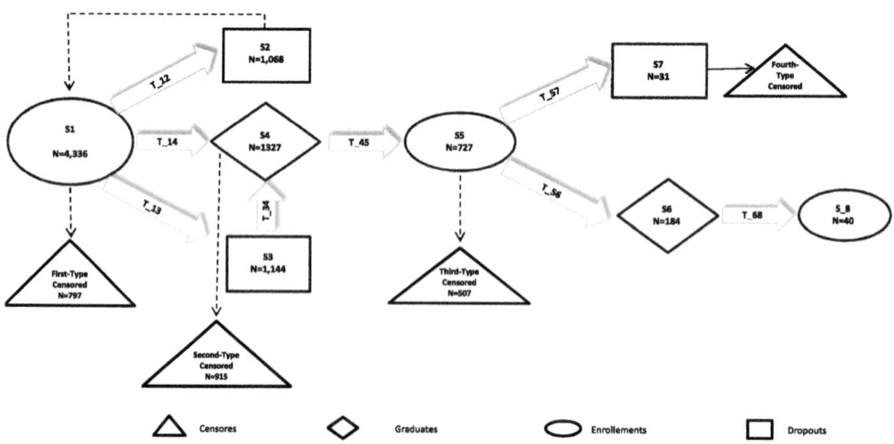

Fig. 1 Path diagram of students' careers

In the first level degree we consider two kinds of dropouts, S_2 and S_3. It is important to differentiate between these two kinds of dropouts since S_2 is treated in the modelling approach as an absorbing state, whereas S_3 as a transition state (students can reenrol and continue their studies in another degree programme).

3.2 *Definition of Time and Censored Units*

The dataset contains the information on the exact time (month/day/year) on which students enter in any of the states experienced during their careers. However, for making the analysis in line with the main scheduled events in the a.y. (e.g., terms, graduation and exam sections) the time has been divided in to trimesters. This makes the total exposure time of the cohort equal to 26 trimesters (interval of time) in which observed events have been classified.

Furthermore the conversion of the time from a.y. to trimesters allows us also to take into account that students may experience more competing or noncompeting events in the same a.y. such as (i) graduation in the first level degree and enrollment to the second level degree, (ii) graduation in the second level degree and enrolment to a postgraduate programme, (iii) enrollment at the first level or second level degree and dropout. In the case of implicit dropout, the expected time at which the event occurred has been imputed equal to the 30 of October of the year after the last one in which student paid the academic fees (first trimester of each a.y.). Different right-censored units are in theory observable in the analysis of students'careers: (i) students who did not experience any transition in at least one of the listed states and are still enrolled at the first-level degree programme (first-type censored units); (ii) students who after the graduation at the first level did not transit to any of the states linked to S_4 (second-type censored units); (iii) students who are still enrolled to the second level degree programmes (third-type censored units); (iv) students who formal dropout and who can still continue their studies (Fourth-Type censored units).

3.3 *Multistate Models*

In order to simplify the possible pathways a multistate model without recurrent events was adopted, allowing students to move just in one direction.

In this way a restriction has been advanced on the possible events that units can experience. However, this hypothesis can be easily relaxed in further analysis.

A model with eight transitions was set up: T_{12}) enrollment $\rightarrow$ dropout ($S_1 \rightarrow S_2$); T_{13}) enrollment $\rightarrow$ formal dropout ($S_1 \rightarrow S_3$); T_{14}) enrollment $\rightarrow$ first level graduate ($S_1 \rightarrow S_4$); T_{34}) formal dropout $\rightarrow$ first level graduate ($S_3 \rightarrow S_4$); T_{45}) graduate $\rightarrow$ enrollment at the second level ($S_4 \rightarrow S_5$); T_{56}) enrollment at second level $\rightarrow$ graduate at second level ($S_4 \rightarrow S_6$); T_{57}) enrollment at second level $\rightarrow$ dropout during the second level ($S_5 \rightarrow S_7$); T_{68}) graduate at the second level → postgraduate courses ($S_6 \rightarrow S_8$) (see Fig. 1).

The specified pathways allow students enrolled at the first level degree to formal dropout and reenroll. We allowed the transition from S_3 to S_4 in order to follow those students who reenroll to a degree programme and got a degree after dropping out, whereas we are not interested (in this phase) in following those who did not get any degree after dropping out (e.g., those who drop repeatedly and did not get any academic title). The transition matrix defined in the model is showed in Table 1, transitions towards positive states are highlighted in bold.

The risk of transition across the states S_1–S_8 is thus described by a set of eight hazards which inform on how the hazard to experience any transition varies along the time. The information about students available in the archive has been used to fit a discrete-time Cox's proportional-hazards model and to single-out profiles of students more at risk to experience specific pathways. Specifically, the following covariates have been included in order to analyze the risk of transition from a state to another: (i) students details—sex (SEX), age (AGE), residence (RES); (ii) students' educational background—type of secondary school attended (LICEO, 1=Liceo, 0=Others); delay in school graduation expressed as difference from the age of 19 (DELSCH), delay in enrollment at the university expressed in terms of difference in years between the year in which students enroll to the university and the year in which they ended the secondary school (DELEN); final mark in the secondary schools (MARKSCH); if student changes faculty or degree programme at least once (CHANGEP). The effects of covariates on each transition have been tested using the likelihood ratio test for comparing nested models. The following predictors were considered in the final model: SEX, LICEO, MARKSCH, DELSCH, DELEN, CHANGP.

The results of this first explorative model (see Table 2) show that males have a hazard higher than females to implicit dropout (experience transition T_{12}), whereas females have a hazard higher to graduate after dropping out (to experience T_{34}). Students who come from a LICEO have lower hazards to dropout (experience T_{12} and T_{13}) and higher hazards to graduate at first and second degree level. As the final mark in the secondary school increases, the hazards to dropout during the first or second level degree (T_{12},T_{13}, T_{57}) decrease and the hazards to get a first or second degree (T_{14}, T_{56}) increase. As expected, students who did not have a regular academic pathway (because changed at least one time their first level degree programme) have a lower hazard to take the first level degree (T_{14}) and a higher hazard to dropout during the second level. For any year of delay in getting the secondary school degree (DELSCH) the hazards to graduate decrease and the hazards to dropout increase. Finally, the delay in the enrollment at the university has a significant effect on the hazard to dropout and to graduate, but the effect on the first is higher.

Table 1 Transition matrix: possible transitions and number of students who experience each transition

From	To									
		S_2	S_3	S_4	S_5	S_6	S_7	S_8	No events	Total
S_1		$T_{12} = 1068$	$T_{13} = 1144$	$T_{14} = 1327$	0	0	0	0	797	4336
S_2	0	0	0	0	0	0	0	0	0	0
S_3	0	0	0	$T_{34} = 76$	0	0	0	0	1068	1144
S_4	0	0	0	0	$T_{45} = 722$	0	0	0	681	1403
S_5	0	0	0	0	0	$T_{56} = 184$	$T_{57} = 31$	0	507	722
S_6	0	0	0	0	0	0	0	$T_{68} = 40$	144	184
S_7	0	0	0	0	0	0	0	0	0	0
S_8	0	0	0	0	0	0	0	0	0	0

Table 2 Multistate-model results

Trans	Sex		LICEO		DELEN		MARKSCH		DELSCH		CHANGP	
	$\hat{\beta}$	*p*-value	$\hat{\beta}$	*p*-value	$\hat{\beta}$	*p*-value	$\hat{\beta}$	*p*-value	$\hat{\beta}$	*p*-value	$\hat{\beta}$	*p*-value
T_{12}	0.174	**	−0.680	***	0.026	***	−3.015	***	0.045	***	−1.633	***
T_{13}	0.091	0.130	−0.396	***	−0.021	*	−2.240	***	0.004	0.631	1.418	***
T_{14}	−0.166	**	0.362	***	0.020	**	3.502	***	−0.037	*	−0.503	***
T_{34}	−0.913	***	0.718	**	−0.087	0.308	0.125	0.897	−0.508	**	−0.068	0.793
T_{45}	−0.049	0.759	0.411	**	−0.062	0.197	3.996	***	0.008	0.788	−0.156	0.730
T_{56}	−0.163	0.681	0.086	0.817	0.035	0.301	−1.840	0.207	−0.140	0.488	0.675	0.360
T_{57}	0.084	0.514	−0.264	0.034 *	−0.067	*	−2.098	***	−0.081	0.151	1.427	***
T_{78}	0.240	0.469	0.418	0.226			1.680	0.322	0.057	0.316		

Significance codes: ***0.001, **0.01, *0.05, 0.1

4 Conclusion

This work makes just a first attempt to show the usefulness of multistate models in assessing the effect of students' features on the probability to experience positive and negative states. Moving from the estimated transition probabilities is possible to compare the probability to experience the different pathways for students with different profiles. In this first explorative analysis we considered just few information regarding the educational background of students to show the informative power of the methodology. The research is still in progress and requires more administrative information to improve the definition of the possible states that the students can experience. This first exercise would just be an attempt to show as the approach can be used as a screening tool to evaluate faculties or degree Programmes' inefficiencies adjusting for students' educational and socioeconomic background. We are also aware that transitions after the first level graduation are affected by two kinds of censored units: the loss of follow-up imputable to students who migrate in other universities at the end of the first level degree (about 20 % of the students who enrolled to the second level) and the administrative censoring (the end of the study) [7]. The analysis in progress aims also to adjust the bias in the transition by extending the exposure time.

References

1. Adelfio, G., Boscaino, G., Capursi, V.: A new indicator for higher education student performance. High. Educ. **68**(5), 653–668 (2014). doi:10.1007/s10734-014-9737-x
2. Anderson, K.P., Keiding, N.: Multi-state models for event history analysis. Stat. Methods Med. Res. **11**, 91–115 (2002)
3. Attanasio, M., Boscaino, G., Capursi, V., Plaia, A.: Indicators and measures for the assessment of University students' careers. In: Proceedings of the 8th Scientific Meeting of the Classification and Data Analysis Group of the Italian Statistical Society, pp. 1–4, University of Pavia, September 7–9 2011
4. Cox, D.R.: Regression models and life tables (with Discussion). J. R. Stat. Soc. Ser. B **34**, 187–220 (1972)
5. de Wreede, L.C., Fiocco, M., Putter, H.: mstate: an R package for the analysis of competing risks and multi-state models. J. Stat. Softw. **38**(7), 1–30 (2011)
6. Meira-Machado, L., de Uña-Àlvarez J., Cadarso-Suárez, C., Anderson, P.: Multi-state models for the analysis of time-to-event data. Stat. Methods Med. Res. **11**, 195–222 (2009)
7. Putter, H., Fiocco, M., Geskus., R.B.: Tutorial in biostatistics: competing risks and multi-state models. Stat. Med. **26**, 2389–2340 (2007)
8. Ortiz, E.A., Dehon, C.: Roads to success in the Belgian French community's higher education system: predictors of dropout and degree completion at the Université Libre de Bruxelles. Res. High. Educ. **54**(6), 693–723 (2013)
9. Porcu, M., Sulis, I.: The credit accumulation process to assess the performances of degree programs: an adjusted indicator based on the result of entrance tests. In: Giudici, P., Ingrassia, S., Vichi, M. (eds.) Statistical Models for Data Analysis. Studies in Classification, Data Analysis, and Knowledge Organization. Springer, Heidelberg (2013)

10. Scott, M., Kennedy, B.: Pitfalls in pathways: some perspectives on competing risks event history analysis in education research. J. Educ. Behav. Stat. **30**(4), 413–442 (2005)
11. Singer, J., Willett, J.: It's about time: using discrete-time survival analysis to study duration and the timing of events. J. Educ. Stat. **18**(2), 155–195 (1993)
12. Singer, J., Willett, J.: Applied Longitudinal Data Analysis: Modelling Changes and Event Occurrance. Oxford University Press, London (2003)
13. Tedesco, N.: Un approccio Multilivello sulla ricerca delle determinanti del richio di laurea e di abbandono dell'Ateneo di Cagliari. In: Fabbris, L. (ed.) Laid-out: Scoprire i rischi con l'analisi di segmentazione. Collana determinazione e prevenzione di rischi sociali e sanitari, vol. 3. CLEUP, Padova (2003)

Printed by Printforce, the Netherlands